LONDON MATHEMATICAL SOCIETY LECTURE NOTE SERIES

T0269198

Managing Editor: Professor I.M.James,
Mathematical Institute, 24-29 St Giles, Oxford

1. General cohomology theory and K-theory, P.HILTON
4. Algebraic topology: a student's guide, J.F.ADAMS
5. Commutative algebra, J.T.KNIGHT
8. Integration and harmonic analysis on compact groups, R.E.EDWARDS
9. Elliptic functions and elliptic curves, P.DU VAL
10. Numerical ranges II, F.F.BONSALL & J.DUNCAN
11. New developments in topology, G.SEGAL (ed.)
12. Symposium on complex analysis, Canterbury, 1973, J.CLUNIE & W.K.HAYMAN (eds.)
13. Combinatorics: Proceedings of the British combinatorial conference 1973, T.P.McDONOUGH & V.C.MAVRON (eds.)
14. Analytic theory of abelian varieties, H.P.F.SWINNERTON-DYER
15. An introduction to topological groups, P.J.HIGGINS
16. Topics in finite groups, T.M.GAGEN
17. Differentiable germs and catastrophes, Th.BROCKER & L.LANDER
18. A geometric approach to homology theory, S.BUONCRISTIANO, C.P.ROURKE & B.J.SANDERSON
20. Sheaf theory, B.R.TENNISON
21. Automatic continuity of linear operators, A.M.SINCLAIR
23. Parallelisms of complete designs, P.J.CAMERON
24. The topology of Stiefel manifolds, I.M.JAMES
25. Lie groups and compact groups, J.F.PRICE
26. Transformation groups: Proceedings of the conference in the University of Newcastle upon Tyne, August 1976, C.KOSNIOWSKI
27. Skew field constructions, P.M.COHN
28. Brownian motion, Hardy spaces and bounded mean oscillation, K.E.PETERSEN
29. Pontryagin duality and the structure of locally compact abelian groups, S.A.MORRIS
30. Interaction models, N.L.BIGGS
31. Continuous crossed products and type III von Neumann algebras, A.VAN DAELE
32. Uniform algebras and Jensen measures, T.W.GAMELIN
33. Permutation groups and combinatorial structures, N.L.BIGGS & A.T.WHITE
34. Representation theory of Lie groups, M.F.ATIYAH et al.
35. Trace ideals and their applications, B.SIMON
36. Homological group theory, C.T.C.WALL (ed.)
37. Partially ordered rings and semi-algebraic geometry, G.W.BRUMFIEL
38. Surveys in combinatorics, B.BOLLOBAS (ed.)
39. Affine sets and affine groups, D.G.NORTHCOTT
40. Introduction to H_p spaces, P.J.KOOSIS
41. Theory and applications of Hopf bifurcation, B.D.HASSARD, N.D.KAZARINOFF & Y-H.WAN
42. Topics in the theory of group presentations, D.L.JOHNSON
43. Graphs, codes and designs, P.J.CAMERON & J.H.VAN LINT
44. Z/2-homotopy theory, M.C.CRABB
45. Recursion theory: its generalisations and applications, F.R.DRAKE & S.S.WAINER (eds.)
46. p-adic analysis: a short course on recent work, N.KOBLITZ
47. Coding the Universe, A. BELLER, R. JENSEN & P. WELCH
48. Low-dimensional topology, R. BROWN & T.L. THICKSTUN (eds.)
49. Finite geometries and designs, P. CAMERON, J.W.P. HIRSCHFELD & D.R. HUGHES (eds.)
50. Commutator Calculus and groups of homotopy classes, H.J. BAUES
51. Synthetic differential geometry, A. KOCK
52. Combinatorics, H.N.V. TEMPERLEY (ed.)
53. Singularity theory, V.I. ARNOLD
54. Markov processes and related problems of analysis, E.B. DYNKIN
55. Ordered permutation groups, A.M.W. GLASS
56. Journees arithmetiques 1980, J.V. ARMITAGE (ed.)
57. Techniques of geometric topology, R.A. FENN
58. Singularities of differentiable functions, J. MARTINET
59. Applicable differential geometry, F.A.E. PIRANI and M. CRAMPIN
60. Integrable systems, S.P. NOVIKOV et al.

London Mathematical Society Lecture Note Series. 53

Singularity Theory

Selected Papers

V.I. ARNOLD
Professor of Mathematics, Moscow University

CAMBRIDGE UNIVERSITY PRESS
CAMBRIDGE
LONDON NEW YORK NEW ROCHELLE
MELBOURNE SYDNEY

CAMBRIDGE UNIVERSITY PRESS
Cambridge, New York, Melbourne, Madrid, Cape Town, Singapore, São Paulo

Cambridge University Press
The Edinburgh Building, Cambridge CB2 8RU, UK

Published in the United States of America by Cambridge University Press, New York

www.cambridge.org
Information on this title: www.cambridge.org/9780521285117

First published 1981
Re-issued in this digitally printed version 2008

A catalogue record for this publication is available from the British Library

ISBN 978-0-521-28511-7 paperback

CONTENTS

Intoduction (C.T.C. WALL) 1

Singularities of smooth mappings (Volume 23, 1968) 3

On matrices depending on parameters (Volume 26, 1971) 46

Remarks on the stationary phase method and Coxeter numbers (Volume 28, 1973) 61

Normal forms of functions in neighbourhoods of degenerate critical points (Volume 29, 1974) 91

Critical points of smooth functions and their normal forms (Volume 30, 1975) 132

Critical points of functions on a manifold with boundary, the simple Lie groups B_k, C_k and F_4 and singularities of evolutes (Volume 33, 1978) 207

Indices of singular points of 1-forms on a manifold with boundary, convolution of invariants of reflection groups and singular projections of smooth surfaces (Volume 34, 1979) 225

INTRODUCTION

C. T. C. Wall

Professor Arnold is well known for his researches on a variety of topics in pure and applied mathematics, but perhaps no field owes more to him than singularity theory. In this volume are collected 7 survey articles of his on singularity theory that have appeared over the last decade. The first of these, written at a time (1968) when the subject was rapidly opening up, remains an excellent general introduction to the field as a whole.

However the core of the volume, consisting of 3 articles which appeared in 1973–75, consists of an account of the classification of critical points of smooth functions, and of the reinterpretations of a key class of functions (those with normal form depending on at most one parameter) – in relation to Lie groups, spherical and hyperbolic triangles, and definiteness of the intersection form – obtained by Arnold and his students during that period. Together, these results constitute one of the most beautiful discoveries in mathematics in recent years: and the further detailed study of these classes of singularities has revealed at each stage unexpected and rich structure.

Although Arnold does not shrink from describing the detailed calculations from which these lists are derived (the articles contain extensive – though not complete – sections of proofs) the surveys are far from being dry lists. He shows how a problem on estimating oscillatory integrals led him to start classifying functions; and by defining and computing an invariant (the 'Arnold index') applies the classification back to the original problem. In another paper, he relates the singularities of functions to those of projections of Lagrangian (and Legendre) submanifolds, and to the structure of caustics. The analysis of singularities of evolutes led to an extension of the work, in which singularities on the boundary of a manifold are investigated: an extension which in many ways completes the pattern set by the original.

Since catastrophe theory as such is not discussed in this volume, it is perhaps worth emphasizing here that Arnold's classification of simple singularities (A_k, D_k, E_k) contains and supersedes Thom's list of elementary catastrophes ($A_k: k \geqslant 5, D_4, D_5$). Moreover, the theory of Lagrangian maps coincides with the so called "catastrophe map". But Arnold goes further, and the relation of the theory of oscillating integrals to singularity theory, which he developed, has been aptly termed 'quantum catastrophe theory'.

The reader of this volume should not expect completeness: the results in these papers have stimulated much further work, and much yet remains to be discovered. But these surveys do contain Arnold's own analysis and synthesis of a decade's work on a fascinating topic, and it is with great pleasure that I introduce them to the reader.

SINGULARITIES OF SMOOTH MAPPINGS

V. I. ARNOL'D

The paper is based on a course of lectures on the local theory of singularities delivered in 1966 at a Summer School in Katsiveli and at the Moscow State University.

Contents

Chapter 1. The structure of singularities
§1. Examples 3
§2. The classes Σ^I 9
§3. The quadratic differential 14
§4. The local ring of a singularity and the Weierstrass preparation theorem 19
Appendix. A proof of the Weierstrass preparation theorem 22
Chapter 2. Deformations of singularities
§5. "Infinite-dimensional Lie groups" acting on "infinite-dimensional manifolds" 26
§6. The stability theorem 30
§7. Proof of convergence 36
§8. In the neighbourhood of an isolated critical point every analytic function is equivalent to a polynomial 40
References 44

Chapter I

THE STRUCTURE OF SINGULARITIES

§1. Examples

The theory of singularities of smooth[2] mappings is concerned with local properties of differentiable mappings of differentiable manifolds,

---

[1] The author is grateful to B. Malgrange, Yu.I. Manin, B. Morin, V.P. Palamodov and R. Thom for fruitful discussions and to S.M. Vishik, A.G. Kushnirenko and A.M. Leontovich for their assistance in preparing these lectures for the printer.

[2] Here and in the sequel "smooth" and "differentiable" mean "infinitely differentiable". The tangent space at a point x in a manifold M is denoted by TM_x. The differential at x of a mapping $f:M \to N$ is denoted by $f_x: TM_x \to TN_{f\,x})$.

$$f\colon M^m \longrightarrow N^n,$$

invariant under diffeomorphisms

$$h\colon M \longrightarrow M, \qquad k\colon N \longrightarrow N.$$

EXAMPLE 1. Consider the mapping $f\colon \mathbf{R}^1 \to \mathbf{R}^1$ given by the formula $y = f(x) = x^2$ (fig. 1). In the neighbourhood of every $x \neq 0$, f is a diffeomorphism.

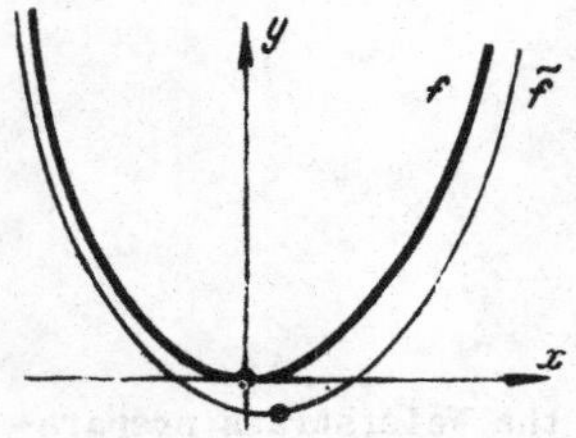

Fig. 1.

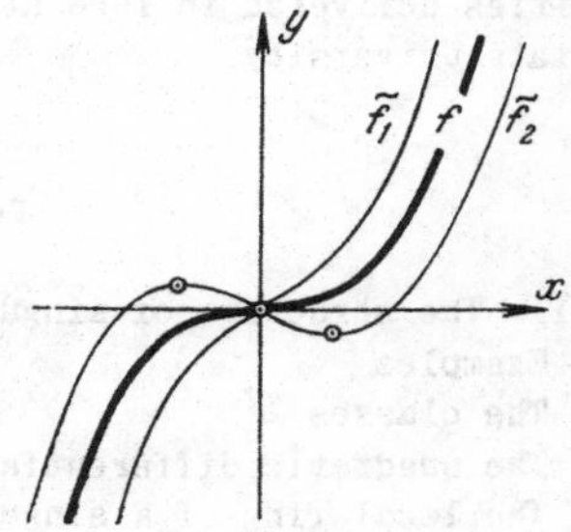

Fig. 2.

This is not the case in the neighbourhood of $x = 0$; f has a *singularity*, the differential of f degenerates at 0. This singularity is *stable*: every mapping $\tilde{f}$ near f whose derivatives are near to those of f has a similar singularity.

EXAMPLE 2. The mapping $y = x^3$ (fig. 2) also has a singularity, but it is *unstable*: for small deformations the singularity can vanish ($\tilde{f}_1$) or split into two ($\tilde{f}_2$).

We give now a general

DEFINITION 1. A differentiable mapping $f\colon M \to N$ is said to be *stable* if for any differentiable mapping $\tilde{f}\colon M \to N$ sufficiently close[1] to f there are diffeomorphisms $h\colon M \to M$, $k\colon N \to N$ close to the identity "converting" f to $\tilde{f}$, that is, such that the diagram

$$\begin{array}{ccc} M & \xrightarrow{f} & N \\ {\scriptstyle h}\downarrow & & \downarrow{\scriptstyle k} \\ M & \xrightarrow{\tilde{f}} & N \end{array}$$

commutes. If there are *homeomorphisms* h and k that are near to 1_M and 1_N and make the diagram commute, then the mapping is said to be topologically stable.

---

[1] The topology on the space of differentiable mappings is that defined by these neighbourhoods of zero in the space of differentiable functions of local coordinates:

$$U(k, \varepsilon) = \left\{ \varphi(x)\colon \max_{|\alpha| \leq k} \left| \frac{\partial^{|\alpha|}\varphi}{\partial x^\alpha} \right| < \varepsilon \right\}$$

Slightly more complicated is the definition of stability at a point.

DEFINITION 2. A smooth mapping $f\colon U \to N^n$ defined on a neighbourhood U of a point x_0 in M^m is said to be *stable at* x_0 if for every mapping $\tilde{f}\colon U \to N^n$ sufficiently near to f there exist neighbourhoods $V \subset U \subset M^m$ of x_0 and $W \subset N^n$ of $y_0 = f(x_0)$, and diffeomorphic embeddings $h\colon V \to U$, $k\colon W \to N^n$ close to the identity and such that the diagram

$$\begin{array}{ccc} M^m \supset U \supset V & \xrightarrow{f} & W \subset N^n \\ h\downarrow & & \downarrow k \\ M^m \supset U & \xrightarrow{\tilde{f}} & N^n \end{array}$$

commutes. If, in addition, the h and k can be taken as homeomorphisms, we get the definition of *topological stability* at x_0; if f, $\tilde{f}$, h and k can be taken as real (complex) analytic functions, we get the definition of real (complex) analytic stability at x_0.

EXAMPLE 3. The implicit function theorem asserts that a mapping whose rank at x_0 is maximal is stable at x_0.

EXAMPLE 4. "Morse's Lemma" (see [1], p. 14) states that a mapping $\mathbf{R}^n \to \mathbf{R}^1$ given by a non-degenerate quadratic form

$$f(x) = x_1^2 + \ldots + x_k^2 - x_{k+1}^2 - \ldots - x_n^2,$$

is stable at 0.

The ideal to which the theory of singularities strives is achieved in the special case of mappings $\mathbf{R}^n \to \mathbf{R}^1$ (Morse theory). The results of this theory that interest us can be stated as follows:

THEOREM 1. 1) *The stable mappings* $f\colon M^m \to \mathbf{R}^1$ *of a compact manifold* M^m *to the line form an everywhere dense set in the space of all smooth mappings.*

2) *A mapping f is stable if and only if the following two conditions are satisfied:*

M_1. *f is stable at every point (that is, every critical point of the function f is non-degenerate).*

M_2. *All critical values of f are distinct.*

3) *The mapping* $f\colon M^m \to \mathbf{R}^1$ *is stable at* x_0 *if and only if coordinates* $x_1, x_2, \ldots, x_m$, y *can be introduced in neighbourhoods of* x_0 *in* M^m *and* $y_0 = f(x_0)$ *in* $\mathbf{R}^1$ *in such a way that f can be written in one of the* $m + 2$ *forms*:

$$\text{M I.}\quad y = x_1,$$
$$\text{M II}_k.\quad y = x_1^2 + \ldots + x_k^2 - x_{k+1}^2 - \ldots - x_m^2 \quad (k = 0, 1, \ldots, m).$$

The following important case – the theory of two-dimensional manifolds – has received exhaustive study (Whitney [2]).

THEOREM 2. *The mapping* $f\colon M^2 \to N^2$ *is stable at the point* x_0 *if and only if it is equivalent in some neighbourhood of* x_0 *to one of three mappings* (fig. 3):

WI. $y_1 = x_1$, $y_2 = x_2$ *(regular point)*,

WII. $y_1 = x_1$, $y_2 = x_2^2$ *(fold)*,

WIII. $y_1 = x_1$, $y_2 = x_1x_2 - \frac{1}{3}x_2^3$ *(cusp) of a neighbourhood of* 0 *in the*

(x_1, x_2)-plane into a neighbourhood of 0 in the (y_1, y_2)-plane.

The stable mappings f: $M^2 \to \mathbf{R}^2$ of a compact surface into the plane form an everywhere dense set in the space of all smooth mappings.

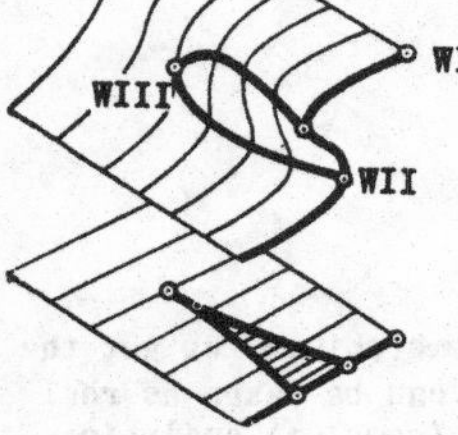

Fig. 3.

The smooth mapping f: $M^2 \to \mathbf{R}^2$ is stable if and only if the following two conditions are satisfied:

WI. *The mapping is stable at every point in M^2.*

WII. *The images of folds intersect only pairwise and at non-zero angles, whereas images of folds and cusps do not intersect.*

EXAMPLE 5. The Whitney mapping W. III. is stable (see fig. 4). Let us examine the structure more closely. Each line $x_1 = C$ goes into the line $y_1 = C$. But if $C < 0$, the image is monotone, while if $C > 0$ it is like $\tilde{f}_2$ in Example 2.

The differential f_* has the matrix

$$\left\| \begin{matrix} 1 & 0 \\ x_2 & x_1 - x_2^2 \end{matrix} \right\|.$$

The rank of f_* is 2 everywhere except on the parabolic fold $x_1 = x_2^2$. On the fold f_* has a kernel parallel to the x_2-axis. The image of the fold is the semi-cubic parabola $y_1 = x_2^2$, $y_2 = \frac{2}{3}x_2^3$. Every point inside the angle has three pre-images,

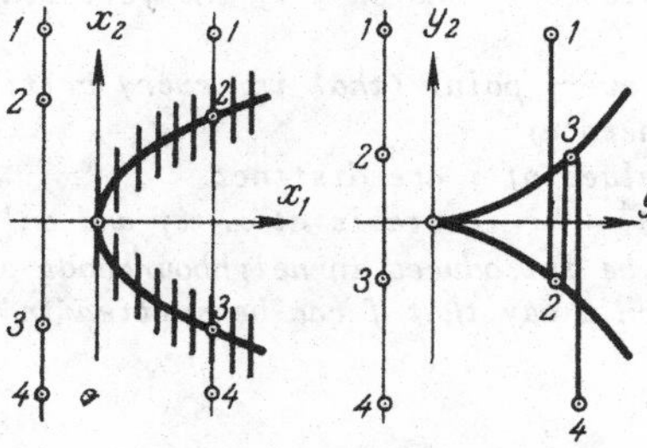

Fig. 4.

each point outside it has one. Note that the fold in the x-plane has no singularity, whereas its image in the y-plane has a singularity at 0. This is explained by the fact that the restriction of f to the parabola $x_1 = x_2^2$ has rank 0 at 0 (the kernel of f_* is tangent to the parabola at 0).

EXAMPLE 6. Consider the mapping $z \to z^2$ of the complex plane onto itself. The singularity at 0 is clearly not one of the types I, II, III. By Whitney's theorem, the mapping is unstable and small deformations can make it into a stable mapping having no singularities apart from folds and cusps. It turns out that it is enough to consider the nearby mappings $z \to z^2 + 2\varepsilon\bar{z}$.

The branch-point splits into three cusps, the circumference into a fold (Fig. 5). The image of the fold is a hypocycloid with three angles;

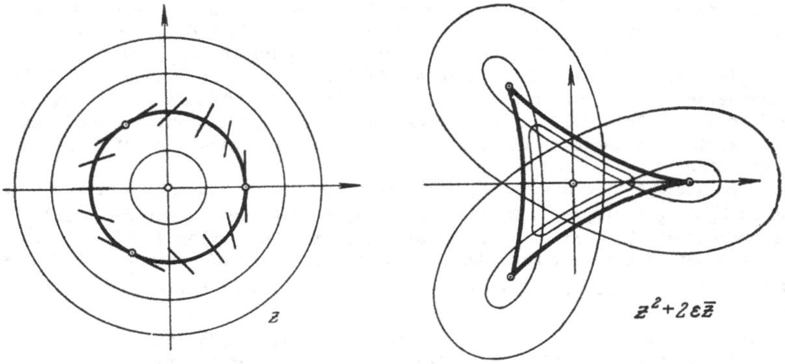

Fig. 5.

the points of the inside have four pre-images, those outside have two. The images of circles of large radius make two turns around the origin.

These examples might give rise to the hope that in higher dimensions every singularity is almost stable, and that stable singularities are easily classifiable. Indeed such a classification exists in dimensions less than six (Whitney [26]). However, the position is completely different in higher dimensions.

THEOREM 3 (Thom [3]). *For $n \geqslant 9$ there exist mappings $\mathbf{R}^n \to \mathbf{R}^n$ unstable at 0 but such that every sufficiently close mapping is unstable.*

Thus, for $n \geqslant 9$ the stable mappings do not form a dense set in the space of all mappings $M^n \to N^n$; there exist *continuous invariants* of smooth mappings f with respect to diffeomorphisms h and k. We give the proof of this theorem in §3.

Naturally the topological classification is coarser.

THEOREM 4 (Thom [4]). *There exists a topological classification of germs of smooth mappings f: $\mathbf{R}^m \to \mathbf{R}^n$ if a set of infinite codimension in the space of all germs is neglected.*[1]

More accurately this means the following. Denote by S the space of all germs of smooth mappings f: $\mathbf{R}^m \to \mathbf{R}^n$ at 0. Then

$$S = \bigcup_{\gamma\in\Gamma} S_\gamma + S_\infty,$$

where Γ is a countable set, each S_γ has finite codimension in S and

[1] The proof has not yet been published.

consists of topologically equivalent[1] germs; and S_∞ has infinite codimension.

We give here the definitions of some of the terms arising. The *germ* of the mapping $f: M^m \to N^n$ at x_0 is the equivalence class of mappings of neighbourhoods of x_0, $\varphi: U \to N^n$, $x_0 \in U \subset M^m$, where we say that two mappings $\varphi_1: U_1 \to N^n$, $\varphi_2: U_2 \to N^n$ are *equivalent* if they coincide on some neighbourhood of x_0 contained in $U_1 \cap U_2$. The *codimension* of a submanifold $A^k \subset B^n$ is $n-k$; that is, "the number of equations giving A locally". This definition carries over to the infinite-dimensional manifold S of germs when one uses the finite-dimensional approximations of S by the spaces of jets.

Two mappings $\varphi_1: U_1 \to N^n$, $\varphi_2: U_2 \to N^n$ are said to have *tangency of order k* at a point in $U_1 \cap U_2$ if $|\varphi_1(x) - \varphi_2(x)| = o(|x|^k)$ in some (and therefore in all) local coordinates. The *jet of order k* of the germ of a mapping $f: M^m \to N^n$ at x in M^m is the set of germs of mappings at x having tangency of order k with f at x. We denote this jet by $j_x^k(f)$. Clearly, $j_x^0(f) = f(x)$, $j_x^1(f)$ is defined by the differential $f_x: TM_x \to TN_{f(x)}$ of f at x.

Let $x_1, \ldots, x_m$ and $y_1, \ldots, y_n$ be choices of local coordinates in M and N, respectively. Then the jet $j_0^k(f)$ is defined by a segment of the Taylor series for f,

$$j_0^k(f) \sim f|_0 + \left.\frac{\partial f}{\partial x}\right|_0 x + \frac{1}{2}\left(\left.\frac{\partial^2 f}{\partial x^2}\right|_0 x, x\right) + \ldots + \frac{1}{k!}\left(\left.\frac{\partial^k f}{\partial x^k}\right|_0 x, \ldots, x\right).$$

Let $J_x^k(M^m, N^n)$ stand for the space of jets of order k of germs $f: M^m \to N^n$ at x. The preceding formula introduces into this space the structure of a finite-dimensional manifold of dimension

$$\dim J_x^k(M^m, N^n) = n + mn + \frac{m(m+1)}{2} n + \ldots + \binom{k+m-1}{m-1} n.$$

There is a natural mapping

$$\pi^k: S \to J_x^k,$$

associating with each germ at x its jet $j_x^k(f)$.

A set $S' \subset S$ is of *finite codimension l*, if for some k,

$$S' = (\pi^k)^{-1} J',$$

where J' is a submanifold of codimension l in J_x^k. In other words, *the set S' of germs has finite codimension l if it is given by l conditions on the Taylor coefficients of fixed order*. Further, S' has *infinite codimension if it lies in the intersection of a sequence of sets of increasing codimensions*.

The set of all jets of order k of germs of mappings $f: M^m \to N^n$ at different points forms a fibre bundle $J^k(M^m, N^n)$ with base $J^0(M, N) = M \times N$. It can be regarded as a vector bundle with the same base, and then we call it the *bundle of k-jets of mappings of M into N*. Finally, proceeding to the limit, as $k \to \infty$, in the sequence of projections $J^{k+1} \to J^k \to \ldots \to J^0$, we get the bundle of jets $J^\infty = J(M, N)$. The notation $J_x^\infty(M, N) = J_x(M, N)$ and $j_x^\infty(f) = j_x(f)$ has a similar meaning.

---

1 Two germs $f_1, f_2: M^m \to N^n$ at x in M^m are topologically equivalent if there is a germ of a homeomorphism $h: M^m \to M^m$ fixing x and a germ of a homeomorphism $k: N^n \to N^n$ "taking $f_1(x)$ to $f_2(x)$", that is, such that the diagram

$$\begin{array}{ccc} M^m & \xrightarrow{f_1} & N^n \\ h\downarrow & & \downarrow k \\ M^m & \xrightarrow{f_2} & N^n \end{array}$$

commutes.

§2. The classes Σ^I

Let f: $M^m \to N^n$ be a smooth mapping and i an integer $\geqslant 0$.

DEFINITION 1. A point x of M^m lies in the set $\Sigma^i(f) \subset M$ if the kernel of the differential f_*: $TM_x^m \to TN_{f(x)}^n$ has dimension i. We say that f has singularity Σ^i at x or a singularity of class Σ^i.

EXAMPLE 1. For the Whitney singularity (fig. 6)

$$y_1 = x_1, \quad y_2 = x_1 x_2 - \frac{1}{3} x_2^3$$

we have

$$f_x = \begin{Vmatrix} 1 & 0 \\ x_2 & x_1 - x_2^2 \end{Vmatrix}, \quad \dim \operatorname{Ker} f_x = \begin{cases} 0 & \text{if} \quad x_1 \neq x_2^2, \\ 1 & \text{if} \quad x_1 = x_2^2. \end{cases}$$

Thus, Σ^0 is the whole plane excluding the parabola $x_1 = x_2^2$, Σ^1 is the parabola, the kernel is parallel to the x_2-axis; and for $i \geqslant 2$, Σ^i is empty.

We remark that the parabola Σ^1 is a smooth manifold which includes the cusp point 0. This point is distinguished by the fact that the kernel of f_0 is tangent to Σ^1 at it. In other words, f *restricted to* Σ^1, has rank 1 at all points other than 0. Thus, 0 lies in $\Sigma^1(f|\Sigma^1(f))$, and all other points on the parabola lie in $\Sigma^0(f\Sigma^1(f))$. We write $\Sigma^{i_2}(f|\Sigma^{i_1}(f))$ as $\Sigma^{i_1 i_2}(f)$. With this notation the Whitney cusp 0 lies in $\Sigma^{11}(f) \subset \Sigma^1(f)$.

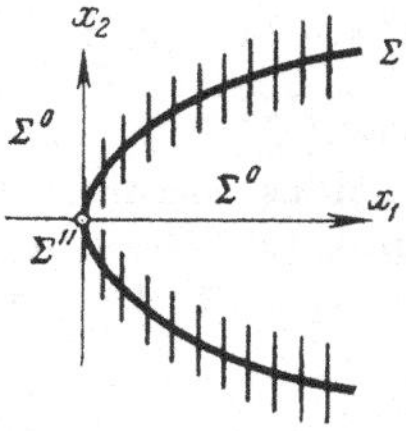

Fig. 6.

For any set $I = i_1, i_2, \ldots, i_n$ of integers the set $\Sigma^I(f)$ is defined by induction as follows.

DEFINITION 2. Let $\Sigma^I(f) = \Sigma^{i_1, \ldots, i_k}(f) \subset M$, M a smooth manifold. Then

$$\Sigma^{i_1, i_2, \ldots, i_k, i_{k+1}}(f) = \Sigma^{i_{k+1}}(f \mid \Sigma^I(f))$$

is the set of all points where the kernel of the differential of the restriction of f to $\Sigma^I(f)$ has dimension i_{k+1}.

REMARK. By the definition, the manifolds

$$M \supset \Sigma^{i_1} \supset \Sigma^{i_1 i_2} \supset \Sigma^{i_1 i_2 i_3} \supset \ldots$$

are embedded in one another. Thus, the kernels of the restrictions of f to these embedded submanifolds are also embedded in one another. So the *sequence of numbers* $i_1, i_2, \ldots$ comprising I must be non-decreasing, $m \geqslant i_1 \geqslant i_2 \geqslant i_3 \geqslant \ldots \geqslant 0$. If one of these inequalities is violated, then Σ^I is empty.

The set $\Sigma^I(f)$ is not necessarily a manifold, therefore the definition given above (due to Thom) does not give a definition of $\Sigma^I(f)$ for all f.

Boardman [5] has proposed a definition of $\Sigma^I(f)$ in terms of the space of jets. For any set $I = i_1, \ldots i_k$ of integers he defines a subset Σ^I of the space of k-jets $J^k(M^m, N^n)$, not depending on any mapping f (see below, p. 11). He has proved:

THEOREM 1. *For any* $I = i_1, \ldots, i_k$ *the set* Σ^I *is a (not necessarily closed) submanifold of codimension* $\nu_I(m, n)$ *in* $J^k(M^m, N^n)$. *(The formula for* ν_I *is given below, p.9).*

The significance of Σ^I is that a "good" mapping f has singularity $\Sigma^I(f)$ at x in the sense of the preceding definition if and only if the jet of f at x lies in Σ^I.

DEFINITION 3. Let $f\colon M^m \to N^n$ be a smooth mapping. The *induced mapping* $\bar{f}\colon M^m \to J^k(M^m, N^n)$ associates with each point x of M^m the jet of f at x:

$$\bar{f}(x) = j_x^k(f).$$

A mapping is called "good" if its induced mapping $\bar{f}$ is transversal[1] on Σ^I.

Boardman has proved

THEOREM 2. 1) *If* f *is good, then* $\Sigma^I(f) = \bar{f}^{-1}(\Sigma^I)$; *that is,* $\Sigma^I(f)$ *is a manifold of codimension* $\nu_I(m, n)$ *in* M^m, *and* $x \in \Sigma^I(f)$ *if and only if the jet of* f *at* x *lies in* Σ^I.

2) *Every smooth mapping can be approximated, together with an arbitrary number of its derivatives, as closely as desired by a good mapping.*

Assertion 2) follows from 1) and Thom's transversality lemma. For $k = 1$ these results were established by Thom [3] and for $k = 2$ by Levine [3].

Let us consider the case $k = 1$ in more detail. A 1-jet of a smooth mapping taking x_0 in M^m to y_0 in N^n is given in local coordinates $x_1, \ldots, x_m$ in M^m and $y_1, \ldots, y_n$ in N^n by the matrix of the differential

$$f_{x_0} = \left\| \frac{\partial y_i}{\partial x_j}\bigg|_{x_0} \right\| \qquad (i = 1, \ldots, n;\ j = 1, \ldots, m).$$

The $m \times n$ matrices form an mn-dimensional linear space L. Consider the set L_r of matrices of rank r. The numbers $k = m - r$ and $l = n - r$ can be called the coranks.

LEMMA 1. *The matrices of rank* r *form a smooth (non-closed) submanifold* L_r *in the space* L *of all* $m \times n$ *matrices, and the codimension of* L_r *is the product of the coranks:*

$$\dim L_r = mn - kl = mn - (m - r)(n - r).$$

PROOF. Since $GL(n, \mathbb{R}) \times GL(m, \mathbb{R})$ acts transitively on L_r, it is sufficient to consider the neighbourhood of the following matrix in L_r:

[1] Let A, B, C be smooth manifolds, $f\colon A \to B$ and $g\colon C \to B$ smooth mappings. Then f and g are said to be transversal if for every pair $a \in A$, $c \in C$ of points for which $f(a) = g(c) = b$ we have $f_*(TA_a) + f_*(TC_c) = TB_b$. If g is an *embedding*, then we speak of the transversality of f on the manifold C. Here $f^{-1}(C)$ is a submanifold of A and its codimension in A is the codimension of C in B (implicit function theorem).

Thom's "Transversality Lemma" asserts that *the set of mappings* $f\colon A \to B$ *transversal to a given mapping* $g\colon C \to B$ *is everywhere dense in the space of all differentiable mappings*. Further, *the set of mappings* f *such that* f *is transversal on an arbitrary submanifold of the space of jets is everywhere dense (and clearly open).*

$$A_0 = \begin{array}{c|c|c|l} & r & l & \\ \hline r & \begin{matrix} 1_1 & & \\ & \ddots & \\ & & 1 \end{matrix} & 0 & \\ \cline{1-3} k & 0 & 0 & \\ \hline \end{array} \begin{matrix} \\ \} m \\ \\ \end{matrix} \quad \text{(n columns)}$$

For a nearby matrix A to lie in L_r it is necessary and sufficient that the kl bordering minors are 0. The resulting equations have the form

$$a_{ij}+O(a^2)=0,\quad r<i\leqslant n,\quad r<j\leqslant m,$$

and are therefore independent for $a = 0$, as required.

EXAMPLE 2. The manifold of square matrices of corank 1 has codimension 1; the manifold of square matrices of corank 2 has codimension 4, and in general, the codimension is k^2 for corank k.

COROLLARY 1. *Suppose that the mapping $f: M^m \to \mathbf{R}^n$ is "good" in the sense that the induced mapping $x \to$ (matrix of the differential at x) is transversal on the manifold L_r of matrices of rank r. Then the set $\Sigma^i(f)$, $i = m - r$, defined at the beginning of this section, is a submanifold of M^m of codimension $(m - r)(n - r) = i(n - m + i)$.*

EXAMPLE 3. Let us consider mappings of manifolds of the same dimension ($m = n$). The codimension of $\Sigma^k(f)$, where the rank of a "good" mapping splits into k units, is k^2.

On the other hand, a mapping "of general type" is good, because every mapping can be approximated by a good one (Thom's Lemma [3]).

It follows from this, for example, that the set Σ^2 of singularities has codimension 4 and so cannot arise for mappings of the plane to the plane (see Example 5 in §1.). But it can even be *non-removable* for mappings $\mathbf{R}^n \to \mathbf{R}^n$ when n is greater than 3.

EXAMPLE 4. The two good mappings $f_{\pm}$ given by the formulae

$$\begin{aligned} y_1 &= x_1, \\ y_2 &= x_2, \\ y_3 &= x_3^2 \pm x_4^2 + x_1x_3 + x_2x_4, \\ y_4 &= x_3x_4, \end{aligned}$$

both have a non-removable point Σ^{20} at zero.

In the next section it will be shown that the germs of $f_{\pm}$ at 0 are *not equivalent*. This means that the classification of singularities into the classes Σ^I is not complete.

Boardman's formula for the codimension of Σ^I, $I = i_1, i_2, \ldots, i_k$ is

$$\nu_I(m, n) = (n - m + i_1)\,\mu(i_1, i_2, \ldots, i_k) - (i_1 - i_2)\,\mu(i_2, i_3, \ldots, i_k) - \ldots \\ \ldots - (i_{k-1} - i_k)\,\mu(i_k),$$

where $\mu(i_1, i_2, \ldots, i_k)$ is the number of sequences $j_1, j_2, \ldots, j_k$ of integers satisfying the conditions:

a) $j_1 \geqslant j_2 \geqslant \ldots \geqslant j_k$.

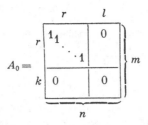

b) $i_r \geqslant j_r \geqslant 0$ for all r $(1 < r \leqslant k)$ when $j_1 > 0$.

EXAMPLE 5. For $k = 1$ we have $I = i$, $\mu(i) = i$, so that $\nu_I(m, n) = (n - m + i)i$, that is, formula (*) for the product of coranks.

EXAMPLE 6. For $I = \underbrace{1, 1, \ldots, 1}_{k}$ we have $\mu(1, 1, \ldots, 1) = k$, $\nu_I(m, n) = (n - m + 1)k$.

In particular, the Whitney singularity of class $\Sigma^{1,1}$ has codimension 2 when $m = n$, and is therefore non-removable for mappings of the plane to the plane at discrete points, for mappings $R^3 \to R^3$ on curves, and so on.

B. Morin [6] has considered good mappings of class $\Sigma^{1,1,\ldots,1,0}$ for all m, n. It turns out that they are always stable if $m \geqslant \nu_I = (n - m + 1)k$, and completely characterized by their own class. For instance, the following assertions are equivalent for the mapping $f\colon R^n \to R^n$:

a) $x \in \Sigma^{\overbrace{1,1,\ldots,1,0}^{n}}$

b) The germ of f at x is equivalent to that of the "generalized Whitney mapping"

$$y_1 = x_1,$$
$$\cdots$$
$$y_{n-1} = x_{n-1},$$
$$y_n = x_1 x_n + x_2 x_n^2 + \ldots + x_{n-1} x_n^{n-1} + x_n^{n+1}.$$

EXAMPLE 7. For $I = i, j$ we have $\mu(i, j) = i(1 + j) - \dfrac{j(j-1)}{2}$, from which we get the formula found earlier by Levine,

$$\nu_{i,j}(m, n) = (n - m + i)\,i + \frac{j}{2}\left[(n - m + i)(2i - j + 1) - 2i + 2j\right].$$

In particular, for $m = n$,

$$\nu_{i,j}(n, n) = i^2 + \frac{j}{2}\left[2i^2 - ij + 2j - i\right] = i^2 + j\left[i^2 - \frac{j(i-1)}{2}\right].$$

Hence it follows that *a singularity of class* $\Sigma^{i,j}$ *first appears as non-removable for* $m = n$ *if*

i,j	1, 0	1, 1	2, 0	2, 1	2, 2	3, 0	3, 1	3, 2	3, 3	4, 0	4, 1	4, 2	4, 3	4, 4
$\nu_{i,j} = n$	1	2	4	7	10	9	16	22	27	16	29	40	49	56

EXAMPLE 8. For $m = n \leqslant 16$ the following classes are realized as points (for good mappings):

$n=\nu_I$	n	4	7	9	10	13	15	16
I	1_n	2	2, 1	3	2, 2 2, 1_2	2, 1_3	2, 2, 1	3, 1 4 2, 1_4

where 1_n denotes $\underbrace{1, 1, \ldots, 1}_{n}$.

BOARDMAN'S DEFINITION OF Σ^I. We first give the definition in non-invariant terms, using local coordinates $x_1, \ldots, x_m$ in M^m and $y_1, \ldots, y_n$ in N^n.

DEFINITION 4. Let B be an ideal in the ring A of germs of infinitely differentiable functions $\varphi(x_1, \ldots, x_m)$. The ideal $\Delta_k(B)$ of A generated by B and the Jacobians $\det\left|\frac{\partial\varphi_i}{\partial x_j}\right|$, consisting of partial derivatives of functions in B, is called a *Jacobian extension of* B.

REMARK. The Jacobian extension is invariant, that is, it does not depend on the coordinate system $x_1, \ldots, x_m$. This is because $\frac{\partial\varphi}{\partial x'} = \frac{\partial\varphi}{\partial x}\frac{\partial x}{\partial x'}$, and the determinant is multilinear.

LEMMA 2. $\Delta_{k+1}(B) \subset \Delta_k(B)$ *for* $k = 1, 2, \ldots$.

This follows on expanding determinants by rows.

DEFINITION 5. A Jacobian extension $\Delta_k(B)$ is called *critical* if $\Delta_k(B) \neq A$, $\Delta_{k-1}(B) = A$.

In other words, for a critical extension the order of the adjoint minors is the least order for which the extension does not coincide with the whole ring.

EXAMPLE 9. Take $m = 4$, and let B be the ideal generated by x_1, x_2, x_3^2, x_4^2. Then the critical Jacobian extension $\Delta_3(B)$ is generated by x_1, x_2, x_3, x_4. For this ideal $\Delta_3(B)$ the critical extension is the fifth: $\Delta_5\Delta_3(B) = \Delta_3(B)$.

EXAMPLE 10. Take $m = 1$ and let B be the ideal generated by x^3. We have the sequence of critical extensions: $\Delta_1(B) = Ax^2$, $\Delta_1\Delta_1(B) = Ax$, $\Delta_2\Delta_1\Delta_1(B) = Ax$.

For convenience we use a different numbering for extensions in what follows.

Notation: $\Delta^k = \Delta_{m-k+1}$. Thus, in Example 9 the critical extensions are Δ^2 and $\Delta^0\Delta^2$; in Example 10 they are Δ^1, $\Delta^1\Delta^1$ and $\Delta^0\Delta^1\Delta^1$.

Let I be a set of integers $i_1 \geqslant i_2 \geqslant \ldots \geqslant i_k$.

DEFINITION 6. Suppose that the germ of the mapping $f\colon M^m \to N^n$ is given in coordinates x, y by the formulae $y_i = f_i(x)$, $f(0) = 0$. *We say that f has a singularity of class Σ^I at 0 if the successive critical Jacobian extensions of the ideal generated by the functions $f_i(x)$, $i = 1, \ldots, n$, are the $\Delta^{i_k}\Delta^{i_{k-1}} \ldots \Delta^{i_1}$.*

EXAMPLE 11. The mapping $y = x^3$ has a singularity of class Σ^{110} at 0; $y = x^{k+1}$ has a singularity of class $\Sigma^{1_k,0}$. The mapping $y_1 = x_1x_2$, $y_2 = x_1^2 - x_2^2$ and the mapping of Example 4 have singularity of class $\Sigma^{2,0}$.

REMARK. It is clear that the definition given above imposes a restriction only on the coefficients of the Taylor expansion of orders up to k inclusive. It is easy to check that, in fact, *the conditions do not depend on the coordinate system* and are imposed only on the k-jet $j_0^k(f)$. The set of all k-jets satisfying these conditions also defines the intersection of $\Sigma^I \subset J^k(M, N)$ with the fibre bundle $J^k(M, N) \to M \times N$.

§3. The quadratic differential

The rank of the first differential gives rise to the singularity classes Σ^I. An investigation of the quadratic part of the mapping yields a finer classification: with each singularity we associate, in an invariant manner, its family of quadratic forms. The second differential is defined uniquely only on the kernel of the first differential, and only to within the image of the first. Therefore we define the quadratic differential of the mapping $f: M^m \to N^n$ at x in M^m to be a quadratic[1] mapping

$$f_x\colon \operatorname{Ker}(f_x) \longrightarrow \operatorname{Coker}(f_x),$$

where $\operatorname{Ker} f_x \subset TM_x$ is the kernel of the first differential $f_x: TM_x \to TN_{f(x)}$, and $\operatorname{Coker} f_x = TN_{f(x)}/f_x TM_x$ is its cokernel.

Firstly we define f_{xx} by means of local coordinates

$$X\colon TM_x^m \longrightarrow M^m, \qquad Y\colon TN_{f(x)}^n \longrightarrow N^n,$$

where

$$X(0)=x,\ Y(0)=f(x),\ \frac{d}{dt}\Big|_{t=0} X(\xi t)=\xi, \qquad \frac{d}{dt}\Big|_{t=0} Y(\xi t)=\xi.$$

In these coordinates f takes the form

$$\varphi\colon TM_x \longrightarrow TN_{f(x)}, \quad \text{where} \quad \varphi = Y^{-1}\circ f \circ X.$$

DEFINITION 1. The value of f_{xx} at ξ in $\operatorname{Ker} f_x$ is

$$f_{xx}(\xi) = \lim_{t\to 0} \frac{\varphi(t\xi)}{t^2} \Big/ f_* TM_x \in \operatorname{Coker} f_x.$$

LEMMA 1. *The quadratic differential f_{xx} is independent of the choice of local coordinates X, Y.*

The proof is clear from Taylor's formula (Fig. 7),

$$\varphi(\eta) = \frac{\partial\varphi}{\partial\xi}\Big|_0 \eta + \frac{1}{2}\left(\frac{\partial^2\varphi}{\partial\xi^2}\Big|_0 \eta, \eta\right) + O(\eta^3).$$

If $\eta \in \operatorname{Ker} f_x$, $\eta = O(t)$, $\zeta = O(t^2)$, then

$$\varphi(\eta+\zeta) = f_x(\eta+\zeta) + \frac{1}{2}\left(\frac{\partial^2\varphi}{\partial\xi^2}\Big|_0 \eta, \eta\right) + \left(\frac{\partial^2\varphi}{\partial\xi^2}\Big|_0 \eta, \zeta\right) + \frac{1}{2}\left(\frac{\partial^2\varphi}{\partial\xi^2}\Big|_0 \zeta, \zeta\right) + \ldots =$$

$$= f_x\zeta + \frac{1}{2}\left(\frac{\partial^2\varphi}{\partial\xi^2}\Big|_0 \eta, \eta\right) + O(t^3)$$

---

1 A mapping $a\colon A \to B$ of linear space is said to be *quadratic* if there is symmetric bilinear mapping $a'\colon A + A \to B$ such that $a = a' \circ \Delta$, where Δ is the diagonal mapping $\Delta\colon A \to A + A$, $\Delta(x) = (x, x)$.

and so on. At the same time we have proved that, for a local system of coordinates in which $\eta_1, \ldots, \eta_k$ are the coordinates of $\eta \in \operatorname{Ker} f_x$ and $\varphi_1, \ldots, \varphi_l$ the coordinates in Coker f_x, the quadratic differential is given by the formula

$$(f_{xx}(\eta))_i = \frac{1}{2} \sum_{j,k} \frac{\partial^2 \varphi_i}{\partial \xi_j \, \partial \xi_k} \eta_j \eta_k.$$

EXAMPLE 1. For the mapping in Example 4 of §2, the quadratic differential is given by f_{xx}: $x_3, x_4 \to x_3^2 \pm x_4^2, x_3 x_4$.

REMARK. The *cubic* differential cannot be defined in a similar way. To give a definition of invariant differentials of higher orders we have to iterate a construction of Porteous, which associates with each mapping g: $F_1 \to F_2$ of vector bundles over M an invariantly defined "inner derivative" at x in M, (dg): $TM_x \to \operatorname{Hom}(\operatorname{Ker} g|_x, \operatorname{Coker} g|_x)$.

In particular, with the differential f_x we associate the mapping $TM \to f^*TN$ of the tangent bundle of M to the inverse image of the tangent bundle of N. The derivative (dg) defines a bilinear mapping

$$TM_x \times \operatorname{Ker} g|_x \to \operatorname{Coker} g|_x,$$
$$(\xi, \eta) \to (dg)(\xi)\eta,$$

which reduces to the quadratic differential when $\xi = \eta$:

$$dg(\eta, \eta) = f^* 2 f_{xx}(\eta).$$

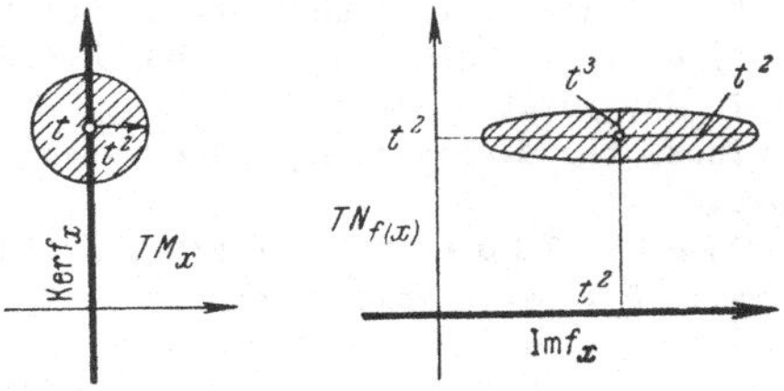

Fig. 7.

The quadratic differential f_{xx} is associated with a bundle L of quadratic forms, in an invariant fashion. Denote by F the linear space of all real quadratic forms on Ker f_x, and by C' the dual space of Coker f_x. To every form

$$\alpha: \operatorname{Coker} f_x \to \mathbf{R}^1, \qquad \alpha \in C'$$

there corresponds a quadratic form $\alpha \circ f_{xx} \in F$.

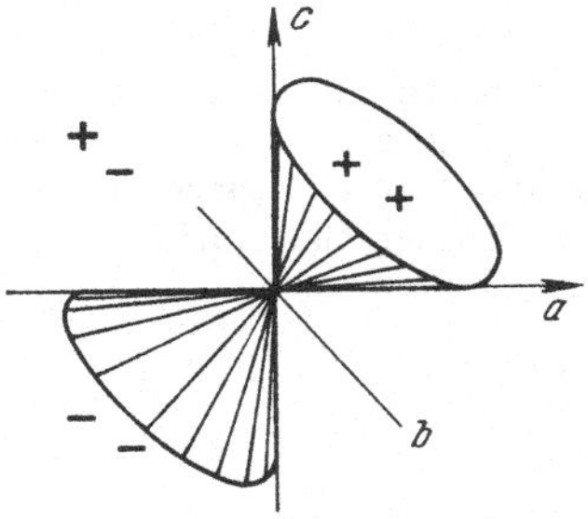

Fig. 8.

DEFINITION 2. *The linear mapping of the cokernel of f_x into the space of quadratic forms on the kernel of f_x defined by $L(\alpha) = \alpha \circ f_{xx}$, L: $C' \to F$, is called the bundle of quadratic forms corresponding to f_{xx}.*

EXAMPLE 2. Suppose that Ker f_x and Coker f_x both have dimension 2. The space F of quadratic forms in two variables is 3-dimensional, and L is a mapping of the two-dimensional plane C' into the 3-dimensional space F. This space F has an additional structure defined by classifying forms according to their indices of inertia. Let x_1, x_2 be coordinates in Ker f_x. We take coordinates in F so that the point with coordinates a, b, c means the form

$$ax_1^2 + 2bx_1x_2 + cx_2^2.$$

Forms of rank 1 (parabolic) form the cone $b^2 = ac$ (Fig. 8). The vertex is the null form. The insides of the two halves of the cone contain elliptic forms of type + + and − −, and the outside contains hyperbolic forms of type + −.

The bundle L is represented by the subspace $L(C')$ of F, and this subspace can be situated in one of seven ways:

1. A plane lying entirely outside the cone (all forms hyperbolic). Example: $\alpha_1(x_1^2 - x_2^2) + \alpha_2 x_1 x_2$.
2. A plane intersecting the cone (two parabolic forms in the bundle). Example: $\alpha_1 x_1^2 + \alpha_2 x_2^2$.
3. A plane tangent to the cone (one parabolic form). Example: $\alpha_1 x_1^2 + \alpha_2 x_1 x_2$.
4. A line inside the cone. Example: $\alpha_1(x_1^2 + x_2^2) + \alpha_2 \cdot 0$.
5. A line outside the cone. Example: $\alpha_1 x_1 x_2 + \alpha_2 \cdot 0$.
6. A line tangent to the cone. Example: $\alpha_1 x_1^2 + \alpha_2 \cdot 0$.
7. The point 0 $(\alpha_1 \cdot 0 + \alpha_2 \cdot 0)$.

Thus, *the disposition of the subspace $L(C') \subset F$ is an invariant of the singularity.*

DEFINITION 3. Let $F(\mathbf{R}^k)$ stand for the space of all quadratic forms in k indeterminates, and let

$$H(c, k) = \mathrm{Hom}(\mathbf{R}^c, F(\mathbf{R}^k))$$

be the linear space of all bundles of c quadratic forms in k variables.

The group $GL(c, \mathbf{R}) \times GL(k, \mathbf{R})$ acts on $H(c, k)$ in a natural way, according to the formula

$$g_c \times g_k L(\xi_c)\,\xi_k = L(g_c^{-1}\xi_c)\, g_k^{-1}\xi_k,$$
$$\xi_c \in \mathbf{R}^c, \quad \xi_k \in \mathbf{R}^k, \quad g_c \in GL(c, \mathbf{R}), \quad g_k \in GL(k, \mathbf{R}).$$

EXAMPLE 3. $H(2, 2)$ is 6-dimensional and splits into 7 orbits under the action of $GL(2, \mathbf{R}) \times GL(2, \mathbf{R})$, as listed in Example 2.

We recall now that with every germ of a mapping $f\colon M \to N$ there is associated a quadratic differential, and with the latter there is associated a bundle of quadratic forms

$$L_f \in \mathrm{Hom}(C', F(K)), \quad C' = \mathrm{Coker}\, f_x, \quad K = \mathrm{Ker}\, f_x.$$

If we *choose* the identifications

$$C' \approx \mathbf{R}^c, \ c = \dim \mathrm{Coker}\, f_x, \quad K \approx \mathbf{R}^k, \quad k = \dim \mathrm{Ker}\, f_x,$$

then L_f corresponds to an element of $H(c, k)$. A change in these identifications of linear spaces gives rise to an operation of $GL(c, \mathbf{R}) \times GL(k, \mathbf{R})$ on $H(c, k)$, as described above. So we have proved:

LEMMA 2. *The construction described associates with each germ of a mapping $f\colon M \to N$ at x an orbit of the action of $GL(c, \mathbf{R}) \times GL(k, \mathbf{R})$ on the space $H(c, k)$, where $c = \dim \mathrm{Coker}\, f_x$, $k = \dim \mathrm{Ker}\, f_x$; and the correspondence is invariant under diffeomorphisms of M and N.*

COROLLARY 1. *Consider the two mappings $f_\pm\colon \mathbf{R}^4 \to \mathbf{R}^4$ given by the* *formulae of Example 4 in* §2:

$$\begin{pmatrix} x_1 \\ x_2 \\ x_3 \\ x_4 \end{pmatrix} \to \begin{pmatrix} x_1 \\ x_2 \\ x_3^2 \pm x_4^2 + x_1x_3 + x_2x_4 \\ x_3x_4 \end{pmatrix}.$$

Their germs at 0 *are not equivalent, that is, there do not exist germs of diffeomorphisms* h, k *with* $h(0) = 0$, $k(0) = 0$ *for which the diagram*

$$\begin{array}{ccc} \mathbf{R}^n & \xrightarrow{f_+} & \mathbf{R}^n \\ h\downarrow & & \downarrow k \\ \mathbf{R}^n & \xrightarrow{f_-} & \mathbf{R}^n \end{array}$$

is commutative.

PROOF. In this case $c = k = 2$, and we have the conditions of Example 2. The bundles $L_\pm$ are given by the formulae

$$\alpha_1(x_3^2 \pm x_4^2) + \alpha_2 x_3 x_4$$

and define a plane in the (a, b, c) space. In the case of f_- this plane lies outside the cone, hence f_- corresponds to the first of the orbits in Example 2. For f_+ the plane intersects the cone, so that f_+ corresponds to the second orbit in that Example. These orbits are different, hence the germs of f_+ and f_- are not equivalent, as we wanted to show.

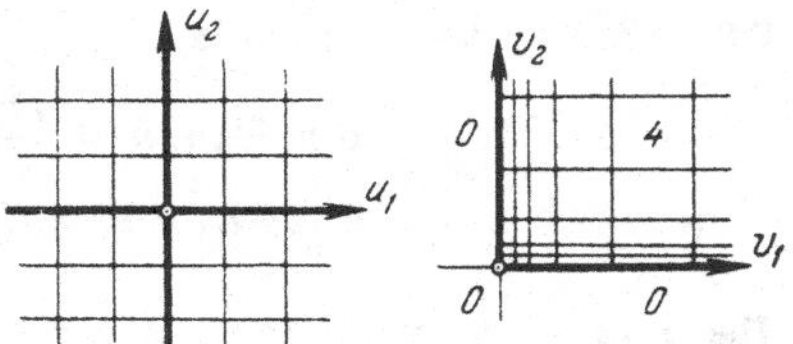

Fig. 9.

REMARK 1. It is easily checked that f_+ and f_- have an unstable singularity of type Σ^{20} at 0. Therefore the classification by the Σ^I is incomplete.

REMARK 2. It is easy to check that f_+ and f_- are distinct in the topological as well as the differentiable sense. To understand their structure better we may regard them as mappings of the plane

$$\begin{matrix} x_3 \\ x_4 \end{matrix} \to \begin{matrix} x_3^2 \pm x_4^2 + x_1x_2 + x_2x_4, \\ x_3x_4, \end{matrix}$$

depending on parameters x_1, x_2. For $x_1 = x_2 = 0$ we get either the complex mapping $z \to z^2$ (for f_-) or a "corner" (for f_+) equivalent to the mapping

$$\begin{matrix} u_1 \\ u_2 \end{matrix} \to \begin{matrix} u_1^2 \\ u_2^2 \end{matrix} = \begin{matrix} v_1, \\ v_2 \end{matrix} \qquad \text{(see fig. 9).}$$

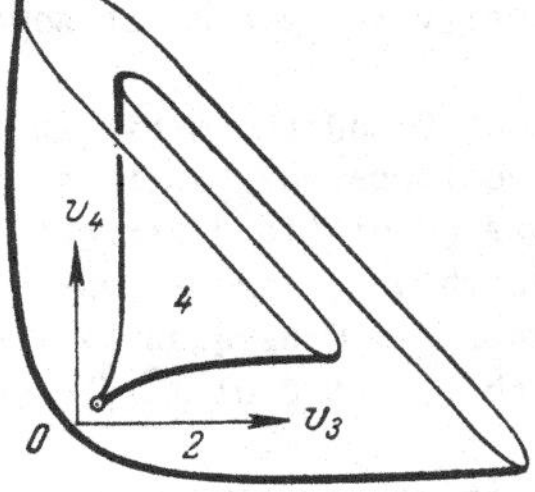

Fig. 10.

Even from this it is clear that f_+ and f_- are topologically inequivalent; the image of f_- covers $\mathbf{R}^4$, whereas that of f_+ does not. For small x_1, x_2 we get nearby mappings (i) in the case of f_- as in Example 6 of §1, (ii) in the case of f_+ as described in Fig. 10.

REMARK 3. Later (§6) we shall prove that *the germs of f_+ and f_- are analytically stable; it can be shown that every mapping $M^4 \to N^4$ may be approximated by a mapping whose germ at each point is equivalent to one of the 7 stable germs given by the formulae:*

$$\Sigma^0:\ y_i = x_i,\quad i = 1, \ldots, 4,$$
$$\Sigma^{10}:\ y_i = x_i,\quad i = 1, \ldots, 3,\quad y_4 = x_4^2,$$
$$\Sigma^{110}:\ y_i = x_i,\quad i = 1, \ldots, 3,\quad y_4 = x_1x_4 + x_4^3,$$
$$\Sigma^{1110}:\ y_i = x_i,\quad i = 1, \ldots, 3,\ y_4 = x_1x_4 + x_2x_4^2 + x_4^4,$$
$$\Sigma^{11110}:\ y_i = x_i,\quad i = 1, \ldots, 3,\quad y_4 = x_1x_4 + x_2x_4^2 + x_3x_4^3 + x_4^5,$$
$$f_+ = \Sigma_+^{20}:\ y_i = x_i,\quad i = 1, 2,\quad y_3 = x_3^2 + x_4^2 + x_1x_3 + x_2x_4,\quad y_4 = x_3x_4,$$
$$f_- = \Sigma_-^{20}:\ y_i = x_i,\quad i = 1, 2,\quad y_3 = x_3^2 - x_4^2 + x_1x_3 + x_2x_4,\quad y_4 = x_3x_4.$$

We note another consequence of the lemma:

THEOREM 1 (Thom [3]). *The set of stable mappings $M^{n^2} \to N^{n^2}$ is not everywhere dense in the space of all smooth mappings for $n \geqslant 3$.*

The proof is based on the following remark.

LEMMA 3. *The codimension of every orbit of $GL(n, \mathbf{R}) \times GL(n, \mathbf{R})$ in the space $H(n, n)$ of bundles of n quadratic forms in n variables is positive for $n \geqslant 3$.*

PROOF. We have, clearly,

$$\dim H(n, n) = \frac{n^2(n+1)}{2},$$
$$\dim GL(n, \mathbf{R}) \times GL(n, \mathbf{R}) = 2n^2.$$

There is a one-dimensional subgroup (the scalars) leaving all points of H fixed. Therefore the codimension of each orbit is not less than

$$\frac{n^2(n+1)}{2} - (2n^2 - 1) \geqslant 1 \quad \text{for} \quad n \geqslant 3.$$

This proves the lemma.

Consider now a mapping $f\colon M^{n^2} \to N^{n^2}$ having a transversal singularity at 0 of type Σ^n. By the formula for the product of coranks, $\Sigma^n(f)$ has codimension n^2 and every nearby mapping has a singularity of type Σ^n at some nearby point.

Next we examine the quadratic differential f_{xx} at 0 and the orbit in $H(n, n)$ corresponding to it. Since this orbit has codimension $\geqslant 1$, there is a mapping $\tilde{f}$ in an arbitrary neighbourhood of f whose quadratic differential at a point of $\Sigma^n(\tilde{f})$ corresponds to *another* orbit (such an $\tilde{f}$ can be constructed easily, by changing f only on a jet of order 2). Consequently, the germ of every mapping $f\colon M^{n^2} \to N^{n^2}$ is unstable at the point 0 of $\Sigma^n(f)$, and this proves the theorem.

REMARK 4. Let us consider the "plane" of pairs of natural numbers (m, n). Then (m, n) lies in the "region of stability" if every mapping $\mathbf{R}^m \to \mathbf{R}^n$ can be approximated by a stable mapping.

Thus all points $(m, 1)$ lie in the region of stability (Morse, Theorem 1 of §1) as do the point $(2, 2)$ (Whitney, Example 5) and all points (m, n) with $n > 2m$ (Whitney's embedding theorem). On the other hand, as we have seen, (9, 9) lies in the region of instability. Mather announced in the autumn of 1965 that the boundary of the region of stability is as shown in Fig. 11.

REMARK 5. We can restate the theorem mentioned above as the assertion that the differentiable singularities of mappings $M^{n^2} \to N^{n^2}$ for large n have "moduli" (that is, invariants changing continuously with the mapping). For example, it can be seen from our proof that there is at least one modulus for $n \geqslant 3$.

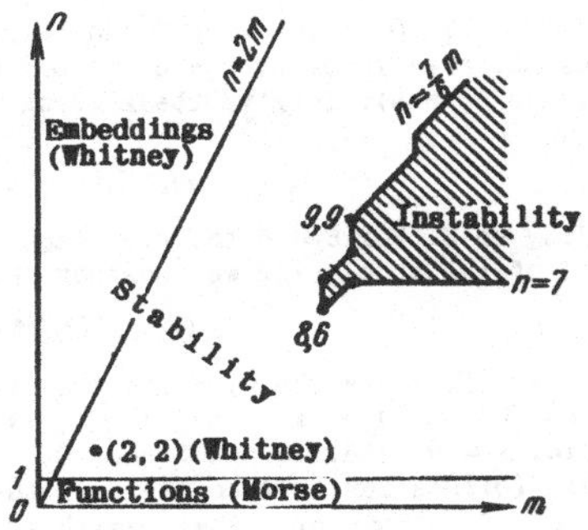

Fig. 11.

Thom has shown that for sufficiently large n the number of moduli is infinite, that is, the space of non-equivalent differentiable singularities is infinite-dimensional.

§4. The local ring of a singularity and the Weierstrass preparation theorem

The local ring of a singularity is a very powerful invariant. This ring is "a ring of functions on infinitesimal pre-images of a point" and is defined as follows.

Let f be the germ at 0 of a smooth map

$$f: \mathbf{R}^m \to \mathbf{R}^n, \quad y = f(x), \quad x \in \mathbf{R}^m, \quad y \in \mathbf{R}^n, \quad f(0) = 0.$$

DEFINITION 1. *The local ring* Q of the germ of f at 0 is the ring

$$Q = \mathscr{E}(x) / f^* [\mathfrak{M}\mathscr{E}(y)] \mathscr{E}(x),$$

where

$\mathscr{E}(x)$ is the ring of germs of smooth functions $\varphi(x)$ at 0,

$\mathscr{E}(y)$ is the ring of germs of smooth functions $\varphi(y)$ at 0,

$\mathfrak{M}\mathscr{E}(y) \subset \mathscr{E}(y)$ is the maximal ideal of $\mathscr{E}(y)$, consisting of germs $\varphi(y)$ vanishing at 0, $\varphi(0) = 0$,

$f^* [\mathfrak{M}\mathscr{E}(y)] \subset \mathscr{E}(x)$ is the image of the maximal ideal in the ring of functions of x, that is, the set of germs of the form $\varphi(f(x))$, $\varphi \in \mathfrak{M}\mathscr{E}(y)$,

$f^* [\mathfrak{M}\mathscr{E}(y)] \mathscr{E}(x)$ is the ideal generated by this last set in $\mathscr{E}(x)$.

EXAMPLE 1. Consider the mapping $\mathbf{R}^1 \to \mathbf{R}^1$ given by $y = x^2$. In this case

$$\mathfrak{M}\mathscr{E}(y) = y\mathscr{E}(y), \quad f^* [\mathfrak{M}\mathscr{E}(y)] = x^2 \mathscr{E}(x^2), \quad f^* [\mathfrak{M}\mathscr{E}(y)] \mathscr{E}(x) = x^2 \mathscr{E}(x).$$

Thus, Q is the two-dimensional local ring

$$Q = \mathscr{E}(x) / x^2 \mathscr{E}(x) = \mathbf{R}[x] / x^2 \mathbf{R}[x]$$

of linear functions $a_0 + a_1 x$ (with multiplication according to the rule $x^2 = 0$).

MOTIVATION. The ring Q can be regarded as the ring of functions on infinitesimal pre-images of $y = 0$ (in our example, the ring of functions on two coincident points). To give these words an exact meaning we consider the family of mappings

$$f_\varepsilon\colon \mathbf{R}^1 \to \mathbf{R}^1, \quad y = x^2 - \varepsilon^2.$$

depending on ε. For $\varepsilon \neq 0$ the pre-images of $y = 0$ consist of two points $x = \pm\,\varepsilon$. Instead of the local ring we consider the analogous ring

$$Q_\varepsilon = F(x)/f_\varepsilon^*[\mathfrak{M}F(y)]\,F(x),$$

where $F(x)$, $F(y)$ are the rings of infinitely differentiable functions of x and y.

PROPOSITION 1. *The ring Q_ε is a ring of functions on the pre-images of the point $y = 0$, that is, on ε and $-\varepsilon$.*

This follows easily from the fact that the ring $f_\varepsilon^*[\mathfrak{M}F(y)]\,F(x)$ is the ring of all smooth functions of x that vanish at the points $x = \pm\,\varepsilon$. It follows from Proposition 1 that every element of Q_ε can be regarded as a linear function $a_0 + a_1 x$. As we have seen, the local ring of Example 1 also consists of linear functions of x.

PROPOSITION 2. *As $\varepsilon \to 0$, the multiplication in Q_ε tends to that in the local ring Q.*

This follows from the fact that both pre-images $x = \pm\,\varepsilon$ tend to 0 as $\varepsilon \to 0$. Thus, *the local ring Q is the limit of the rings Q_ε of functions on the inverse image of $f_\varepsilon^{-1}(0)$, where the pre-images merge as $\varepsilon \to 0$.* In particular, it is clear from this that it is no accident that the dimension of Q is the number of pre-images in the more general situation.

EXAMPLE 2. Consider the mapping $\mathbf{R}^1 \to \mathbf{R}^1$, $y = x^n$. Here

$$\mathfrak{M}\mathscr{E}(y) = y\mathscr{E}(y), \quad f^*[\mathfrak{M}\mathscr{E}(y)] = x^n\mathscr{E}(x^n), \quad f^*[\mathfrak{M}\mathscr{E}(y)]\,\mathscr{E}(x) = x^n\mathscr{E}(x).$$

Therefore Q is the n-dimensional ring

$$Q = \mathscr{E}(x)/x^n\mathscr{E}(x) = \mathbf{R}[x]/x^n\mathbf{R}[x]$$

of polynomials $a_0 + a_1 x + \ldots + a_{n-1}x^{n-1}$, with multiplication determined by $x^n = 0$. (This is called the ring of *truncated* polynomials). The dimension n is the multiplicity of the pre-image of 0.

EXAMPLE 3. This ring has a generalized Whitney singularity:

$$f\colon \mathbf{R}^{n-1} \to \mathbf{R}^{n-1}, \quad y_i = x_i \ (i = 1, 2, \ldots, n-2),$$
$$y_n = x_1 x_{n-1} + x_2 x_{n-1}^2 + \ldots + x_{n-2}x_{n-1}^{n-2} + x_{n-1}^n.$$

EXAMPLE 4. The two mappings $f_\pm\colon \mathbf{R}^4 \to \mathbf{R}^4$ of class Σ^{20} (Example 4 of §2),

$$\begin{matrix} x_1 & & x_1, \\ x_2 & \longrightarrow & x_2, \\ x_3 & & x_3^2 \pm x_4^2 + x_4 x_3 + x_2 x_4, \\ x_4 & & x_3 x_4 \end{matrix}$$

have *non-isomorphic* 4-dimensional rings

$$Q = \mathbf{R}[x_3, x_4]/\{x_3, x_4, x_3^2 \pm x_4^2\}\,\mathbf{R}[x_3, x_4].$$

The equivalence classes of the monomials 1, x_3, x_4, x_3^2, for instance, can be taken as four linear generators for Q. The multiplication in Q then takes the form $x_3 x_4 = 0$, $x_4^2 = \mp\, x_3^2$, $x_3^3 = 0$, $x_4^3 = 0$.

REMARK. If the germs of f_1 and f_2 at 0 are differentiably equivalent, then the rings $Q_1 = \mathscr{E}_1/I_1$ and $Q_2 = \mathscr{E}_2/I_2$ are equivalent in the sense that there is a commutative diagram

$$\begin{array}{ccccccccc} 0 & \longrightarrow & I_1 & \longrightarrow & \mathscr{E}_1 & \longrightarrow & Q_1 & \longrightarrow & 0 \\ & & \downarrow\uparrow & & \downarrow\uparrow & & \downarrow\uparrow & & \\ 0 & \longrightarrow & I_2 & \longrightarrow & \mathscr{E}_2 & \longrightarrow & Q_2 & \longrightarrow & 0. \end{array}$$

Up to equivalence the ring Q represents the strongest invariant of a singularity.

Mather has announced the following results:

1. *A stable singularity is uniquely determined by its ring; that is, if the germs f_1 and f_2 are stable and the rings Q_1 and Q_2 are equivalent, then f_1 and f_2 are differentiably equivalent:*

$$\begin{array}{ccc} M & \xrightarrow{f_1} & N \\ h\downarrow & & \downarrow k \\ M & \xrightarrow{f_2} & N \end{array}$$

where h and k are germs of diffeomorphisms.

2. *If the germ f: $\mathbf{R}^m \to \mathbf{R}^n$, $n > m$, lies in Σ^{20} at 0, then the local ring Q is equivalent to one of the following:*

$\mathrm{I}_{a,b}$ $\mathbf{R}[[x, y]]/\{xy, x^a + y^b\}$, $b \geqslant a \geqslant 2$,
$\mathrm{II}_{a,b}$ $\mathbf{R}[[x, y]]/\{xy, x^a - y^b\}$, $b \geqslant a \geqslant 2$ even,
$\mathrm{III}_{a,b}$ $\mathbf{R}[[x, y]]/\{x^a, y^b, xy\}$, $n > m$,
IV_a $\mathbf{R}[[x, y]]/\{x^2 + y^2, x^a\}$, $a \geqslant 3$,
V_a $\mathbf{R}[[x, y]]/\{x^a, y^b, x_y\}$, $a \geqslant 3$, $n > m$.

The codimensions of the corresponding singularities can also be computed:

$$\operatorname{codim} \mathrm{I}_{a,b} = \operatorname{codim} \mathrm{II}_{a,b} = (a+b-1)(m-n+1)+1,$$
$$\operatorname{codim} \mathrm{III}_{a,b} = (a+b-2)(n-m+2)+2,$$
$$\operatorname{codim} \mathrm{IV}_a = (2a+1)(n-m+1)+1,$$
$$\operatorname{codim} \mathrm{V}_a = (2a+2)(n-m+2)+2.$$

EXAMPLES 5. A stable singularity of the mapping f: $\mathbf{R}^n \to \mathbf{R}^n$, $y = f(x)$, with local ring of type I_{ab} (II_{ab}), $n = a + b$, is given by the formulae

$$y_i = x_i \ (i = 1, \ldots, a-1), \ y_i' = x_i' \ (i = 1, \ldots, b-1), \ y_a = x_a x_b',$$

$$y_b' = x_1 x_a + x_2 x_a^2 + \ldots + x_{a-1} x_a^{a-1} + x_a^a + x_1' x_b' + x_2' x_b'^2 + \ldots + x_{b-1}' x_b'^{b-1} \pm x_b'^b$$

for a suitable choice of coordinates

$$x = x_1, \ldots, x_{a-1}, x_a, x_1', \ldots, x_{b-1}', x_b'; \ y = y_1, \ldots, y_{a-1}, y_a, y_1', \ldots, y_{b-1}', y_b'.$$

The stability of this singularity will be proved later, in §6.

The fundamental technical tool for working with singularities of smooth mappings is the Weierstrass "preparation theorem" and its generalizations. As before, let f: $\mathbf{R}^m \to \mathbf{R}^n$, $y = f(x)$, $f(0) = 0$, be the germ of a smooth mapping.

THEOREM 1. *Suppose that the local ring Q is finite-dimensional as a real linear space, and let $e_1(x), \ldots, e_r(x)$ in $\mathscr{E}(x)$ be representatives of generators of Q. Then every germ $\varphi(x)$ in $\mathscr{E}(x)$ has a representation of the form*

$$\varphi(x)=\varphi_1(y)e_1(x)+\ldots+\varphi_r(y)e_r(x),\ y=f(x),\ \varphi_k\in\mathscr{E}(y),$$

that is, $\mathscr{E}(x)$ *is a finite-dimensional* $\mathscr{E}(y)$ *-module.*

This is the form in which Malgrange [7], [8] proved the theorem. This proof, which is very complicated, was recently simplified by Mather [9]. An analytic version of the theorem (see [10]) preceded Malgrange's paper.

Let $f\colon \mathbf{C}^m \to \mathbf{C}^n$, $y = f(x)$, $f(0) = 0$, be the germ of an analytic mapping at 0, $\mathscr{H}(x)$ the ring of germs of functions of x analytic at 0, $\mathscr{H}(y)$ the same thing for y.

THEOREM 2. *If the local ring*

$$Q=\mathscr{H}(x)/(f^*[\mathfrak{M}\mathscr{H}(y)])\mathscr{H}(x)$$

is finite-dimensional over $\mathbf{C}$ *and* $e_1(x), \ldots, e_r(x)$ *denote representatives of generators of* Q *in* $\mathscr{H}(x)$, *then every germ* $\varphi(x)$ *in* $\mathscr{H}(x)$ *has a representation of the form*

$$\varphi(x)=\varphi_1(y)e_1(x)+\ldots+\varphi_r(y)e_r(x),\ y=f(x),\ \varphi_k\in\mathscr{H}(y),$$

that is, $\mathscr{H}(x)$ *is a finite-dimensional* $\mathscr{H}(y)$*-module.*

This theorem will be useful in the sequel; we give a proof below.

APPENDIX

A PROOF OF THE PREPARATION THEOREM

1. First of all we make some preliminary remarks. Let Q be an r-dimensional local ring of mappings $y = f(x)$, $f(0) = 0$.

LEMMA 1. *The r-th power of every element of the maximal ideal of* Q *is 0:* $\mathfrak{M}^r=0$.

PROOF. Take a in $\mathfrak{M}$; for $\lambda \neq 0$, $(a-\lambda e)^{-1}$ *exists*. We interpret multiplication by a as an operator $A\colon Q \to Q$, and show that all eigenvalues of A are zero. Let ξ be an eigenvector, $a\xi = \lambda\xi$, $\lambda \neq 0$. Then $(a - \lambda e)\xi = 0$. Multiplying by $(a - \lambda e)^{-1}$ we get $\xi = 0$, contrary to assumption. Therefore $\lambda = 0$, hence $A^r = 0$, so that $a^r = 0$ and the lemma is proved.

COROLLARY 1. *Consider the germs* $x_1, \ldots, x_m$ *of local coordinates in* $\mathbf{C}^m$ *at 0. Then there exist m decompositions*

$$x_i^r=\sum_{j=1}^{n} y_j\lambda_{ji}(x),\ \lambda_{ji}(x)\in\mathscr{H}(x). \tag{1}$$

PROOF. The class of x_i in Q lies in the maximal ideal. By the lemma, the class of x_i^r is zero, as required.

To (1) we add the obvious decompositions

$$x^{\mathbf{k}}=\sum_{i=1}^{r} a_{\mathbf{k}i}e_i(x)+\sum_{j=1}^{n} y_j\lambda_{j\mathbf{k}}(x),\ \lambda_{j\mathbf{k}}\in\mathscr{H}(x),\ a_{\mathbf{k}i}\in\mathbf{C}, \tag{2}$$

where $x^{\mathbf{k}}$ denotes the monomial $x_1^{k_1}x_2^{k_2}\ldots x_m^{k_m}$ and all $k_i < r$. There are only finitely many such monomials.

2. Formal preparation theorem. Every germ $\varphi \in \mathscr{H}(x)$ is given by a power series

$$\varphi(x) = \sum_{\mathbf{k} \geqslant 0} \varphi_{\mathbf{k}} x^{\mathbf{k}}, \quad x^{\mathbf{k}} = x_1^{k_1} \ldots x_m^{k_m}. \tag{3}$$

Our immediate aim is to construct power series for the $\varphi_i(y)$ such that, formally[1]

$$\varphi(x) = \varphi_1(y) e_1(x) + \ldots + \varphi_r(y) e_r(x), \text{ where } y = f(x). \tag{4}$$

With each power series (3) we associate the polynomial

$$\rho[\varphi] = \sum_{\substack{k_i < r \\ i=1, \ldots, m}} \varphi_{\mathbf{k}} x^{\mathbf{k}}. \tag{5}$$

We divide the remainder of the series by x_i^r. To this end we use the notation

$$\tau_i[\varphi] = \sum_{k_i \geqslant r} \varphi_{\mathbf{k}} \frac{x^{\mathbf{k}}}{x_i^r} \quad (i = 1, \ldots, m), \tag{6}$$

and divide φ by x_1^r, the remainder by x_2^r, and so on:

$$\sigma_i[\varphi] = \tau_i\left[\varphi - \sum_{j=1}^{i-1} x_j^r \sigma_j[\varphi]\right], \; \sigma_1 = \tau_1 \; (i = 1, \ldots, m). \tag{7}$$

Then

$$\varphi = \rho[\varphi] + \sum_{i=1}^{m} \sigma_i[\varphi] x_i^r. \tag{8}$$

Every monomial $\varphi_{\mathbf{k}} x^{\mathbf{k}}$ for which $k_1, \ldots, k_{i-1} < r$, $k_i \geqslant r$, gives a contribution $\varphi_{\mathbf{k}} \dfrac{x^{\mathbf{k}}}{x_i^r}$ to $\sigma_i[\varphi]$; if all $k_i < r$, then $\varphi_{\mathbf{k}} x^{\mathbf{k}}$ occurs in $\rho[\varphi]$.

We transform $\rho[\varphi]$ and x_i^r by means of (1) and (2). We get

$$\varphi = p[\varphi] + \sum_{j=1}^{n} y_j s_j[\varphi], \tag{9}$$

where

$$p[\varphi] = \sum_{i=1}^{r} p_i[\varphi] e_i, \quad p_i[\varphi] = \sum_{\substack{k_l < r \\ l=1, \ldots, m}} \varphi_{\mathbf{k}} a_{\mathbf{k}i} \tag{10}$$

---

1 Here $f(x)$ can be understood as a formal power series and the $e_i(x)$ as formal power series in x representing the generators of the ring

$$\hat{Q} = \Phi(x) / f^*[\mathfrak{M}\Phi(y)]\,\Phi(x),$$

where $\Phi(x)$ ($\Phi(y)$) is the ring of formal power series in x (in y). The formal Weierstrass theorem will then be proved; *if a formal mapping f has a finitely generated ring $\hat{Q}$ and $e_i(x)$ are representatives of the generators, then (4) is valid for any $\varphi \in \Phi(x)$.*

and

$$s_j[\varphi]=\sum_{\substack{k_l<r\\ l=1,\ldots,m}}\varphi_{\mathbf{k}}\lambda_{j\mathbf{k}}(x)+\sum_{i=1}^{m}\sigma_i[\varphi]\lambda_{jl}(x)\ (j=1,\ \ldots,\ n). \tag{11}$$

Iterating (9) we arrive

$$\varphi=p[\varphi]+\sum_{j=1}^{n}y_jps_j[\varphi]+\sum_{j_1,\,j_2=1}^{n}y_{j_1}y_{j_2}s_{j_2}s_{j_1}[\varphi]$$

and generally for each l,

$$\varphi=p[\varphi]+\sum_{j_1=1}^{n}y_{j_1}ps_{j_1}[\varphi]+\sum_{j_1,\,j_2=1}^{n}y_{j_1}y_{j_2}ps_{j_2}s_{j_1}[\varphi]+\ldots$$
$$\ldots+\sum_{j_1,\,\ldots,\,j_l=1}^{n}y_{j_1}\ldots y_{j_l}ps_{j_l}\ldots s_{j_1}[\varphi]+$$
$$+\sum_{j_1,\,\ldots,\,j_l,\,j_{l+1}=1}^{n}y_{j_1}\ldots y_{j_l}y_{l+1}s_{j_{l+1}}s_{j_l}\ldots s_{j_1}[\varphi]. \tag{12}$$

Collecting the coefficients of e_i we get finally the required expression (4) in the form $\varphi=\sum_{i=1}^{r}\varphi_i(y)e_i$, where the power series

$$\varphi_i(y)=\sum_{l=0}^{\infty}\sum_{j_1,\,\ldots,\,j_l=1}^{n}\varphi_{ij_1\ldots j_l}y_{j_1}\ldots y_{j_l} \tag{13}$$

has coefficients

$$\varphi_{ij_1\ldots j_l}=p_is_{j_l}\ldots s_{j_1}[\varphi]. \tag{14}$$

This proves the formal theorem.

3. Investigation of convergence. If φ is the germ of an analytic function, then by Cauchy's inequality there exist $C>0$, $N>0$ such that

$$|\varphi_{\mathbf{k}}|<CN^{|\mathbf{k}|},\ |\mathbf{k}|=k_1+\ldots+k_m. \tag{15}$$

This is expressed by the symbol $\varphi\prec(C,N)$. The inequality (15) is satisfied by some $C=C(N)$ if $N>1/R$, where R is the radius of convergence. The expressions (1) and (2) are fixed and finite in number.

Suppose that

$$|a_{\mathbf{k}i}|<A,\ A>0. \tag{16}$$

Further we may assume that

$$\lambda_{j\mathbf{k}}\prec\left(1,\ \frac{1}{2}\right),\ \lambda_{jl}(x)\prec\left(1,\ \frac{1}{2}\right) \tag{17}$$

(this can be achieved by a suitable choice of the scales of x and y). From From (15), (16), (7) we find that

$$\varphi\prec(C,N)\Rightarrow\sigma_i[\varphi]\prec(CN^r,\ N), \tag{18}$$

while from (15), (16), (5), (10),

$$\varphi\prec(C,N)\Rightarrow|p_i[\varphi]|<A_1\cdot ACN^r, \tag{19}$$

where $A_1=A_1(r,\ m)>0$. Furthermore it is easy to prove:

LEMMA 2. *Suppose that* $\varphi \prec (1, \theta)$, $\psi \prec (C, N)$, $0 < \theta < 1 < N$. *Then*

$$\varphi\psi < (A_2C, N), \text{ where } A_2 = A_2(\theta, m) > 0. \tag{20}$$

PROOF OF LEMMA. If

$$\varphi = \sum_{\mathbf{k}\geqslant 0} \varphi_{\mathbf{k}} x^{\mathbf{k}}, \ |\varphi_{\mathbf{k}}| < \theta^{|\mathbf{k}|}, \quad \psi = \sum_{\mathbf{k}\geqslant 0} \psi_{\mathbf{k}} x^{\mathbf{k}}, \ |\psi_{\mathbf{k}}| < CN^{|\mathbf{k}|},$$

then

$$\varphi\psi = \sum_{\mathbf{k}\geqslant 0} (\varphi\psi)_{\mathbf{k}} x^{\mathbf{k}},$$

where

$$|(\varphi\psi)_{\mathbf{k}}| = \Big|\sum_{0\leqslant \varkappa \leqslant \mathbf{k}} \varphi_{\varkappa}\psi_{\mathbf{k}-\varkappa}\Big| \leqslant \sum_{0\leqslant \varkappa \leqslant \mathbf{k}} \theta^{|\varkappa|} CN^{|\mathbf{k}-\varkappa|} \leqslant$$

$$\leqslant \sum_{0\leqslant l\leqslant |\mathbf{k}|} \theta^l CN^{|\mathbf{k}|}(l+1)^m \leqslant CN^{|\mathbf{k}|} \sum_{0\leqslant l\leqslant |\mathbf{k}|} \theta^l (l+1)^m \leqslant CN^{|\mathbf{k}|} A_2(\theta, m),$$

where this time

$$A_2 = \sum_{0\leqslant l\leqslant\infty} \theta^l (l+1)^m < \infty,$$

as we wanted to show.

Next, we find from (11), (17), (18), (20) that

$$|(s_j[\varphi])_{\mathbf{k}}| \leqslant r^m CN^r + mA_2CN^rN^{|\mathbf{k}|},$$

that is,

$$\varphi \prec (C, N) \Rightarrow s_j[\varphi] \prec (A_3CN^r, N), \text{ where } A_3(r, m) = mA_2\left(\tfrac{1}{2}, {}^r m\right) + r^m. \tag{21}$$

Repeating (21) we get

$$s_{j_l} \ldots s_{j_1}[\varphi] \prec (C(A_3N^r)^l, N). \tag{22}$$

It follows from (14), (22) and (19) that

$$|\varphi_{ij_1 \ldots j_l}| \leqslant A_1 \cdot A \cdot N^r (A_3N^r)^l C. \tag{23}$$

Rewrite (13) in the form

$$\varphi_i(y) = \sum_{\lambda} \varphi_{i\lambda} y^{\lambda}, \ y^{\lambda} = y_1^{\lambda_1} \ldots y_n^{\lambda_n}, \ |\lambda| = l. \tag{24}$$

Every monomial y^{λ} can be written in the form $y_{j_1} \ldots y_{j_l}$ in not more than n^l ways. It is then clear from (13) and (23) that, for any $i = 1, \ldots, r$,

$$|\varphi_{i\lambda}| \leqslant n^l A_1 A N^r (A_3N^r)^l C < DM,$$

where $D = A_1AN^rC$, $M = nA_3N^r$. Thus the series (24) represents the germ of an analytic function $\varphi_i(y) \in \mathscr{H}(y)$, as required.

Chapter 2

DEFORMATIONS OF SINGULARITIES

§5. "Infinite-dimensional Lie groups" acting on "infinite-dimensional manifolds"

We want to discuss the infinite-dimensional analogue of the following situation. Let G be a Lie group, M a smooth manifold, and let $G \times M \to M$ be a given operation of G on M. The point m_* of M is called *stable* if every point m of M sufficiently near it is the image of m_* under some element of G, that is, if the orbit Gm_* of m_* contains a neighbourhood of m_* in M. Further, m_* is said to be *infinitesimally stable* if every point m of M sufficiently near m_* is the image of m_* under an infinitely small displacement in G. In other words, the mapping L of the tangent space of G at the identity into the tangent space of M at m_* given by

$$L\colon TG_e \to TM_{m_*}, \quad L = \left.\frac{\partial gm}{\partial g}\right|_{e,\, m_*} \tag{1}$$

must be a mapping *onto*. We could also say that m_* is infinitesimally stable if the operator L has a (one-sided) inverse

$$\overline{L}\colon TM_{m_*} \to TG_e, \quad L\overline{L} = E. \tag{2}$$

It follows quickly from the implicit function theorem that

THEOREM 1 (on stability). *Every infinitesimally stable point is stable.*

Here I have in mind a proof not depending on the implicit function theorem that lends itself better to a generalization to the infinite-dimensional case.

Suppose that coordinate systems have been introduced in a neighbourhood of the identity of G and in a neighbourhood of m_* in M. Then we can identity the points g in G and m in M in our neighbourhoods with elements of the linear spaces TG_e and TM_{m_*}. The action of G on M is then given by the formulae

$$gm = m + Lg + R(g, m), \tag{3}$$

where $L = L(m_*)$ is the differential of the action given by (1) and R is a remainder term of second order of smallness:

$$|R(g, m)| \leqslant C_1 |g| (|g| + |m|). \tag{4}$$

In these terms stability of m_* means the solubility of the equation $gm = 0$ in g for sufficiently small m.

We solve this equation by successive approximations

$$g_s = -\overline{L}m_{s-1}, \quad m_s = g_s m_{s-1} = R(g_s, m_{s-1}) \quad (s = 1, 2, \ldots), \ m_0 = m. \tag{5}$$

The solution is given by the infinite product

$$g = \lim_{s\to\infty} g_s \ldots g_2 g_1. \tag{6}$$

The convergence of the approximations follows for sufficiently small m_0 from the estimates

$$|g_s| < C_2 |m_{s-1}|, \quad |m_{s+1}| < C_3 |m_s|^2 \quad (C_2 = \|\overline{L}\|, \quad C_3 = C_1 C_2 (1 + C_2)), \tag{7}$$

which come from (4) and (5). It is clear from (7) that the sequence $|m_s|$ decreases faster than any geometric progression for sufficiently small $|m_0|$, so that the product (6) converges to a solution of the equation $gm = 0$.

The advantage of applying these arguments (over applying the implicit function theorem) to the solution of $m + Lg + R(g, m) = 0$ lies in the fact that *we use the value of the inverse operator L only at the single point m_** (whereas the value of L is needed in a neighbourhood of m_* for the implicit function theorem). Of course, in the finite-dimensional case the difference is inessential; on the other hand, in infinite-dimensional cases it turns out to be impossible, or very difficult, to evaluate L in a neighbourhood of m_*.

The stability theorem admits of a generalization to the infinite-dimensional case when the g and m in (3) are elements of function spaces. A similar generalization has essentially been used in the paper [11] of H. Cartan, and particularly in Kolmogorov [12]. An abstract formulation of the conditions of this generalization would be too cumbersome[1] because algebraic and topological notions turn up which have not yet been given names. For this reason I have restricted myself to an indication of the character of the restrictions imposed, and to a detailed discussion of an example in §6 and §7.

The most essential condition is that the operator L must have *finite order*. Suppose, for example, that we are considering spaces of analytic functions. Here L will be of finite order ν if it takes functions m analytic in a neighbourhood of a domain U to functions analytic in a domain V such that for any $\delta > 0$,

$$|\bar{L}m|_V \leqslant C|m|_{U+\delta}\delta^{-\nu}, \tag{8}$$

where $U + \delta$ denotes the δ-neighbourhood[2] of U.

It turns out that *if the operator L in* (3) *has finite order* (and satisfies certain other conditions whose formulation would take up too much space, but which are always satisfied in natural questions of analysis) *then the successive approximations converge for sufficiently small $m = m_0$ to a solution of the equation $gm = 0$.*

We mention now some analytic questions whose solutions arise as special cases of the infinite-dimensional stability theorem not stated above.

EXAMPLE 1. *The stability of the germ of an analytic mapping from one complex space to another.*

Let

$$y = f(x), \quad f(0) = 0, \quad x \in X, \quad y \in Y, \tag{9}$$

---

1 And would certainly not lie on the Procrustean bed of the theory of infinite-dimensional manifolds with Banach charts.

2 For instance, the operator $\frac{d^\nu}{dz^\nu}$ has order ν, as can be seen from Cauchy's formula

$$\frac{d^\nu}{dz^\nu} m(z) = \frac{\nu!}{2\pi i} \oint_{|\zeta - z| = \delta} \frac{m(\zeta)\, d\zeta}{(\zeta - z)^{\nu+1}}.$$

be an analytic mapping of a domain in an n-dimensional complex affine space X into another such space Y. The germ of f at 0 can be regarded as a point in the "infinite-dimensional manifold" M of germs of mappings. An "infinite-dimensional local Lie group G" acts on M, namely the direct product of local Lie groups of analytic mappings of domains of X into X, and of domains of Y into Y. The elements $h \times k$ of G operate on M as change of coordinates in image and in pre-image:

$$(h \times k) f = k \circ f \circ h^{-1}.$$

The question of the stability of (9) is that of the stability of f in M under the action of the "group" G.

In connection with the general stability theorem it is sufficient to have infinitesimal stability and the estimate (2) for the inverse operator. In this case it turns out that it is enough for infinitesimal stability to know that the following n linear equations in $2n$ germs at 0 of the analytic $n \times n$ matrices $H_i(x)$, $K_i(y)$ have a solution:

$$x_i E = \frac{\partial f}{\partial x} H_i(x) + K_i(f(x)) \quad (E \text{ the } n \times n \text{ unit matrix}. \tag{10}$$

Here the inverse operator $\overline{L}$ has finite order and we obtain

COROLLARY 1. *If the n equations* (10) *are soluble, then f is stable at* 0.

A similar proof is given in §§6 and 7. Mather proved an analogous theorem in [13] for infinitely differentiable functions.

EXAMPLE 2. *Every analytic function is equivalent to a polynomial in the neighbourhood of an isolated critical point.*

Let

$$f\colon \ \mathbf{C}^n \to \mathbf{C}^1, \quad f(0) = 0, \quad df|_0 = 0,$$

be the germ of an analytic function having 0 as an isolated critical point. The group of germs at 0 of bi-analytic diffeomorphisms $g\colon \mathbf{C}^n \to \mathbf{C}^n$, $g(0) = 0$, operates on the space of functions $f\colon \mathbf{C}^n \to \mathbf{C}^1$ as a group of change of variables:

$$gf = f \circ g.$$

Let M be the "manifold" consisting of functions f having a fixed jet of order k at 0:

$$M = \{f\colon j_0^k(f) = j\}.$$

Consider the subgroup of the group of germs of diffeomorphisms leaving fixed a k-jet of f at 0:

$$j_0^k(gf) = j_0^k(f).$$

This subgroup G contains, for example, all germs that coincide with the identity up to order k: $g(x) - x = o(|x|^k)$. The "infinite-dimensional Lie group G" acts on the "infinite-dimensional manifold M".

Suppose now that 0 is an *isolated* critical point. Then, for sufficiently large k, the point f in M is infinitesimally stable. Furthermore, the inverse operator $\overline{L}$ turns out to have finite order, and we get from the general theorem:

COROLLARY 2. *If 0 is an isolated critical point of a complex analytic function f, then an analytic system of coordinates can be chosen in a neighbourhood of 0 such that f is a polynomial in those coordinates.*

See §8 for the proof (and the dissertation [14] of Tougeron for another proof).

EXAMPLE 3. *A normal form for the germ of an analytic mapping of $\mathbf{C}^n$ into itself at a fixed point.*

Let

$$m\colon z \to \mu z + \ldots, \quad \mu = dm|_{z=0}, \tag{11}$$

be the germ of an analytic mapping of complex n-dimensional space $\mathbf{C}^n$ into itself. Clearly m leaves 0 fixed. We ask whether it is possible to "change coordinates"

$$g\colon z \to w = z + \ldots \tag{12}$$

in the neighbourhood of 0 in such a way that m is brought to linear normal form

$$g \circ m \circ g^{-1}\colon w \to \mu w.$$

This is a question of the stability of the point m of the "infinite-dimensional manifold" M of germs of type (11); the "infinite-dimensional group" G of germs of the form (12) acts on M as a group of change of variables:

$$gm = g \circ m \circ g^{-1}.$$

Straightforward calculation shows that the point μ is infinitesimally stable and the inverse operator $\bar{L}$ has finite order if the eigenvalues μ_i of the endomorphism μ satisfy the condition

$$\Big|\mu_i - \prod_{j=1}^{n} \mu_j^{k_j}\Big| > C|k|^{-\nu} \text{ for all integers } k_j \geqslant 0, \quad |k| = \sum_{j=1}^{n} k_j \geqslant 2. \tag{13}$$

From the general theorem we get:

COROLLARY 3. *If the eigenvalues of μ satisfy* (13), *then* (11) *reduces to linear normal form under a bi-analytic change of coordinates.*

This result is contained in C.L. Siegel's book [15] and is very non-trivial even for $n = 1$. The same method can be used to give his results on normal linear form for systems of analytic ordinary differential equations in the neighbourhood of a fixed point, or of periodic motion (see [16]); and also some of the results of [17] on normal forms for analytic differential equations on the torus.

EXAMPLE 4. *Decomposition of analytic matrix functions.*

Let $D = D_1 \cap D_2$ be the intersection of two sufficiently good bounded domains in the complex z-plane. We consider the "infinite-dimensional manifold" M of non-degenerate matrix functions analytic in D, $m\colon D \to GL(n, \mathbf{C})$. The "infinite-dimensional Lie groups" of analytic non-degenerate matrix functions in D_1 and in D_2 act on M by left and right multiplication:

$$(g_1 \times g_2)\, m = g_1 m g_2^{-1}, \qquad g_1\colon D_1 \longrightarrow GL(n, \mathbf{C}), \qquad g_2\colon D_2 \longrightarrow GL(n, \mathbf{C}).$$

It turns out that the unit matrix $m = E$ is infinitesimally stable with respect to such an operation; from the general theorem we get

COROLLARY 4. *For any two closed domains $F_1 \subset D_1$, $F_2 \subset D_2$, every matrix function $m(z)$ sufficiently near a constant in D can be represented on the intersection $F_1 \cap F_2$ in the form of a product $m(z) = g_1(z) g_2^{-1}(z)$ of matrices analytic in both domains: $g_1\colon F_1 \to GL(n, \mathbf{C})$, $g_2\colon F_2 \to GL(n, \mathbf{C})$.*

This result is due to H. Cartan [11].

§6. The stability theorem

We consider an analytic mapping f of a neighbourhood of 0 in an n-dimensional complex affine space X into another such space Y:

$$y=f(x),\quad x=x_1,\ \dots,\ x_n;\quad y=y_1,\ \dots,\ y_n,\quad f=f_1,\ \dots,\ f_n;\quad f(0)=0. \quad (1)$$

We recall that f is said to be *(analytically) stable* at 0 if every nearby mapping f_0 is equivalent to it. To be precise, if for every neighbourhood U of 0 in X and every mapping f_0: $U \to Y$ sufficiently near f there exist neighbourhoods U', W of 0, $0 \in U' \subset U \subset X$, $0 \in W \subset f(U) \subset Y$ and analytic diffeomorphic embeddings h: $U' \to U$, k: $W \to Y$ near to the identity and carrying f_0 to f so that in U',

$$k^{-1}\circ f_0\circ h=f, \qquad \begin{array}{ccc} X\supset U\supset U' & \xrightarrow{f} & W\subset Y \\ h\downarrow & & \downarrow k \\ U & \xrightarrow{f_0} & Y \end{array} \quad (2)$$

Here nearness is to be understood relative to the topology given by these neighbourhoods of zero

$$U(\varepsilon,\ r)=\{\varphi\colon \max_{|x|\leq r}|\varphi(x)|<\varepsilon\}$$

in the space of analytic functions.

The main result of this chapter is the following.

THEOREM 1 (on stability[1]). *The mapping* (1) *is stable at* 0 *if in some neighbourhood of* $x = 0$ *there exist* n *decompositions*

$$x_iE=f'H_i+K_i \qquad (i=1,\ \dots,\ n), \quad (3)$$

where E *is the* $n\times n$ *unit matrix;* $f'(x) = \dfrac{\partial f}{\partial x}$ *is the Jacobian matrix of* f*;* $H_i(x)$, $K_i(y)$ *are* $n \times n$ *matrices analytic in neighbourhoods of* $x = 0$ *(of* $y = 0$*); and* $y = f(x)$.

EXAMPLE 1. The mapping $y = x^2$ of the line is stable at 0, because for any $\varphi(x)$ there exists a decomposition of the form

$$\varphi(x)=2xh(x)+k(x^2).$$

On the other hand, the mapping $y = x^3$ is unstable at 0, because there is no decomposition $x = 3x^2h(x) + k(x^3)$ of the type (3).

EXAMPLE 2. Consider the Whitney cusp Σ^{110}:

$$y_1=x_1x_2-\frac{1}{3}x_1^3,\quad y_2=x_2,\quad n=2, \quad (4)$$

where to get decompositions (3) we may take

$$H_1=\begin{vmatrix}0 & 1\\ 1 & x_1\end{vmatrix},\quad -K_1=\begin{vmatrix}0 & y_2\\ 1 & 0\end{vmatrix},\quad H_2=0,\quad K_2=y_2E.$$

---

[1] The infinite-dimensional analogue of this theorem has been proved recently by Mather [13].

COROLLARY 1. *The mapping* (4) *is stable at* 0 (Whitney's Theorem [2]).

EXAMPLE 3. Consider the more general singularity of class $\Sigma^{1_{n},0}$: $C^n \to C^n$:

$$\left.\begin{array}{l} y_1 = x_1x_2 - x_1^2x_3 + x_1^3x_4 - \ldots + (-1)^n x_1^{n-1}x_n + \frac{1}{n+1}(-x_1)^{n+1}, \\ y_2 = x_2, \ \ldots, \ y_n = x_n. \end{array}\right\} \qquad (5)$$

To obtain decompositions (3) in this case it is sufficient to take $H_i = 0$,

$$H_i = 0, \quad K_i = y_iE \quad \text{for} \quad i > 1,$$

$$H_1 = \begin{vmatrix} 0 & & & & 1 \\ 1 & x_1 & & & 2x_3 \\ & 1 & x_1 & & 3x_4 \\ \ldots & \ldots & \ldots & \ldots & \ldots \\ & & 1 & x_1 & (n-1)x_n \\ & & & 1 & x_1 \end{vmatrix} \qquad -K_1 = \begin{vmatrix} 0 & & & & y_2 \\ 1 & 0 & & & 2y_3 \\ & 1 & 0 & & 3y_4 \\ \ldots & \ldots & \ldots & \ldots & \ldots \\ & & 1 & 0 & (n-1)y_n \\ & & & 1 & 0 \end{vmatrix}.$$

COROLLARY 2. *The mapping* (5) *is stable at* 0 (theorem of B. Morin [6] and Jo Ging-tzung [18]).

EXAMPLE 4. Consider a non-parabolic mapping of class $\Sigma^{2 0}$, $C^4 \to C^4$:

$$y_1 = x_1, \quad y_2 = x_2, \quad y_3 = x_3x_4, \quad y_4 = \frac{1}{2}(x_3^2 \pm x_4^2) + x_1x_3 + x_2x_4. \qquad (6)$$

For the matrices in (3) we can take

$$H_i = 0, \quad K_i = y_iF \qquad (i = 1, 2),$$

$$H_3 = \begin{vmatrix} x_1 + x_3 & 0 & 0 & 1 \\ 0 & x_3 & \mp 1 & 0 \\ -x_3 & 0 & 0 & 0 \\ 0 & 0 & 1 & 0 \end{vmatrix}, \quad K_3 = \begin{vmatrix} -y_1 & 0 & 0 & -1 \\ 0 & 0 & \pm 1 & 0 \\ y_3 & 0 & 0 & 0 \\ 0 & -y_3 & -y_2 & 0 \end{vmatrix},$$

and similar expressions for H_4, K_4.

COROLLARY 3. *The singularity* (6) *is stable at* 0 (see Whitney [26]).

REMARK 1. If f and the decompositions (3) are real, then f is real analytically stable. In particular, this applies to the mappings (4), (5), (6) of the preceding examples.

REMARK 2. In fact, we prove a stronger assertion than that of the stability theorem. For example, every mapping f_0 having tangency of sufficiently high order with f at 0 is already sufficiently near to f for the existence of a commutative diagram (2). Thus, in particular, such a mapping is equivalent to a polynomial, namely to a segment of its Taylor series.

The proof of the stability theorem is given below. Condition (3) is nothing more than the condition of infinitesimal stability of f at 0 (see §5).

INFINITESIMAL STABILITY. We recall that a mapping f is said to be infinitesimally stable if every mapping f_0 "infinitely near" to f is the image of f under "infinitely small" diffeomorphisms h, k. To write down the corresponding formulae, we consider three arbitrary mappings

$$f_0\colon\ x \to f(x)+\boldsymbol{\varphi}(x), \quad h\colon\ x \to x+\mathbf{h}(x), \quad k\colon\ y \to y+\mathbf{k}(y).$$

Here $\boldsymbol{\varphi}(x)$, $\mathbf{h}(x)$, $\mathbf{k}(y)$ are sets of germs of n analytic functions, which we shall (illegally) call "vector fields".

LEMMA 1. *The mapping $k \circ f \circ h$ is given by the formula*

$$\left.\begin{aligned} &k\circ f_0\circ h\colon x \to f(x)+\overline{\boldsymbol{\varphi}}(x), \ \text{где}\ \overline{\boldsymbol{\varphi}}(x)=\Sigma_0+\Sigma_1+\Sigma_2+\Sigma_3,\\ &\Sigma_0=\boldsymbol{\varphi}(x)+f'(x)\mathbf{h}(x)+\mathbf{k}(f(x)),\\ &\Sigma_1=f(x+\mathbf{h}(x))-f(x)-f'(x)\mathbf{h}(x),\\ &\Sigma_2=\boldsymbol{\varphi}(x+\mathbf{h}(x))-\boldsymbol{\varphi}(x),\\ &\Sigma_3=\mathbf{k}(f(x+\mathbf{h}(x))+\boldsymbol{\varphi}(x+\mathbf{h}(x)))-\mathbf{k}(f(x)).\end{aligned}\right\} \tag{7}$$

The proof is by an obvious substitution.

If we take $\boldsymbol{\varphi}$, $\mathbf{h}$, $\mathbf{k}$ to have values of the first order of smallness then, as is clear from (7), Σ_0 is also of the first order of smallness, and Σ_1, Σ_2, Σ_3 of the second order. Thus for the infinitesimal stability of f the equation $\Sigma_0 = 0$ must have a solution in $\mathbf{h}$, $\mathbf{k}$.

DEFINITION 1. *A mapping f is said to be infinitesimally stable at 0 if for every vector field $\boldsymbol{\varphi}(x)$ analytic at 0 there exist analytic vector fields $\mathbf{h}(x)$ and $\mathbf{k}(y)$ at 0 such that $\boldsymbol{\varphi}$ is expressed by the formula*

$$-\boldsymbol{\varphi}(x)=f'(x)\mathbf{h}(x)+\mathbf{k}(f(x)) \tag{8}$$

in some neighbourhood of $x = 0$.

We denote by V a sufficiently small neighbourhood of 0 in Y and by U the component of 0 in the pre-image $f^{-1}V$. Let $\mathcal{L}$ be an operator from the space of functions of x analytic in U into the space of functions analytic in x (or in y) in the closure of U (or of V).

DEFINITION 2. *The operator $\mathcal{L}$ has degree p and order ν if for any function φ,*

$$\|\mathcal{L}\boldsymbol{\varphi}\|_{U\ (or\ V)} < C(\|\varphi\|_{U+\delta})^{p}\,\delta^{-\nu} \tag{9}$$

for arbitrary δ, $0 < \delta < \delta_0$.

Here $\|\ \|_D = \sup_{z\in D}$, and $U + \delta$ is the δ-neighbourhood of U.

Now we state a theorem concerning the solubility of equation (8).

THEOREM 2 (on infinitesimal stability). *If there exist n decompositions* (3), *then f is infinitesimally stable. Furthermore, there exist linear operators $\mathbf{H}$ and $\mathbf{K}$ of degree 1 and of finite order ν with vector values and vector arguments,*

$$\mathbf{H}[\boldsymbol{\varphi}(x)]=\mathbf{h}(x), \quad \mathbf{K}[\boldsymbol{\varphi}(x)]=\mathbf{k}(y),$$

satisfying (8):

$$-\boldsymbol{\varphi}=f'\mathbf{H}[\boldsymbol{\varphi}]+K[\boldsymbol{\varphi}].$$

Here the order ν does not exceed $r - 1$, where r is the multiplicity of f at 0.

The proof of this theorem is given on pages 31 – 33. Using the theorem on infinitesimal stability it is easy to prove the theorem of stability by the successive approximations that are described in abstract form in §5. In our case these approximations come to the following.

PROOF OF THEOREM 1 (on stability). Suppose that $y = f_0(x)$, $f_0(x) = f(+) + \varphi(x)$, is a mapping near $y = f(x)$. We want to establish the equivalence of f and f_0 and introduce the following notation for the remainder term in (7) for $\Sigma_0 = 0$:

$$R[\varphi, \mathbf{h}, \mathbf{k}] = \Sigma_1 + \Sigma_2 + \Sigma_3, \tag{10}$$

where the expressions for Σ_1, Σ_2, Σ_3 are indicated in (7). We set

$$\mathbf{R}[\varphi] = R[\varphi, H[\varphi], K[\varphi]], \tag{11}$$

where H and K are the operators in the statement of the theorem on infinitesimal stability.

Starting with φ_0, we define inductively vector fields

$$\varphi_s = \mathbf{R}[\varphi_{s-1}], \qquad \mathbf{h}_s = \mathbf{H}[\varphi_{s-1}], \qquad \mathbf{k}_s = \mathbf{K}[\varphi_{s-1}], \tag{12}$$

and mappings

$$h_s\colon x \longrightarrow x + \mathbf{h}_s(x), \qquad k_s\colon y \longrightarrow y + \mathbf{k}_s(y), \qquad f_s\colon x \longrightarrow f(x) + \varphi_s(x), \tag{13}$$

and further mappings

$$h^s = h_1 \circ h_2 \circ \ldots \circ h_s, \qquad k^s = k_s \circ \ldots \circ k_1. \tag{14}$$

Finally we consider mappings whose convergence will be proved in §7:

$$h = \lim_{s \to \infty} h^s, \quad k = \lim_{s \to \infty} k^s. \tag{15}$$

Note that by Lemma 1

$$k_s \circ f_{s-1} \circ h_s = f_s, \tag{16}$$

(because $\Sigma_0 = 0$ by definition of H and K). And so, by (14), (16),

$$k^s \circ f_0 \circ h^s = f_s. \tag{17}$$

Since $\lim\limits_{s \to \infty} f_s = f$, we get from (15), (17) that

$$k \circ f_0 \circ h = f, \tag{18}$$

which proves the stability theorem.

PROOF OF THE THEOREM ON INFINITESIMAL STABILITY. We shall prove that the existence of n decompositions as in (3) is equivalent to infinitesimal stability.

LEMMA 1. *The mapping $y = f(x)$ is infinitesimally stable at 0 if and only if for every function $\varphi(x)$ analytic at 0 there exist $n \times n$ matrices $H(x)$ and $K(y)$ analytic at 0 such that*

$$\varphi(x) E = f'H + K(f(x)), \text{ E the unit $n \times n$ matrix.} \tag{1}$$

PROOF. Suppose that the decompositions (1) exist, and that

$$\varphi(x) = \sum_{i=1}^{n} \varphi_i(x) \mathbf{e}_i.$$

is the coordinate representation of the field φ. Then

$$-\varphi = f'\mathbf{h} + \mathbf{k}, \quad \text{where } \mathbf{h} = -\sum_{i=1}^{n} \mathbf{h}_i, \quad \mathbf{k} = -\sum_{i=1}^{n} \mathbf{k}_i, \tag{2}$$

where $\mathbf{h}_i$ and $\mathbf{k}_i$ are the *i*-th columns of the matrices *H*, *K*, respectively, that arise in $\varphi_i E$. Conversely, suppose that a decomposition (8) exists. Then

$$-\varphi(x)\mathbf{e}_i = f'\mathbf{h}_i + \mathbf{k}_i \qquad (i = 1, \ldots, n).$$

Therefore the matrices *H*, *K* with the columns $-\mathbf{h}_i$, $-\mathbf{k}_i$ satisfy (1), as we wanted to show.

In particular, if *f* is infinitesimally stable at 0, then there exist *n* decompositions (3). The converse is also true: having *n* such decompositions, one can construct an expression (1) for any function φ(*x*) and so, by Lemma 1, all the decompositions (8).

LEMMA 2. *If the functions φ_1 and φ_2 have decompositions like* **(1),**

$$\varphi_1 E = f'H_1 + K_1, \quad \varphi_2 E = f'H_2 + K_2,$$

then their product also has such a decomposition:

$$\varphi_1\varphi_2 E = f'H_{12} + K_{12},$$

where

$$H_{12} = \varphi_1 H_2 + H_1 K_2, \quad K_{12} = K_1 K_2.$$

The proof of Lemma 2 is straightforward. According to the lemma we derive from (3) that there is a decomposition (1) for any polynomial φ(*x*). On passage to the limit it then follows that we can construct a decomposition (1) for any analytic function φ(*x*); however, insofar as we need estimates for *H* and *K* in terms of φ, it is convenient to derive (1) by other means, namely by using the Weierstrass preparation theorem.

LEMMA 3. *If decompositions* **(3)** *exist, then f has finite multiplicity at* **0.**

LEMMA 4. *If f is an analytic function of finite multiplicity r at* **0,** *then every function φ(x) analytic at* **0** *can be written as a sum of r terms,*

$$\varphi(x) = \sum_{k=1}^{r} \psi_k(y)\, e_k(x), \quad y = f(x), \tag{3}$$

where $e_1(x), \ldots, e_r(x)$ are monomials in $x_1, \ldots, x_n$ not depending on φ, and $\psi_1(y), \ldots, \psi_r(y)$ are functions of y analytic at **0.** *Further, there exist (linear) operators $\mathbf{W}_k$ of degree* **1** *and finite order* **r − 1** *for which*

$$\psi_k = \mathbf{W}_k[\varphi] \quad (k = 1, \ldots, r). \tag{4}$$

The proof of Lemma 3 is given below. Lemma 4 represents an analytic

refinement of the Weierstrass preparation theorem and was proved[1] in [19].

Theorem 2 of §6 on infinitesimal stability follows easily from Lemmas 1 – 4. By Lemma 3, the multiplicity r is finite. By Lemma 4 there exist monomials e_k and operators $\mathbf{W}_k$. Using Lemma 2 and starting from the decompositions (3), we decompose the monomials e_k like this:

$$e_k E = f' \mathscr{H}_k(x) + \mathscr{K}_k(y) \qquad (k = 1, \ldots, r). \tag{5}$$

Next we define operators $\mathscr{H}$ and $\mathscr{K}$, acting on functions $\varphi(x)$ and taking values that are $n \times n$ matrices analytic in x (or y), by the following relations:

$$\mathscr{H}[\varphi] = \sum_{k=1}^{r} \mathbf{W}_k[\varphi]\mathscr{H}_k, \quad \mathscr{K}[\varphi] = \sum_{k=1}^{r} \mathbf{W}_k[\varphi]\mathscr{K}_k.$$

Then in view of (3) – (5),

$$\varphi(x)E = f'\mathscr{H}[\varphi](x) + \mathscr{K}[\varphi](y). \tag{6}$$

Relation (6) reduces to (1) with $\mathscr{H}[\varphi] = H$, $\mathscr{K}[\varphi] = K$. By Lemma 1 the mapping f is infinitesimally stable at 0.

Furthermore, $\mathscr{H}$ and $\mathscr{K}$ are linear operators of order ν, because the $\mathbf{W}_k$ are. The operators $\mathbf{H}$: $\varphi \to \mathbf{h}$, $\mathbf{K}$: $\varphi \to \mathbf{k}$ are linear of order ν, because $\mathbf{h}$ and $\mathbf{k}$ are obtained from $\mathscr{H}[\varphi]$, and $\mathscr{K}[\varphi]$ by using (2). This proves the theorem on infinitesimal stability.

PROOF OF LEMMA 3. Suppose that f has infinite multiplicity at 0, that is, for each r every neighbourhood contains r different points with the same value $f(x)$. Then 0 is not an isolated point of $f^{-1}(0)$, and $f^{-1}(0)$ contains a curve γ passing through 0 (see for instance Hervé [21], Chapter IV, Theorem 11). Clearly, $\det|f'| = 0$ around this curve. Take a sequence of points $\xi_i \to 0$ on γ, and consider the $n + 1$ linear subspaces

$$f'|_\xi(TX_\xi) = T_\xi \subseteq TY_0 \qquad (\xi = \xi_i, \xi_{i+1}, \ldots, \xi_{i+n}).$$

Since $\det|f'|_\xi = 0$, we must have $\dim T_\xi < n$. Now let $\varphi(x)$ be a "vector field". If the $n + 1$ planes $\varphi(\xi) + T_\xi$, $\xi = \xi_i, \ldots, \xi_{i+n}$ in the n-dimensional space TY_0 do not intersect in a single point, then the field $\varphi(x)$ cannot be represented in the form

$$-\varphi = f'\mathbf{h} + \mathbf{k}(f(x))$$

for any $\xi_i, \ldots, \xi_{i+n}$ contained in a neighbourhood of 0.

But for fixed i and almost all φ, the $n + 1$ planes $\varphi(\xi) + T_\xi$, $\xi = \xi_i, \ldots, \xi_{i+n}$ do not intersect in a single point in the space of vector polynomials of sufficiently high degree. Thus, for almost all vector polynomials φ, the same must be true for all i. So for almost all φ a decomposition $-\varphi = f'\mathbf{h} + \mathbf{k}$ is impossible in every region containing $\xi_i, \ldots, \xi_{i+n}$ for any i, that is, in any neighbourhood of 0. On the other hand, it follows from Lemmas 1 and 2 that φ admits a decomposition $-\varphi = f'\mathbf{h} + \mathbf{k}$. This contradiction proves Lemma 3.

---

1 We remark that this lemma is the only difficult part of our proof of the stability theorem, and the only place where analyticity is used in an essential way. If an estimate similar to Lemma 4 could be found in the infinitely differential case, then the successive approximations in §5 combined with the Nash-Moser smoothing [20] would lead immediately to an infinite-dimensional stability theorem (and such a theorem has been proved by Mather, using a completely different method).

§7. Proof of convergence

Notation. To ease the discussion we introduce some abbreviated notation. Let V be a sufficiently small closed neighbourhood of $y = 0$, $V \subset V_0$. Then we denote the component of 0 in $f^{-1}(V)$ by U. If φ is a function[1] of y analytic in V, then we write

$$|\varphi|_V = \max_{y \in V} |\varphi(y)|.$$

If φ is a function of x analytic in U, then

$$|\varphi|_V = \max_{x \in U} |\varphi(x)|.$$

For $\delta > 0$, the symbol $V - \delta$ denotes the set of points having a δ-neighbourhood lying entirely within V. In what follows the following obvious fact will often be used:

LEMMA 1. *If* $C < \left\| \frac{\partial f}{\partial x} \right\|^{-1}$, *then* $f^{-1}(V - \delta) \subset U - C\delta$.

PROOF. For any x_0 in $f^{-1}(V - \delta)$, $f(x_0) \in V - \delta$. Let us consider the segment $f(x_0 + \theta h)$, $0 \leqslant \theta \leqslant 1$, $|h| \leqslant C\delta$. We have
$|f(x_0 + \theta h) - f(x_0)| \leqslant \left\| \frac{\partial f}{\partial x} \right\|_V |h| \leqslant \delta$, so that $x_0 + \theta h \in U$, as required.

Let L be an operator taking functions φ analytic in one domain to functions $L[\varphi]$ analytic in another, and let $p \geqslant 0$, $\nu \geqslant 0$, $\nu_0 \geqslant 0$. We write

$$L[\varphi] \prec \varphi^p (\nu \,|\, \nu_0)$$

and say that L is an operator of degree p if, for some $\delta_0 > 0$, we have

$$|L[\varphi]|_{V-\delta} \leqslant \frac{|\varphi|_V^p}{\delta^\nu}, \quad \text{whenever} \quad |\varphi|_V < \delta^{\nu_0}, \ 0 < \delta < \delta_0, \ 0 \in V \subset V_0.$$

A similar definition holds for operators $L[\varphi_1, \ldots, \varphi_k]$ of several arguments. The relation

$$L[\varphi_1, \ldots, \varphi_k] \prec \Phi(\varphi_1, \ldots, \varphi_k) \quad (\nu \,|\, \nu_{01}, \ldots, \nu_{0k})$$

means that

$$|L[\varphi]_{V-\delta}| \leqslant \frac{\Phi(|\varphi_1|_V, \ldots, |\varphi_k|_V)}{\delta^\nu}, \quad \text{whenever} \quad |\varphi_i|_V < \delta^{\nu_{0i}},$$

$$i = 1, \ldots, k, \quad 0 < \delta < \delta_0.$$

Examples. EXAMPLE 1. *Put* $L = \frac{\partial^\alpha}{\partial x^\alpha}$. *Then* $L[\varphi] \prec \varphi(\alpha + 1 \,|\, 0)$.

This is because by Cauchy's formula and Lemma 1,

$$\left| \frac{\partial^\alpha \varphi}{\partial x^\alpha} \right|_{V-\delta} \leqslant \frac{\alpha! \, |\varphi|_V}{(C\delta)^\alpha} < \frac{|\varphi|_V}{\delta^{\alpha+1}} \text{ when } \delta_0 < \left(\alpha! \left\| \frac{\partial f}{\partial x} \right\|_V^\alpha \right)^{-1}.$$

EXAMPLE 2. *Put* $L[\varphi] = f(x + \varphi) - f(x)$. *Then* $L[\varphi] \prec \varphi(1 \,|\, 2)$.

For if $x \in f^{-1}(V - \delta)$, $|\varphi| < \delta^2$, $\delta < \delta_0 < \left\| \frac{\partial f}{\partial x} \right\|_V^{-1}$, then

1 With values in a normed space of numbers, vectors or matrices.

$$|L\varphi|_{V-\delta} \leqslant \left\| \frac{\partial f}{\partial x} \right\|_V |\varphi| \leqslant |\varphi| \delta^{-1}.$$

Similarly, *if* $L[\varphi] = f(x+\varphi) - f(x) - f(x)\varphi$, *then* $L[\varphi] \prec \varphi^2 (1\,|\,2)$.

EXAMPLE 3. *If* $L[\varphi, h] = \varphi(x+h) - \varphi(x)$, *then* $L[\varphi, h] \prec \varphi h (1\,|\,0,2)$. The proof is similar.

EXAMPLE 4. *If* $L[\varphi] \prec \varphi^l (\lambda\,|\,\lambda_0)$, $M[\varphi] \prec \varphi^m (\mu\,|\,\mu_0)$, *then*

$(L+M)[\varphi] \prec \varphi^r (\nu\,|\,\nu_0)$, *where* $r = \min(l, m)$, $\nu = \max(\lambda, \mu) + 1$, $\nu_0 = \max(\lambda_0, \mu_0)$,

$L[\varphi] M[\varphi] \prec \varphi^r (\nu\,|\,\nu_0)$, *where* $r = l+m$, $\nu = \lambda + \mu$, $\nu_0 = \max(\lambda_0, \mu_0)$,

$M[L[\varphi]] \prec \varphi^r (\nu\,|\,\nu_0)$, *where* $r = lm$, $\nu = m\lambda + \mu + 1$, $\nu_0 = \max\left(\lambda_0 + 1, \frac{\lambda + \mu_0 + 1}{l}\right)$.

The proof is obvious.

Estimates for the operators Σ, RH, K, R. We turn now to an estimate of the operators Σ_1, Σ_2, Σ_3 in formula (7) of §6. We find that

$$\Sigma_1[\mathbf{h}] \prec \mathbf{h}^2 (1\,|\,2),$$
$$\Sigma_2[\varphi, \mathbf{h}] \prec \varphi\mathbf{h} (1\,|\,0, 2),$$
$$\Sigma_3[\varphi, \mathbf{h}, \mathbf{k}] \prec \mathbf{k}(\varphi + \mathbf{h}) (2\,|\,3, 3, 0).$$

The estimates of Σ_1 and Σ_2 are contained in Examples 2 and 3 above. To deal with Σ_3 we use the representation $\Sigma_3 = S_3[\mathbf{k}, \Delta[\mathbf{h}, \varphi]]$, where

$$S_3[\mathbf{k}, \Delta] = \mathbf{k}(f(x) + \Delta(x)) - \mathbf{k}(f(x)) \prec \mathbf{k}\Delta (1\,|\,0, 2),$$
$$\Delta[\mathbf{h}, \varphi] = f(x+\mathbf{h}) - f(x) + \varphi(x+\mathbf{h}) \prec \mathbf{h} + \varphi (1\,|\,2, 0).$$

Finally, for $R = \Sigma_1 + \Sigma_2 + \Sigma_3$ we get

$$R[\varphi, \mathbf{h}, \mathbf{k}] \prec \mathbf{h}^2 + \mathbf{h}\varphi + \mathbf{k}(\varphi + \mathbf{h}) (3\,|\,3, 3, 0).$$

Now let H and K be the operators in the theorem on infinitesimal stability. These operators have order not more than $r - 1$, where r is the multiplicity of the mapping at 0. By Lemma 1 we can write in our new notation

$$\mathbf{H}[\varphi] \prec \varphi (r\,|\,0), \quad \mathbf{K}[\varphi] \prec \varphi (r\,|\,0). \tag{1}$$

Combining these estimates with those for $R[\varphi, \mathbf{h}, \mathbf{k}]$ we get

$$\mathbf{R}[\varphi] \prec \varphi^2 (2r+1\,|\,r+4). \tag{2}$$

for the operator $\mathbf{R}[\varphi] = R[\varphi, \mathbf{H}[\varphi], \mathbf{K}[\varphi]]$. This shows that the remainder term has degree higher than one, and this ensures the rapid convergence of the successive approximations.

A convergence lemma. For the proof of convergence we use some obvious assumptions.

LEMMA 2. *Suppose that we are given a sequence* $\delta_1 > \delta_2 > \ldots > 0$ *of numbers, a sequence* $U_0 \supset U_1 \supset \ldots \supset 0$ *of domains and a sequence*

$$h_s\colon\ U_s \to U_{s-1},\quad h_s(x) = x + \mathbf{h}_s(x),$$

of analytic mappings for which the following conditions hold:

A) $U_s \subset U_{s-1} - C\delta_s$, $\| \mathbf{h}_s \| \leqslant C\delta_s$,

B) $\left\| \frac{d\mathbf{h}_s}{dx} \right\|_{U_s} < \delta_s$, $\sum_{s=1}^{\infty} \delta_s < \infty$.

Then the sequence of mappings h^s: $U_s \to U_0$, $h^s = h_1 \circ h_2 \circ \ldots \circ h_s$ *on the intersection* $U_\infty = \bigcap_{s\to\infty} U_s$ *converges to a differentiable mapping* h: $U_\infty \to U_0$, $x \to x + \mathbf{h}(x)$.

PROOF. If $x \in U_{s+1}$, then by A) the segment $x + \theta \mathbf{h}_{s+1}(x)$, $0 < \theta \leqslant 1$, lies in U_s. Therefore

$$|h^{s+1}x - h^s x| = |h^s(x + \mathbf{h}_{s+1}(x)) - h^s(x)| \leqslant \left\| \frac{dh^s}{dx} \right\|_{U_s} \cdot \| \mathbf{h}_{s+1} \|_{U_{s+1}}.$$

But, by B),

$$\left\| \frac{dh^s}{dx} \right\|_{U_s} \leqslant \prod_{i=1}^{s} \left\| \frac{dh_s}{dx} \right\|_{U_s} \leqslant \prod_{i=1}^{s} \left(1 + \left\| \frac{dh_i}{dx} \right\|_{U_i} \right) \leqslant \prod_{i=1}^{s} (1 + \delta_i) < C_1.$$

Thus, $|h^{s+1}x - h^s x|_{U_{s+1}} < C_1 C\delta_{s+1}$, so that the sequence h^s is uniformly convergent in U_∞. Further, $\frac{dh^{s+1}}{dx} = \frac{dh^s}{dx} \frac{dh_{s+1}}{dx}$ so that

$$\left\| \frac{dh^{s+1}}{dx} - \frac{dh^s}{dx} \right\| \leqslant \left\| \frac{dh^s}{dx} \right\| \left\| \frac{dh_{s+1}}{dx} \right\| \leqslant C_1 \delta_{s+1}.$$

Therefore the product converges uniformly in U_∞, and this concludes the proof of the lemma.

LEMMA 3. *Let* A_s *be linear operators such that* $\| A_s \| \leqslant \delta_s \leqslant \frac{1}{2}$, $\sum_{s=1}^{\infty} \delta_s = C < \infty$. *Then* $\lim_{s\to\infty} \prod_{i=1}^{s} (E + A_i) = \Pi$ *exists and* $\| \Pi \| \leqslant e^C$, $\| \Pi^{-1} \| \leqslant e^{2C}$.

PROOF.

$$\| \Pi \| \leqslant \prod_{s=1}^{\infty} (1 + \delta_s) \leqslant e^C, \quad \| \Pi^{-1} \| \leqslant \prod_{s=1}^{\infty} (1 - \delta_s)^{-1} \leqslant \prod_{s=1}^{\infty} (1 + 2\delta_s) \leqslant e^{2C}.$$

LEMMA 4. *Set* $\mathbf{k}'[\mathbf{k}, \varphi, \mathbf{h}] = \mathbf{k}(f(x + \mathbf{h}(x)) + \varphi(x + \mathbf{h}(x)))$, $\mathbf{k}''[\varphi] = \mathbf{k}'[\mathbf{K}[\varphi], \varphi, \mathbf{H}[\varphi]]$. *Then*

$$\mathbf{k}''[\varphi] \prec \varphi(r+1 \,|\, r+3). \tag{3}$$

The proof of (3) is similar to the argument above; first we introduce the estimate

$$\mathbf{k}'[\mathbf{k}, \varphi, \mathbf{h}] \prec \mathbf{k}(0 \,|\, 0, 2, 2),$$

and then use the estimates (1).

Convergence of the sequences of mappings h^s, k^s. Set $V_0 = \{y\colon |y| \leqslant \rho_0\}$, and assume that $|\varphi_0|_{V_0} \leqslant M_0$. We have to show that for sufficiently small M_0, the sequences $h^s = h_1 \circ \ldots \circ h_s$, $k^s = k_s \circ \ldots \circ k_1$ defined by the relations

$$h_s\colon x \to x + \mathbf{h}_s(x), \qquad k_s\colon y \to y + \mathbf{k}_s(y),$$
$$\mathbf{h}_s = \mathbf{H}[\varphi_{s-1}], \quad \mathbf{k}_s = \mathbf{K}[\varphi_{s-1}], \quad \varphi_s = \mathbf{R}[\varphi_{s-1}],$$

(s = 1, 2, ...) converge in some neighbourhood of 0.

We use the estimates (1), (2), (3) already established:

$$\mathbf{H}[\varphi] \prec \varphi(r\,|\,0),\ \mathbf{K}[\varphi] \prec \varphi(r\,|\,0),\ \mathbf{R}[\varphi] \prec \varphi^2(2r+1\,|\,r+4),$$
$$\mathbf{k}''[\varphi] \prec \varphi(r+1\,|\,r+3),$$

and the following immediate consequence of (1):

$$\frac{\partial}{\partial x}\mathbf{H}[\varphi] \prec \varphi(r+2\,|\,0),\quad \frac{\partial}{\partial y}\mathbf{K}[\varphi] \prec \varphi(r+2\,|\,0). \tag{4}$$

Let δ_0 be the constant in (1) - (4), C the constant of Lemma 1, and let δ_1 be a sufficiently small positive number:

$$0 < \delta_1 \leqslant \min(0.1;\ 0{,}1\rho;\ \delta_0;\ 0.5C).$$

We show that, *for the convergence of the sequences h^s, k^s, it is sufficient that*

$$|\varphi_0|_{V_0} \leqslant M_0 = \delta_1^T,\quad T = 4r+2. \tag{5}$$

Define a sequence $\delta_1 > \delta_2 > \ldots > 0$ of numbers and a sequence $V_0 \supset V_1 \supset \ldots \supset 0$, $U_0 \supset U_1 \supset \ldots \supset 0$, of domains by the rules

$$\delta_{s+1} = \delta_s^{1\frac{1}{2}},\ V_s = V_{s-1} - \delta_s,\ U_s = f^{-1}V_s \qquad (s = 1, 2, \ldots). \tag{6}$$

By the choice of δ_1, we have $\sum_{s=1}^{\infty} \delta_s < 0.2\rho$. Therefore $V_\infty = \bigcap_{s \geqslant 1} V_s$ contains a neighbourhood of 0. We use the notation $|\cdot|_s = |\cdot|_{V_s}$.

LEMMA 5. *The functions in U_s and in V_s listed above are well-defined, analytic, and satisfy the following inequalities*:

$$|\varphi_s|_s \leqslant M_s = \delta_{s+1}^T \qquad (s = 1, 2, \ldots), \tag{7}$$

$$\max\left(|\mathbf{h}_s|_s,\ |\mathbf{k}''[\varphi_s]|_s,\ \left|\frac{d}{dx}\mathbf{h}_s\right|_s,\ \left|\frac{d}{dy}\mathbf{k}_s\right|_s\right) \leqslant \delta_s^2. \tag{8}$$

PROOF. For $s = 0$ we have (7) = (5). Suppose that $|\varphi_{s-1}|_{s-1} \leqslant \delta_s^T$. Then $|\varphi_{s-1}|_{s-1} \leqslant \delta_{s+1}^{r+4}$, because $\frac{2}{3}T > r + 4$. So we apply the operator $\mathbf{R}$ and by (2), (6) we see that $|\varphi_s|_s = |\mathbf{R}[\varphi_{s-1}]|_s \leqslant \delta_s^{2T}\delta_s^{-(2r+1)} = \delta_s^{\frac{3}{2}T} = \delta_{s+1}^T$, which proves (7) for all $s = 1, 2, \ldots$. The inequalities (8) follow from (7) by means of (1), (3), (4), (6). This proves the lemma.

CONVERGENCE OF h^s. It follows, in particular, from (8) that the mapping h^s: $x \to x + \mathbf{h}_s(x)$ is defined on U_s. The image h_sU_s lies in U_{s-1}; for $|\mathbf{h}_s|_s \leqslant \delta_s^2 < C\delta_s$ by (8), and so $U_s \subset U_{s+1} - C\delta_s$ by Lemma 1.

Thus, the mappings h_s, $s = 1, 2, \ldots$ satisfy all the requirements of Lemma 2 (inequalities A) and B) follow easily from (8)). By Lemma 2, the sequence of mappings $h^s = h_1 \circ \ldots \circ h_s$ converges in U_∞ to an analytic mapping h: $U_\infty \to U_0$. By Lemma 3, h has a non-degenerate differential; therefore h defines a diffeomorphism in some neighbourhood of $x = 0$.

CONVERGENCE OF k^s. Consider next the mappings k_s: $y \to y + \mathbf{k}_s(y)$. It follows from the estimate for $\mathbf{k}''$ in (8) that k_s is defined and analytic in $f_{s-1} \circ h_s(U_s)$, where f_{s-1}: $x \to f(x) + \varphi_{s-1}(x)$ in U_s. And in U_s we have from (16) of §6 that $f_s = k_s \circ f_{s-1} \circ h_s$. We show that *the mapping* $k^s = k_s \circ \ldots \circ k_1$ *is defined and analytic in* $f_0 \circ h^s(U_s)$. For $s = 1$ this has already been done. Suppose that k^{s-1} is defined and analytic in $f_0 \circ h^{s-1}(U_{s-1})$, and that $f_{s-1} = k^{s-1} \circ f_0 \circ h^{s-1}$ in U_{s-1}. Then we get $f_s = k_s \circ (k^{s-1} \circ f_0 \circ h^{s-1}) \circ h_s$ in U_s, since h_s: $U_s \to U_{s-1}$; and it follows that $k^s = k_s \circ k^{s-1}$ is defined and analytic in $f_0 \circ h^s(U_s)$, and also that $f_s = k^s \circ f_0 \circ h^s$.

Furthermore, using the estimate (8) for $\|\mathbf{k}''\|$, we see that k^s converges uniformly in $V' = f_0 \circ h(U_\infty)$ to the mapping $\mathbf{k}$: $V' \to V_\infty$.

The fact that $\mathbf{k}$ is differentiable and non-degenerate follows from Lemma 3 and the estimate (8) for $\left|\frac{d}{dy}\mathbf{k}\right|$. Consequently, $\mathbf{k}$ defines a diffeomorphism from some neighbourhood of $f_0 \circ h(0)$ to a neighbourhood of $y = 0$. It is easy to derive[1] the limit relation $f = k \circ f_0 \circ h$ from (7), (8) and the fact that $f_s = k^s \circ f_0 \circ h^s$.

Finally, the estimate (8) shows that for sufficiently small M_0 the mappings h and k are arbitrarily near to the identity. This proves the stability theorem.

REMARK. *Every mapping* f_0 *having tangency of sufficiently high order* m *with* f *at* 0 *is equivalent to* f *at* 0.

For suppose that $f_0(x) = f(x) + \varphi_0(x)$, $|\varphi_0(x)| \leqslant C_1|x|^{m+1}$. Since f has finite multiplicity r, we get

$$x^r \in (\mathfrak{M}\mathcal{H}_y)\mathcal{H}_x, \quad |x|^r \leqslant C_2|f(x)|.$$

(see the Appendix to §4). Therefore $|\varphi_0(x)|_{V_0} \leqslant C_3\rho^{\frac{m+1}{r}}$. If $\frac{m+1}{r} > T = 4r + 2$ and ρ is sufficiently small, then (5) is satisfied for $\delta_1 = 0.1\rho$. So, if[2] $m \geqslant 4r^2 + 2r$, then f_0 is equivalent to f in a sufficiently small neighbourhood of $x = 0$. At the same time we have proved:

COROLLARY 1. *The mapping* f *is equivalent to a polynomial at* 0, *namely to its Taylor polynomial of degree* m.

§8. In the neighbourhood of an isolated critical point every analytic function is equivalent to a polynomial

The germs of functions f: $\mathbf{C}^n \to \mathbf{C}$, g: $\mathbf{C}^n \to \mathbf{C}$ at 0 are *equivalent* if there exists a bi-analytic diffeomorphism h of neighbourhoods U, V of 0 in $\mathbf{C}^n$ such that $h(0) = 0$ and the diagram

---

1 Take x in U_∞. Then $h(x)$ and $h^s(x)$ can be connected by a curve in $h^s(U_s)$ whose length is small, as $s \to \infty$. Further, f_0 has derivative in $h^s(U_s)$ bounded by a constant independent of s. Thus, $f_0 \circ h(x)$ and $f_0 \circ h^s(x)$ can also be joined by a short curve in $f_0 \circ h^s(U_s)$. The derivative of k^s is also bounded in $f_0 \circ h^s(U_s)$ independently of s; thus, for large s the points $k^s \circ f_0 \circ h^s(x)$ and $k^s \circ f_0 \circ h(x)$ are near each other. Finally, $k^s \circ f_0 \circ h(x)$ is near to $k \circ f_0 \circ h(x)$ for large s. Thus, $k \circ f_0 \circ h = \lim_{s\to\infty} k^s \circ f_0 \circ h^s = \lim_{s\to\infty} f_s = f$.

2 Actually this is more than we need for our argument.

$$\begin{array}{ccc} U & \xrightarrow{f} & \mathbf{C} \\ {\scriptstyle h}\uparrow & \nearrow {\scriptstyle g} & \\ V & & \end{array}$$

is commutative.

It is easy to see that every analytic function of a single variable is equivalent to a polynomial in the neighbourhood of any critical point. N. Levison [22] has proved the analogous result for functions of two variables. Whitney [23] has given examples of analytic functions of three variables that are not equivalent to a polynomial in the neighbourhood of a critical point. An example of such a function is

$$f(x, y, z) = xy(x-y)(x-yz)(x-ye^z).$$

The proof of the non-equivalence is based on the following remark. A surface transversal to the z-axis at the point z_0 intersects the level set $f = 0$ in five curves passing through the point $x = y = 0$, $z = z_0$. Consider the cross-ratios formed by the tangents at z_0; they are independent of the choice of surface. The cross-ratio α corresponding to the first four factors of f define z_0. Therefore the cross-ratio β corresponding to the last four factors is a function $\beta(\alpha)$. If f were equivalent to a polynomial at 0, then $\beta(\alpha)$ would be algebraic, whereas e^z is transcendental.

Note that the critical point $x = y = z = 0$ in Whitney's example is not isolated.

THEOREM 1. *The germ of an analytic function at an isolated critical point is equivalent to the germ of its Taylor polynomial of sufficiently high degree.*

PROOF[1]. Let $f(x) = f(x_1, \dots, x_n)$ be the given function and suppose that $x = 0$ is an isolated critical point. Let $\mathbf{h}(x) = (h_1(x), \dots, h_n(x))$ be the germ of a "vector field" analytic at zero and $\varphi(x)$ the germ of a function analytic at zero.

LEMMA 1. *The following identity holds:*

$$f(x+\mathbf{h}(x)) + \varphi(x+\mathbf{h}(x)) = f(x) + \Sigma_0 + \Sigma_1 + \Sigma_2,$$

where

$$\Sigma_0 = f'(x)\,\mathbf{h}(x) + \varphi(x),$$
$$\Sigma_1 = f(x+\mathbf{h}(x)) - f(x) - f'(x)\,\mathbf{h}(x),$$
$$\Sigma_2 = \varphi(x+\mathbf{h}(x)) - \varphi(x).$$

Corresponding to the general theme of §5 we must investigate the "infinitesimal" equation $\Sigma_0 = 0$.

LEMMA 2. *Suppose that $\varphi(x)$ has sufficiently high order N at zero. Then there exists a field $\mathbf{h}(x)$ such that*

$$\varphi(x) + f'(x)\,\mathbf{h}(x) = 0.$$

The construction of $\mathbf{h}$ is performed as follows. We consider the mapping

---

[1] Another proof can be found in Tougeron's dissertation [14].

$f'\colon x \to \left(\frac{\partial f}{\partial x_1}, \ldots, \frac{\partial f}{\partial x_n}\right)$. The critical points of f are the inverse images of zero under f'. Since $x = 0$ is an isolated critical point, the analytic function f' has finite multiplicity at zero (the "theorem on zeros"). Consequently the local ring of f' at zero has finite dimension r. It follows that there exist n decompositions

$$x_i^r = f'(x)\,\mathbf{h}_i(x) \qquad (i = 1, \ldots, n),$$

where the $\mathbf{h}_i(x)$ are the germs of vector fields analytic at zero (see the Appendix to §4). We assume that $\mathbf{h}_i(x)$ is analytic in the unit disc with respect to the first coordinate (if not, change the scale).

If N is sufficiently large ($N \geqslant nr$), then the germ of the function $\varphi(x)$, which is analytic at zero and of order N, can be written as

$$\varphi(x) = \sum_{i=1}^{n} x_i^r\,(\varphi_i(x)), \text{ where the } \varphi_i = \sigma_i[\varphi] \qquad (i = 1, \ldots, n)$$

are obtained by the successive division of $\varphi(x)$ by x_i^r (see the Appendix to §4). We introduce the operator

$$\mathbf{H} = -\sum_{i=1}^{n} \mathbf{h}_i \sigma_i. \tag{1}$$

The field $\mathbf{h} = \mathbf{H}[\varphi]$ satisfies the conditions of the lemma, so that the proof of Lemma 2 is complete.

Suppose now that $f_0(x) = f(x) + \varphi_0(x)$ is a mapping close to f. We define by recursion functions φ_s and fields $\mathbf{h}_s$ starting with $\varphi_0(x)$ and continuing with

$$\varphi_s = \mathbf{R}[\varphi_{s-1}], \quad \mathbf{h}_s = \mathbf{H}[\varphi_{s-1}] \qquad (s = 1, 2, \ldots),$$
$$\mathbf{R}[\varphi] = R[\varphi, \mathbf{H}[\varphi]], \quad R[\varphi, \mathbf{h}] = \Sigma_1 + \Sigma_2.$$

We define further mappings:

$$h_s\colon x \to x + \mathbf{h}_s(x), \quad f_s\colon x \to f(x) + \varphi_s(x), \quad h^s = h_1 \circ \ldots \circ h_s.$$

By Lemma 1, $f_{s-1} \circ h_s = f_s$, so that $f_0 \circ h^s = f_s$. Thus, our theorem follows from the following proposition.

LEMMA 3. *If φ_0 has sufficiently high order at 0, then $h = \lim_{s\to\infty} h^s$ is a diffeomorphism in a sufficiently small neighbourhood of $x = 0$, and $f_0 \circ h = f$.*

For the convergence proof we introduce the following notation. Let φ be a function analytic in $|x| \leqslant \rho < 1$ and having order not less than a at 0. Set

$$|\varphi|_{a,\rho} = \max_{|x|\leqslant\rho} \frac{|\varphi(x)|}{|x|^a}.$$

Further, let L be an operator taking analytic functions to analytic functions, and suppose that a, b, ν, $\nu_0 > 0$. The notation

$$|L[\varphi]|_a \prec \Phi(|\varphi|_b)(\nu|\nu_0)$$

means that there exists a constant $\delta_0 > 0$ such that $|L[\varphi]|_{a,\rho-\delta} \leqslant \frac{\Phi(|\varphi|_{b,\rho})}{\delta^\nu}$, if $0<\delta<\left(\delta_0, \frac{\rho}{2}\right)$ and $|\varphi|_{b,\rho} < \delta^{\nu_0}$. The meaning of the symbol $\prec$ for operators of several arguments is defined in a similar way.

EXAMPLE 1. *The following estimate holds:*

$$\left|\frac{d\varphi}{dx}\right|_{a-1} \prec |\varphi|_a \ (2|0).$$

For by using Cauchy's formula for $x \neq 0$, $d=\delta\frac{|x|}{\rho-\delta}$, $0<\delta<\frac{\rho}{2}$, we find $\left|\frac{d\varphi}{dx}\right| \leqslant \frac{|\varphi|_a(|x|+d)^a}{d} \leqslant \frac{2^a}{\delta}|x|^{a-1}$, as required.

In the same way one proves

$$|\Sigma_1[\mathbf{h}]|_a \prec |\mathbf{h}|_a^2(1|1), \ |\Sigma_2[\varphi, \mathbf{h}]|_{a+b-1} \prec |\varphi|_a|\mathbf{h}|_b(2|0, 2). \tag{2}$$

LEMMA 4. *The operator* **H** *satisfies*

$$|\mathbf{H}[\varphi]|_{N-r} \prec |\varphi|_N(n+1|0), \ N \geqslant nr. \tag{3}$$

PROOF. As is clear from (1), it is enough to establish that

$$|\sigma_i[\varphi]|_{N-r} \prec |\varphi|_N(n+1|0).$$

This is done by a simple calculation of the Taylor coefficients. If $|\varphi(x)| \leqslant M|x|^N$ for $|x| \leqslant \rho$, then $\varphi(x) = \sum_{\mathbf{k}} \varphi_{\mathbf{k}} x^{\mathbf{k}}$, where $\mathbf{k} = (k_1, \ldots, k_n)$, and $|\varphi_{\mathbf{k}}| \leqslant M\rho^{N-k}$ $(k = k_1 + \ldots + k_n, \ k \geqslant N)$.

By definition of $\sigma_i[\varphi]$, the coefficients $\varphi_{i,\mathbf{k}}$ of the Taylor expansion

$$\sigma_i[\varphi] = \sum_{\mathbf{k}} \varphi_{i,\mathbf{k}} x^{\mathbf{k}} \qquad (i=1, \ldots, n)$$

are obtained from $\varphi_{\mathbf{k}}$ by displacing the index r positions. Thus,

$$|\varphi_{i,\mathbf{k}}| \leqslant M\rho^{N-k-r} \qquad (k \geqslant N-r).$$

Suppose now that $0 < \delta < 0.5\rho$, so that $\theta = 1 - \delta/r > 0.5$. Then

$$|\sigma_i[\varphi]|_{N-r,\rho-\delta} \leqslant M\sum_{\mathbf{k}}{}' \left|\frac{x}{\rho}\right|^{k-N+r} \leqslant M\theta^{r-N}\sum_{\mathbf{k}}\theta^k \leqslant \frac{M2^N}{(1-\theta)^n} \leqslant \frac{M2^N\rho^n}{\delta^n},$$

as required.

Comparing (2) and (3) we get for $N \geqslant nr$, $n \geqslant 2$,

$$|\mathbf{R}[\varphi]|_N \prec |\varphi|_N^2(2n+4|n+3). \tag{4}$$

PROOF OF LEMMA 3. We prove convergence for

$$|\varphi_0(x)| \leqslant C|x|^m, \ m \geqslant 4n+7+N, \ N=nr, \ n \geqslant 2.$$

Let δ_0 be the constant in (3) and (4). Define sequences δ_s, ρ_s $(s=1,2,\ldots)$ of numbers by the rules $\delta_{s+1} = \delta_s^{1.5}$, $\rho_s = \rho_{s-1} - \delta_s$, $\delta_1 = 0.1\rho_0$.

If ρ_0 is sufficiently small, then $\delta_s < 0.5\rho_{s-1}$, $\delta_s < \delta_0$, $\rho_s > 0.5\rho_0$ and

$$|\varphi_0|_{N,\rho_0} \leqslant \delta_1^T,\ T = 4n + 6.$$

Here we use the notation $|\cdot|_s = |\cdot|_{N,\rho_s}$.

LEMMA 5. *The following estimates hold*:

$$|\varphi_s|_s \leqslant \delta_{s+1}^T \quad (s = 1, 2, \ldots). \tag{5}$$

For $s = 0$, (5) is clearly true. If $|\varphi_{s-1}|_{s-1} \leqslant \delta_s^T$, then by (4),

$$|\varphi_s|_s = |\mathbf{R}[\varphi_{s-1}]|_s \leqslant \delta_s^{2T}\delta_s^{-(2n+4)} = \delta_s^{\frac{3}{2}T} = \delta_{s+1}^T,$$

as required.

It follows from (3) and (5) that

$$|\mathbf{h}_s(x)| < \delta_s^2, \quad \left|\frac{d}{dx}\mathbf{h}_s(x)\right| \leqslant \delta_s^2. \tag{6}$$

for $|x| < \rho_s$. The convergence of the sequence h^s is a consequence of (6) by Lemma 2 of §7 (p.). This proves Lemma 3 and with it the theorem.

REMARK 1. We have shown that the germ of a function is equivalent to its Taylor polynomial of degree $m = nr + 4n + 6$ at zero. It is easy to see that in fact $m = 2r$ will do.

REMARK 2. A.M. Samoilenko has carried the proof over to the case of infinitely differentiable functions, using the Nash-Moser smoothing [20]. Of course, instead of assuming that the critical points are isolated we have to assume that the local ring of mappings $x \to \frac{\partial f}{\partial x}$ is finite-dimensional.

References

[1] J. Milnor, Morse Theory, Princeton 1963.
Translation: *Teoriya Morsa*, Izdat. "Mir", Moscow 1965.

[2] H. Whitney, On singularities of mappings of Euclidean spaces. I. Mappings of the plane into the plane, Ann. of Math. 62 (1955), 374-410.

[3] R. Thom and H. Levine, Singularities of differentiable mappings I, Bonn 1959.

[4] R. Thom, Local topological properties of differentiable mappings, Bombay Colloquium on Differentiable Analysis, Oxford University Press 1964.

[5] J.M. Boardman, Singularities of differentiable maps, University of Warwick (1965), 1-105.

[6] B. Morin, Formes canoniques des singularités d'une application différentiable, Compt. Rendus Acad. Sci. Paris 260 (1965), 5662-5665, 6503-6506.

[7] B. Malgrange, Le théorème de préparation en géométrie différentiable, Séminaire H. Cartan (1962/63), Nos. 11, 12, 13, 22.

[8] B. Malgrange, Ideals of differentiable functions, Oxford University Press, 1966.

[9] J. Mather, On the preparation theorem of Malgrange, Princeton 1966.

[10] C. Houzel, Géométrie analytique locale I, Séminaire H. Cartan (1960/1961), No. 18.

[11] H. Cartan, Sur les matrices holomorphes de *n* variables complexes, J. Math. Pures. Appl. 19 (1940), 1-26.

[12] A.N. Kolmogorov, On the preservation of conditionally periodic motions, Dokl. Akad. Nauk SSSR 98 (1954), 527-530. MR 16, 294.

[13] J. Mather, Structural stability of mappings, Princeton 1966.
[14] J.C. Tougeron, Idéaux des fonctions différentiables, Université de Rennes, 1967.
[15] C.L. Siegel, Vorlesungen über Himmelsmechanik, Springer-Verlag, Berlin 1965. Translation: *Lektsii po necebnoi mekhanik*, Izdat.-Lit, Moscow 1959.
[16] C.L. Siegel. Über die Normal form analytischer Differentialgleichungen in der Nähe einer Gleichgeirichtslösung, Nachr. Akad. Wiss. Göttingen, Math.-Phys. Kl. 1952, 21-30. MR **15**, 222.
[17] V.I. Arnol'd, Small denominators. I. Mappings of a circle onto itself, Izv. Akad. Nauk Ser. Mat. **25** (1961), 21-86. MR **25**, # 4113.
[18] Jo Ging-tzung, Singularities S_{ro}, Sci. Sinica **14** (1965), 816-830.
[19] V.I. Arnol'd, A remark on the Weierstrass preparation theorem, Funktsional. Anal i prilozhen. **1** (1967), 1-8.
[20] J. Moser, A new technique for the construction of solutions of non-linear differential equations, Proc. Acad. Nat. Sci. U.S.A., **47** (1961), 1824-1831.
[21] M. Hervé, Functions of several complex variables, Oxford University Press 1963.
Translation: *Funktsii mnogikh kompleksnykh peremennykh*, Izdat. "Mir", Moscow 1965.
[22] N. Levinson, A polynomial canonical form for certain analytic functions of two variables at a critical point, Bull. Amer. Math. Soc. **66** (1960), 366.
[23] H. Whitney, Local properties of analytic varieties; in the book: Differential and combinatorial topology, Princeton 1965.
[24] H. Whitney, The general type of singularity of a set of $2n$-1 smooth functions of n variables, Duke. Math. J. **10** (1943), 161-172.
[25] H. Whitney, The self-intersections of a smooth n-manifold in $2n$-space, and $(2n-1)$-space, Ann. of Math. Ser. 2, **45** (1944), 220-293.
[26] H. Whitney, On singularities of mappings of euclidean spaces, Symposium Internacional de Topologia Algebraica, Mexico 1958.
[27] R. Thom, Les singularités des applications différentiables, Ann. Inst. Fourier **6** (1956), 43-87.
[28] R. Thom, Remarques sur les problèmes comportant des inéquations différentielles globales, Bull. Soc. Math. France **87** (1959), 455-461.
[29] R. Thom, L'équivalence d'une fonction différentiable et d'un polynome, Topology 3 (1965), 297-307.
[30] A. Haefliger, Sur les self-intersections des applications différentiables, Bull. Soc. Math. France **87** (1959), 351-359.
[31] A. Haefliger, Quelques remarques sur les applications différentiables d'une surface dans le plan, Ann. Inst. Fourier **10** (1960), 47-60.
[32] A. Haefliger, Points multiples d'une application et produit cyclique reduit, Amer. J. Math. **83** (1961), 57-70.
[33] A.W. Tucker, Branched and folded coverings, Bull. Amer. Math. Soc. **42** (1936), 859-862.
[34] Z. Wolfsohn, On differentiable maps of Euclidean n-space into Euclidean m-space, Bull. Amer. Math. Soc. **61** (1955), 171.
[35] N. Levinson, Transformation of an analytic function of several variables to a canonical form, Duke. Math. J. **28** (1961), 345-353.
[36] H.I. Levine, The singularities S_1^q, Illinois, K. Math. **8** (1964), 152-168.
[37] H.I. Levine, Elimination of cusps, Topology 3 (1965), 263.
[38] H.I. Levine, Mappings of manifolds into the plane, Amer. J. Math. **88** (1966), 357-365.
[39] L.S. Pontryagin, Vector fields on manifolds, Mat. Sb. **24** (**66**) (1949), 155-162. = Amer. Math Soc. Translations, no. 13 (1949).

Received by the Editors September 4, 1967.

Translated by J. Wiegold.

To Ivan Georgievich Petrovskii
on his 70th birthday.

ON MATRICES DEPENDING ON PARAMETERS

V. I. Arnold

Given a family of matrices smoothly depending on parameters of endomorphisms of a complex linear space, it is shown that there is a normal form to which the family can be reduced by the choice of a base smoothly depending on the parameters. The formulae obtained are applied to the investigation of bifurcation diagrams of families of matrices.

Contents

Introduction
§1. Versal and universal deformations 47
§2. Versality and transversality 48
§3. An example 50
§4. The construction of transversal deformations 51
§5. Remarks 54
References 59

Introduction

The reduction of a matrix to its Jordan normal form is an unstable operation. For both the normal form itself and the reducing mapping depend discontinuously on the elements of the original matrix. Therefore, if the elements of a matrix are known only approximately, then it is unwise to reduce it to its Jordan form.

Furthermore, when investigating a family of matrices smoothly depending on parameters, then although each individual matrix can be reduced to a Jordan normal form, it is unwise to do so, since in such an operation the smoothness (and also the continuity) relative to the parameters is lost.

Thus, the problem arises of finding the simplest possible normal form to which not only one specific matrix, but an arbitrary family of matrices close to it can be reduced by means of a mapping smoothly depending on the elements of the matrices.

In the present note this problem is solved for a family of matrices depending holomorphically on parameters. In other words, we investigate the problem of a normal form of matrices over the ring of germs of holomorphic functions of several complex variables.

Obviously, it is enough to consider a deformation of a Jordan matrix, that is, a family containing a matrix that reduces to the Jordan normal form when its parameters have the value zero. Theorem 4.4 gives a form to which all close matrices holomorphically depending on parameters can be reduced by a change of base.

This normal form differs from the original Jordan form in that some entries instead of being zero are holomorphic functions of parameters that vanish when the value of the parameters is zero. This is the simplest form in the sense that the number of such entries is as small as possible.

The reduction to this normal form is stable: if the matrix is known only approximately, then both the elements of the normal form and the reducing mapping can be found approximately.

The construction of the normal form is based on a well known theorem on commuting matrices and on standard arguments of the theory of singularities of differentiable mappings [2]. The result obtained has obvious applications in the theory of the bifurcations of non-linear differential and integral equations (see [3]–[7]).

The author is grateful to N. N. Brushlinskaya, D. A. Kazhdan and S. G. Krein for useful discussions.

§ 1. Versal and universal deformations

1.1. Deformations. We consider square complex matrices of order n. Let A_0 be such a matrix. A *deformation* of A_0 is a matrix $A(\lambda)$ of the same order as A_0 with entries that are power series of an arbitrary number k of complex variables λ_i, convergent in a neighbourhood of $\lambda = 0$, with $A(0) = A_0$. A deformation is also called a family, the variables λ_i *parameters*, and the parameter space $\Lambda = \{\lambda\}$ a *base* of the family.

Two deformations $A(\lambda)$ and $B(\lambda)$ of the matrix A_0 are called *equivalent* if there exists a deformation $C(\lambda)$ of the identity matrix with the same base such that $A(\lambda) = C(\lambda)B(\lambda)C^{-1}(\lambda)$.

Let φ be a germ of a mapping $\mathbf{C}^l \to \mathbf{C}^k$ holomorphic at 0, that is, a set of k power series in l complex variables convergent in a neighbourhood of 0. We assume that $\varphi(0) = 0$. The mapping φ of the parameter space $\{\mu\}$ into the base of the deformation $A(\lambda)$ defines a new deformation of A_0 according to the formula

$$(\varphi^*A)(\mu) = A(\varphi(\mu)), \qquad \mu \in \mathbf{C}^l.$$

The deformation φ^*A is said to be *induced* by A under the mapping φ.

1.2. Versal deformations. A deformation $A(\lambda)$ of A_0 is called *versal*[1] if

[1] The term "versal" is obtained from the word universal by discarding the prefix "uni" indicating uniqueness.

every other deformation $B(\mu)$ of A_0 is equivalent to a deformation induced by $A(\lambda)$ under a suitable change of parameters:

$$B(\mu)=C(\mu)A(\varphi(\mu))C^{-1}(\mu), \quad C(0)=E, \quad \varphi(0)=0.$$

A versal deformation $A(\lambda)$ is called *universal* if the change of parameters of φ is determined uniquely by $B(\mu)$.

EXAMPLE. Of the three deformations

$$\begin{pmatrix}\lambda_1 & \lambda_2\\ \lambda_3 & \lambda_4\end{pmatrix}, \quad \begin{pmatrix}\lambda_1 & 1+\lambda_3\\ 0 & \lambda_2\end{pmatrix}, \quad \begin{pmatrix}1+\lambda_1 & 0\\ 0 & \lambda_2\end{pmatrix}$$

the first is versal but not universal, the second is not versal, and the third is not only versal but also universal.

§2. Versality and transversality

2.1. Transversality. Let $N \subset M$ be a smooth submanifold of a manifold M. We consider a smooth mapping A: $\Lambda \to M$ of another manifold Λ into M, and let λ be a point in Λ such that $A(\lambda) \in N$.

The mapping A is called *transversal to N at* λ if the tangent space to M at $A(\lambda)$ is the sum

$$TM_{A(\lambda)}=A_*T\Lambda_\lambda+TN_{A(\lambda)}.$$

2.2. Orbits. We consider the set $M = \mathbf{C}^{n^2}$ of all $n \times n$ matrices and the Lie group $G = GL(n, \mathbf{C})$ of all non-singular $n \times n$ matrices. The group G acts on M according to the formula

$$\text{(1)} \qquad \mathrm{Ad}_g m = gmg^{-1} \quad (m \in M,\ g \in G).$$

Consider the orbit of an arbitrary fixed matrix $A_0 \in M$ under the action of G. This smooth submanifold of M is denoted by N. Thus, the orbit N of A_0 consists of all the matrices similar to A_0.

2.3. The versality condition. A deformation $A(\lambda)$ of a matrix A_0 may be regarded as a mapping A: $\Lambda \to M$ of the base of the family into the space M of matrices. The following lemma is obvious and well known.

LEMMA. *A deformation $A(\lambda)$ is versal if and only if the mapping A is transversal to the orbit of A_0 at $\lambda = 0$.*

We shall prove that the versal deformation A is transversal. Let $B(\mu)$ be any deformation of A_0. Then by the versality of A

$$B(\mu)=C(\mu)A(\varphi(\mu))C^{-1}(\mu),$$

hence

$$B_{*0}=A_*\varphi_*+[C_*, A_0].$$

Consequently, for every tangent to the base of B at the origin of a vector ξ we have $B_*\xi = A_*\varphi_*\xi + [C_*\xi, A_0]$. Hence, any vector in the space TM_{A_0}

is the sum of a vector in the image of A_* and a vector tangent to the orbit. Thus, A is transversal to the orbit.

We precede the proof of the versality of a transversal deformation by the following construction.

2.4. Orbit and centralizer. We consider the mapping α: $G \to M$, given by (1) for a fixed $m = A_0$. The mapping α carries the group G into the orbit of A_0. The stationary subgroup H of A_0 is mapped to A_0; it consists of all the elements of G that commute with A_0.

The mapping α induces a mapping of the tangent space at the unit element of the group into the tangent plane to the orbit

$$\alpha_*: TG_e \to TM_{A_0}, \quad \alpha_* C = [C, A_0].$$

The kernel Ker α_* of α_* is the tangent plane to the stationary subgroup H. The subalgebra Ker α_* of the Lie algebra TG_e consists of all the matrices that commute with A_0 and is called the *centralizer* of A_0.

Note that the dimension of the matrix group G is the same as that of the matrix manifold M (namely n^2). Consequently, *the codimension of the orbit of A_0 is equal to the dimension of its centralizer.*

2.5. Decomposition into a direct product. By means of the unit element of G we introduce a submanifold V, transversal to the centralizer and of complementary dimension (equal to the dimension of the orbit). Let A: $\Lambda \to M$, $A(0) = A_0$, be an arbitrary deformation of A_0 transversal to the orbit of A_0, where the dimension of the base Λ is equal to the codimension of the orbit (that is, has the minimum value for transversal deformations).

Let β: $V \times \Lambda \to M$ be the mapping given by the formula $\beta(v, \lambda) = \mathrm{Ad}_v A(\lambda)$. Note that the dimension of the direct product $V \times \Lambda$ is the same as that of M (namely n^2). From the transversality of A to the orbit and the transversality of V to the centralizer it follows that the derivation $\beta_*|_{(e, 0)}$ is non-singular. By the inverse functions theorem the mapping β determines a holomorphic diffeomorphism between a sufficiently small neighbourhood of $(e, 0)$ in $V \times \Lambda$ and a sufficiently small neighbourhood of A_0 in M. Replacing V and Λ by smaller manifolds, we may assume that a neighbourhood of A_0 in M splits into the direct product $V \times \Lambda$.

2.6. Transversality ⇒ versality. Let A be a transversal deformation. If the dimension of the base is greater than the minimum (that is, the codimension of the orbit), we replace the base by a submanifold whose dimension is equal to the codimension of the orbit and such that the restriction of A to this subspace is still transversal to the orbit. Let us prove that the resulting family is versal. If we obtain more, so much the better.

Let $B(\mu)$ be an arbitrary deformation of A_0. For sufficiently small μ the matrix $B(\mu)$ has a unique representation in the form $B(\mu) = \beta(v, \lambda)$, $v \in V$, $\lambda \in \Lambda$ (see 2.5). Hence, B can be written in the form

$$B(\mu) = C(\mu) A(\varphi(\mu)) C^{-1}(\mu),$$

where $\varphi(\mu) = \pi_2 \beta^{-1} B(\mu)$, $C(\mu) = \pi_1 \beta^{-1} B(\mu)$, and where π_1 and π_2 are the projections of $V \times \Lambda$ onto V and onto Λ. Thus, A is versal, and the lemma of 2.3 is proved.

Note that in spite of the uniqueness of this decomposition, the deformation A need not be universal. The fact is that uniqueness holds only for a fixed submanifold V and the latter can be chosen in different ways.

§ 3. An example

3.1. A Sylvester family. Consider an n-parameter family of $n \times n$ matrices

$$A(\alpha) = \begin{pmatrix} 0 & 1 & & & \\ & 0 & 1 & & \\ \cdot & \cdot & \cdot & \cdot & \cdot \\ & & & 0 & 1 \\ \alpha_1 & \alpha_2 & & \dots & \alpha_n \end{pmatrix}.$$

We call this a *Sylvester* family.[1]

3.2. Versality of a Sylvester family. By a direct computation of the commutators it is easy to verify that a Sylvester family is transversal to the orbit of each of its matrices. Lemma 2.3 therefore has the following consequence.

COROLLARY. *A Sylvester family determines an n-parameter versal deformation of each of its matrices.*

If a matrix A_0 can be reduced to a Sylvester normal form, then also matrices close to A_0 can be reduced to a Sylvester form without losing their smooth dependence on the parameters.

3.3. Universality of Sylvester deformations. Note that the elements α_i of a Sylvester matrix are (to within the sign) the coefficients of its characteristic polynomial. Hence, the elements of the Sylvester normal form of a given matrix are uniquely determined.

COROLLARY. *A Sylvester family defines a universal deformation of each of its matrices.*

Thus, the problem of a normal form of a deformation of a matrix A_0 is solved for the case when A_0 can be reduced to Sylvester form, that is, when only one Jordan block corresponds to each eigenvalue.

3.4. Universal deformation of a Jordan block. In particular, we consider the case when A_0 is a Jordan block with the eigenvalue λ. From Lemma 2.3 and Corollary 3.3 we obtain:

COROLLARY. *A versal (and even a universal) deformation of a Jordan block is an n-parameter deformation*

$$D(\alpha_1, \dots, \alpha_n) = \lambda E + A(\alpha),$$

where $A(\alpha)$ is a Sylvester matrix.

---

[1] This can also be regarded as the family of all ordinary linear differential equations of order n with constant coefficients.

If a matrix A_0 is a Jordan block, then matrices close to A_0 can be reduced to the form just given, and the reducing mapping and the parameters α_i of the Sylvester matrix depend holomorphically on the elements of the original matrix.

3.5. Codimensions of orbits. From Corollary 3.3 we also derive:

COROLLARY. *For any Sylvester matrix, irrespective of the multiplicity of the eigenvalues, the codimension of the orbit (which is equal to the dimension of the centralizer and also to the number of parameters of a universal deformation) is one and the same, namely the order n of the matrix.*

§4. The construction of universal deformations

4.1. The orthogonal complement. We introduce in the space of matrices $M = \mathbf{C}^{n^2}$ a Hermitian scalar production $(A, B) = \mathrm{Tr}\,(AB^*)$, where B^* is the adjoint of B. The corresponding scalar square is simply the sum of the squares of the moduli of the matrix elements.

LEMMA. *The vector $B \in TM_{A_0}$ is perpendicular to the orbit of the matrix A_0 if and only if $[B^*, A_0] = 0$.*

PROOF. The tangent vectors to the orbit of A_0 are the matrices that can be represented in the form $[C, A_0]$. Since B is orthogonal to the orbit, for any matrix C we have $([C, A_0], B) = 0$. In other words, for any C we have $0 = \mathrm{Tr}([C, A_0]B^*) = \mathrm{Tr}(CA_0B^* - A_0CB^*) = \mathrm{Tr}([A_0, B^*]C) = ([A_0, B^*], C^*)$. Since C is arbitrary, this condition is equivalent to $[B^*, A_0] = 0$.

Thus, the lemma is proved: *the orthogonal complement to the orbit of A_0 is the adjoint of its centralizer.*

4.2. Centralizers of Jordan matrices. It is not difficult to compute the centralizers of matrices reduced to Jordan normal forms. We assume at first that the matrix A_0 has only one eigenvalue λ and a sequence of upper Jordan blocks of dimensions $n_1 \geqslant n_2 \geqslant \ldots$

LEMMA.[1] *The matrices that commute with A_0 are precisely those described in Figure 1 below.*

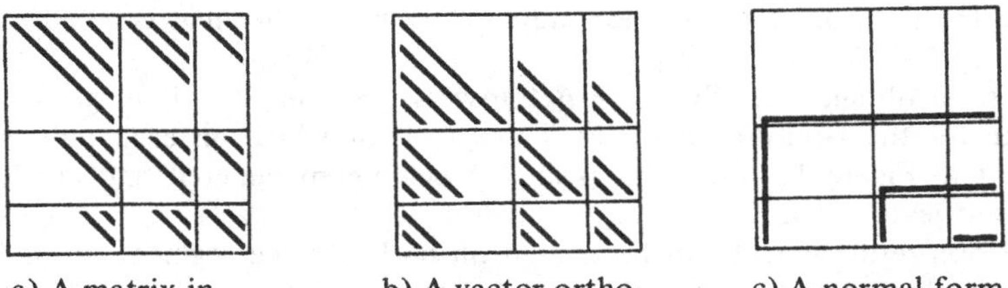

a) A matrix in the centralizer

b) A vector orthogonal to the orbit

c) A normal form

Fig. 1.

[1] The proof is in [8], 199–207.

In Figure 1a each oblique segment denotes a sequence of identical numbers, and the blank entries denote zero. Thus, the number of oblique segments is equal to the dimension of the centralizer.

COROLLARY. *The dimension of the centralizer of the matrix A_0 (which is equal to the codimension of the orbit and to the minimum dimension of a versal deformation) is given by the formula* $d = n_1 + 3n_2 + 5n_3 + \ldots$

If a Jordan matrix A_0 has several eigenvalues, we divide it into blocks corresponding to these eigenvalues. Then matrices commuting with A_0 are block diagonal, where a block corresponding to an eigenvalue λ has the form described in Figure 1a.

The formula for the dimension of the centralizer (the codimension of the orbit, the dimension of a minimal versal family) is obtained from the above by summation over all the distinct eigenvalues.

4.3. Selection of the normal form. For a transversal deformation of a matrix A_0 we may choose the family of the form $A_0 + B$, where B belongs to the orthogonal complement of the orbit of A_0 described above. So we get a versal deformation of A_0 with the minimum number of parameters (equal to the codimension of the orbit).

If A_0 has only one eigenvalue, B has the form described in Figure 1b. Here on each oblique segment is a sequence of identical numbers, which are the parameters of the deformation. The number of parameters is equal to the number of segments and is given by the formula of Corollary 4.2.

The matrix B has many non-zero entries. We can suggest another form of the versal family $A_0 + B$ in which the number of non-zero entries in B is minimal (and equal to the number of parameters).

To this end we choose a base of the centralizer in the following natural way: with each oblique segment in Figure 1a we associate a matrix with 0's and 1's, the 1's being on the given oblique segment.

By writing down the condition of orthogonality in Lemma 4.1, we can verify that the system of independent equations of the tangent space to the orbit consists of the following equations: for each oblique segment of Figure 1b the sum of the corresponding elements of the matrix is equal to zero.

Thus, to obtain a family $A_0 + B$ transversal to the orbit it is sufficient to take for the family of matrices B matrices for which on each oblique segment in Figure 1b one entry is an independent parameter, and all the other entries are zero.

The non-zero element can be chosen on each oblique segment at an arbitrary place. For example, the matrix described in Figure 1c is a suitable choice. So we come to the following result.

4.4. THEOREM. *Every matrix A_0 has a versal deformation, the number of its parameters is equal to the codimension of the orbit and to the dimension of the centralizer of A_0. This number is equal to*

$d = \sum_{\lambda} (n_1 + 3n_2 + 5n_3 + \ldots)$, *where* $n_1 \geqslant n_2 \geqslant \ldots$ *are the orders of Jordan blocks corresponding to the eigenvalue* λ *and the summation is over all the distinct eigenvalues. A versal deformation with fewer parameters does not exist.*

If A_0 *is in the Jordan normal form, then for a versal deformation with the minimum number of parameters we may take a d-parameter "normal form"* $A_0 + B$, *where the blocks of the block-diagonal matrix* B *have the form described in Figure* 1c. *The order of a block is equal to the multiplicity of the corresponding eigenvalue. The entries indicated in Figure* 1c *are independent parameters and the other entries are equal to zero.*

In other words *any complex matrix close to the given matrix can be reduced to the above d-parameter normal form* $A_0 + B$ *(where* A_0 *is the Jordan normal form of the given matrix), so that the reducing mapping and the parameters of the normal form depend holomorphically on the elements of the original matrix.*

Aliter: *each matrix over the ring of germs of functions holomorphic at zero can be reduced to the above d-parameter normal form* $A_0 + B$, *where* A_0 *is the Jordan form of the value of the germ at zero and* B *is a matrix whose only non-zero entries are those indicated in Figure* 1c.

4.5. Examples. If all the eigenvalues of A_0 are distinct, then $d = n$ and our normal form is diagonal.

If A_0 is the null matrix, then $d = n^2$ and our versal deformation is simply the family of all matrices.

If A_0 consist of one Jordan block, then $d = n$ and our normal form is the Sylvester form.

We denote a matrix in Jordan normal form by the product of the determinants of its blocks. For example, $\alpha^3\alpha^2$ denotes a matrix having two Jordan blocks (of order 3 and 2) with the same eigenvalue α, and the matrix $\alpha^2\alpha\beta^2$ has blocks of order 2 and 1 with the eigenvalue α and a block of order 2 with the eigenvalue β.

By Theorem 4.4 *minimal versal deformations of the matrices* $\alpha^3\alpha^2$ *and* $\alpha^3\alpha\beta^2$ *can be chosen in the following form*

$$\begin{vmatrix} \alpha & 1 & 0 & 0 & 0 \\ 0 & \alpha & 1 & 0 & 0 \\ \lambda_1 & \lambda_2 & \alpha+\lambda_3 & \lambda_4 & \lambda_5 \\ \lambda_6 & 0 & 0 & \alpha & 1 \\ \lambda_7 & 0 & 0 & \lambda_8 & \alpha+\lambda_9 \end{vmatrix}, \quad \begin{vmatrix} \alpha & 1 & 0 & 0 & 0 \\ \lambda_1 & \alpha+\lambda_2 & \lambda_3 & 0 & 0 \\ \lambda_4 & 0 & \alpha+\lambda_5 & 0 & 0 \\ 0 & 0 & 0 & \beta & 1 \\ 0 & 0 & 0 & \lambda_6 & \beta+\lambda_7 \end{vmatrix}.$$

The dimensions of the families are, respectively, equal to

$$d = n_1 + 3n_2 = 3 + 3\cdot 2 = 9, \quad d = n_1(\alpha) + 3n_2(\alpha) + n_1(\beta) = 2 + 3\cdot 1 + 2 = 7.$$

Thus, *any family of matrices close to a matrix of order* 5 *with a unique eigenvalue* α *and Jordan blocks of orders* 3 *and* 2 *can be reduced, without loss of smoothness, to the above nine-parameter normal form. If the matrices of the family depend holomorphically on the parameters then the coefficients* $\lambda_1, \lambda_2, \ldots, \lambda_9$ *of the normal form also depend holomorphically on the parameters.*

Similarly, any family of matrices close to $\alpha^2\alpha\beta^2$ can be reduced smoothly to the above seven-parameter normal form.

§ 5. Remarks

5.1. Generalizations. 1. Similar propositions can also be proved for the real case and for differentiable dependence on parameters, but the formulae are tedious.

2. In all the definitions instead of families of matrices we could have talked of fibre sections of fibre endomorphisms of the given vector fibering. This would have made the statements longer, without changing the proofs.

3. Similar (but simpler) results can be obtained for families of rectangular matrices (mappings from one space into another). Here a minimal versal family has the form

$$\begin{pmatrix} E_r & \\ & \begin{matrix} \lambda_{11} \ldots \lambda_{1l} \\ \cdots\cdots \\ \lambda_{k1} \ldots \lambda_{kl} \end{matrix} \end{pmatrix},$$

where r is the rank of A_0, E_r is the identity matrix of order r, and λ_{ij} are independent parameters.

4. In Lemma 2.2 neither the fact that the manifold M consists of matrices, nor that G is a group of matrices are essential: important only is the fact that the Lie group G acts on M.

5.2. Applications. By combining the formulae we have obtained with the transversality theorem [2], we can derive many results on families of matrices "of general form".

For example, from the versality of the family in 5.1.3 we obtain the following corollary.

COROLLARY. *In the space of families of linear mappings* $A : \mathbf{C}^m \to \mathbf{C}^n$ *an open everywhere dense set forms a family of the following kind. The rank of A is everywhere equal to* min (m, n), *except on the "bifurcation diagram", which is formed by a finite number of smooth manifolds* Σ^i *in the parameter space. On* Σ^i *the rank is reduced to i units; the codimension of* Σ^i *is equal to* $(m - r)(n - r)$. *In the neighbourhood of each point the stratification* $\{\Sigma^i\}$ *of the parameter space is diffeomorphic to the direct product of a domain in* $\mathbf{C}^k$ *and the stratification of a neighbourhood of zero in one of the verbal families of* 5.1.3.

EXAMPLE. Let $m = n$ and let the rank of A_0 be $n - 1$. Then the bifurcational diagram of a family of general form in the neighbourhood of

A_0 is a smooth hypersurface Σ^1. If the rank of A_0 is $n - 2$, then the bifurcation diagram consists of a smooth submanifold Σ^2 of codimension 4 and a hypersurface Σ^1 which tends to Σ^2 so that in every plane normal to Σ^2, Σ^1 has close to Σ^2 the form of a quadric cone.

The normal forms of § 3 and § 4 lead to similar results, but to state them we have first to get rid of the continuous parameters – the eigenvalues.

5.3. Orbits of a bundle. We call a *bundle* the set of all matrices whose Jordan normal forms differ only by their eigenvalues, but for which the sets of distinct eigenvalues and the orders of the Jordan blocks are the same. For example, all the diagonal matrices with simple eigenvalues define one bundle.

Each bundle is a semi-algebraic smooth submanifold of the space of matrices, it is a fibre space whose fibres are the orbits. The base of this fibering is the configuration space consisting of sets of coloured eigenvalues (two eigenvalues have the same colour if the corresponding Jordan blocks have the same order).

The splitting into bundles is a finite semi-algebraic stratification of the space $\mathbf{C}^{n^2}$, and we can make use of the transversality theorem.

COROLLARY. *In the space of families of $n \times n$ matrices an open everywhere dense set forms a family transversal to all the bundles of the orbits.*

5.4. Families in general position. Families of matrices that are transversal to all the bundles are said to be in *general position.* Corresponding to the decomposition of the space of matrices into bundles, the parameter space of the family decomposes into submanifolds. In a family in general position almost all the matrices have simple eigenvalues. The exceptional parameter values to which there correspond matrices with multiple eigenvalues define a subset of the parameter space. We call this a *bifurcation diagram.*

The bifurcation diagram of a family in general position is a finite union of smooth manifolds; to each bundle of orbits there corresponds a manifold in the parameter space – a manifold of matrices with fixed orders of the Jordan blocks. The codimension of such a manifold in the parameter space of a family in general position is equal to the codimension of the corresponding bundle in the space of all matrices.

The singularities of the bifurcation diagram of a family in general position are the same as those in bifurcation diagrams of versal deformations.

EXAMPLE. We assume that a family in general position contains a Jordan block of order n. Then the bifurcation diagram (in some neighbourhood of the corresponding value of the parameter) has the form of a

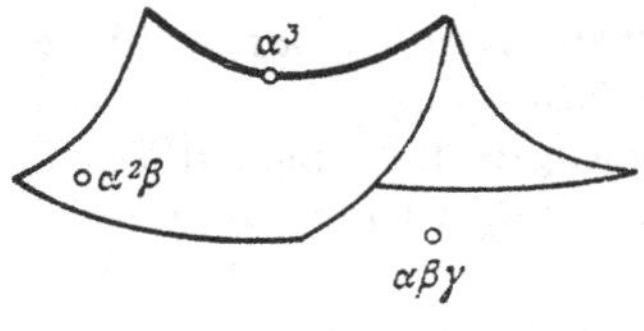

Fig. 2. Bifurcation diagram of a family in general position containing a Jordan block of order 3.

direct product of a swallow tail[1] of the required dimension and a Euclidean space. For example, Figure 2 represents the bifurcation diagram of a family in general position containing a Jordan block of order 3.

5.5. Codimensions of bundles. It is easy to prove:

LEMMA. *The codimension of a bundle can be expressed by the dimensions of Jordan blocks $n_1(\lambda) \geqslant n_2(\lambda) \geqslant \ldots$ according to the formula*

$$c = 2 \sum_{\lambda} (n_2 + 2n_3 + 3n_4 + \ldots) + \sum_{\lambda} (n_1 + n_2 + \ldots - 1).$$

For $c = d - \nu$, where d is the codimension of the orbit, and ν is the number of distinct eigenvalues λ. By Theorem 4.4

$$d = \sum_{\lambda} (n_1 + n_2 + \ldots) + 2 \sum_{\lambda} (n_2 + 2n_2 + 3n_4 + \ldots), \quad \nu = \sum_{\lambda} 1,$$

from which the formula follows.

Note that in the formula for the codimension of a bundle simple eigenvalues make no contribution. In addition, the codimension of a bundle does not depend on the order n of the matrix, but depends only on the sizes of the Jordan blocks corresponding to multiple eigenvalues. Therefore Lemma 5.5 permits us to enumerate quickly singularities of small codimensions in bifurcation diagrams of families in general position irrespective of the order of the matrices.

5.6. Enumeration of singularities of small codimension. The only bundle of codimension 1 in the space of all matrices is that of matrices having a single eigenvalue of multiplicity two. We denote this bundle by α^2, without indicating the simple eigenvalues, which can be arbitrary. For the sake of brevity we say "the codimension of a case" (instead of "the codimension of a bundle in the space of all matrices"). In the case α^2 the bifurcation diagram is a smooth hypersurface.

There are two cases of codimension 2: that of a pair of Jordan blocks each of order 2 with distinct eigenvalues ($\alpha^2 \beta^2$) and that of one Jordan block of order 3 (α^3). The corresponding bifurcation diagrams in the plane are a pair of intersecting straight lines and a semi-cubical parabola.

By Corollary 5.3 *a two-parameter family in general position has a bifurcation diagram with singularities only of these two forms.*

Thus, the versal families of § 4 allow us to investigate the singularities of bifurcation diagrams of families in general position with a large number of

---

[1] A swallow tail of dimension $k - 1$ is a hypersurface in $\mathbf{C}^k$ given by an equation $\Delta(a) = 0$, where Δ is the discriminant of the polynomial $z^{k+1} + a_1 z^{k-1} + \ldots + a_k = 0$. A one-dimensional swallow tail is a semi-cubical parabola; a two-dimensional one is shown on the left of Figure 3.

parameters.[1] With increasing codimension c the number of distinct bundles N grows in the following form:

c	1	2	3	4	5	6	7	8	9
N	1	2	4	7	11	19	30	51	73

All cases of codimension $c < 5$ are enumerated in the following list:

c	1	2	3	4
det	α^2	α^3, $\alpha^2\beta^2$	α^4, $\alpha\alpha$, $\alpha^3\beta^2$, $\alpha^2\beta^2\gamma^2$	α^5, $\alpha^2\alpha$, $\alpha^4\beta^2$, $\alpha\alpha\beta^2$, $\alpha^3\beta^2\gamma^2$, $\alpha^3\beta^3$, $\alpha^2\beta^2\gamma^2\delta^2$.

For example, the column $c = 3$ indicates that *the bifurcation diagrams of three-parameter families in general position have no singularities except those in Fig.* 3.

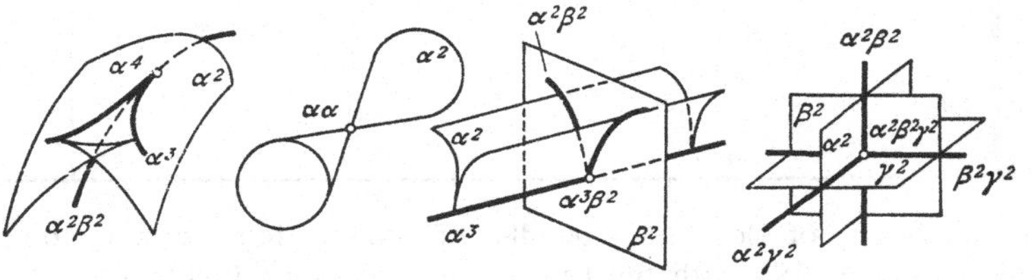

Fig. 3. Singularities of bifurcation diagrams of three-parameter families in general position.

Note that the bifurcation diagram of a family in general position with several distinct multiple eigenvalues is locally diffeomorphic to a direct product of diagrams of families corresponding to cases with one multiple eigenvalue.

We call bundles whose matrices have only one multiple eigenvalue *elementary*. All the other bundles are essentially transversal intersections of elementary ones. In particular, the codimension of a complex singularity is equal to the sum of the codimensions of the elementary components.

---

1 Note that a versal d-parameter deformation of the matrix A_0 is transversal to the orbit and consequently to the bundle containing A_0. The codimension c of this bundle is less than d. Therefore, to construct bifurcation diagrams we can consider deformations with fewer parameters transversal to the bundle but not to the orbit.

Such a deformation is, for example, the c-parameter subfamily of the versal family of Theorem 4.4, which is obtained if one of the diagonal parameters of B is equated to zero (Fig. 1b) for each eigenvalue λ. Another c-parameter deformation, transversal to the bundle, is obtained if in the normal form of Theorem 4.4 we restrict ourselves to matrices B in which all the blocks corresponding to distinct eigenvalues have trace zero.

To prove this consider vectors tangent to the bundle but orthogonal to the orbit. Our assertions follow from the fact that if the matrix A_0 has only one eigenvalue, then such vectors and only they are scalar matrices.

The orders of the Jordan blocks of all the elementary bundles of codimension $c < 16$ are given in the following table:

c	1	2	3	...	7	8	...	11	12	...	15
$\{n_i\}$	2	3	4	...	8	9	...	12	13	...	16
			1, 1	...	5, 1	6, 1	...	9, 1	10, 1	...	13, 1
					2, 2	3, 2	...	6, 2	7, 2	...	10, 2
						1, 1, 1	...	4, 1, 1	5, 1, 1	...	8, 1, 1
								3, 3	4, 3	...	7, 3
									2, 2, 1	...	5, 2, 1
											4, 4
											1, 1, 1, 1

The number a_c of elementary bundles of codimension c can be computed by Euler's method [9] with the help of the generating function

$$\sum_{c=-1}^{\infty} a_c x^{c+1} = \prod_{k=1}^{\infty} (1 - x^{k^2})^{-1}.$$

Here we must take $a_{-1} = a_0 = 1$. The first few coefficients are

c	1	2	3	4	5	6	7	8	9	10	11	12	13	14	15
a_c	1	1	2	2	2	2	3	4	4	4	5	6	6	6	8

ADDED IN PROOF. With each matrix we can associate an interesting family of algebraic manifolds. Namely, we assign to each matrix the set of coefficients of its characteristic polynomial. The space of all matrices then becomes the complete space of families of algebraic manifolds (the fibre of a family is the set of all matrices with a fixed characteristic polynomial).

In exactly the same way the base of a deformation of an arbitrary matrix decomposes into subsets corresponding to matrices with common characteristic polynomials. Our versal deformations are algebraic (linear), therefore their bases are complete spaces of algebraic families of algebraic manifolds.

The explicit formulae of versal deformations make it possible to indicate explicitly the equations defining these families. Computations show that even for the simplest matrices we get highly unusual families of algebraic manifolds. Let us show, for example, how to obtain in this way one of Briescorn's results [12] on rational double points.

The dimension of the manifold of a family constructed for a given matrix is expressed in terms of the orders of the Jordan blocks according to the formula

$$D = d - n = \sum_{\lambda} 2n_2(\lambda) + 4n_3(\lambda) + 6n_4(\lambda) + \ldots$$

(in the notation of 4.2).

Hence, *the family so constructed is that of algebraic surfaces* ($D = 2$) *if and only if to each eigenvalue there corresponds one Jordan block except for a single eigenvalue to which there correspond two Jordan blocks one of which is of order* 1.

By computing the characteristic polynomial of matrices corresponding to a versal family we can write down explicitly the equations of the resulting family of surfaces. After insignificant transformations these can be put into the form

$$xy + z^{k+1} + a_1 z^{k-1} + \ldots + a_k = 0,$$

where $a_1, \ldots, a_k$ are the parameters of the family. But this is the standard versal deformation of a surface with a rational double point of type A_k, which was obtained in detail by M. Artin, Briescorn and Tyurina [13]. Thus, *families of algebraic surfaces arising from versal deformations of matrices, and only these, are versal deformations of surfaces with rational double points of type* A_k.

References

[1] G. Frobenius, Ueber die vertauschbaren Matrizen, S.–B. Preuss. Akad. Wiss. Phys. Math. Kl. (1896), 7–16.

[2] Singularities of differentiable mappings. A collection of translations, Izdat. "Mir", Moscow 1968.

[3] V. I. Arnold, Remarks on singularities of finite codimension in complex dynamical systems, Funktsional Anal. i Prilozhen. **3** (1969), 1–6.
= Functional Anal. Appl. **3** (1969), 1–5.

[4] R. J. Sacker, On invariant surfaces and bifurcation of periodic solutions of ordinary differential equations, New York University, 1964.

[5] N. N. Brushlinskaya, A versal family of vector fields of a Poincaré domain, Funktsional Anal. i Prilozhen. **4** (1970), 6–13.
= Functional Anal. Appl. **4** (1970).

[6] S. G. Krein and V. P. Trofimov, On holomorphic operator functions of several complex variables, Funktsional Anal. i Prilozhen 3 (1969), 85–86.
= Functional. Anal. Appl. 3 (1969), 330–331.
[7] F. V. Atkinson, Multiparameter spectral theory, Bull. Amer. Math. Soc. 74 (1968), 1–27.
[8] F. R. Gantmakher, Teoriya matrits, second ed. Izdat. Nauka, Moscow 1967. Translation: The theory of matrices, Chelsea, New York 1969.
[9] L. Euler, De partitione numerorum, Novi Comment. Acad. Sc. Petrop. 3 (1950/1), 125–135.
[10][1] W. Wasow, On holomorphically similar matrices, J. Math. Anal. Appl. 4 (1962), 202–206.
[11][1] S. Khabbaz and G. Stengle, An application of *K*-theory to global analysis of matrix-valued function, Math. Ann. **179** (1969), 115–122.
[12][1] E. Briescorn, Nice Mathematical Congress, 1970.
[13][1] G. N. Tyurina, The resolution of singularities of plane deformations of rational double points, Funktsional. Anal. i Prilozhen. 4 (1970), 77–83.
= Functional Anal. Appl. 4 (1970).

Translated by H. Freedman.

[1] Added in proof.

To A. N. Kolmogorov on
his seventieth birthday

REMARKS ON THE STATIONARY PHASE METHOD AND COXETER NUMBERS

V. I. Arnol'd

We study integrals of rapidly oscillating functions. Such integrals tend to zero when the length of the oscillation wave tends to zero through wave fronts of constant form. The asymptotic decrease of the integral is determined by the character of the critical points of the function describing the front. If all of these critical points are non-degenerate (Morse), then the integral tends to zero like the wave length raised to the power of half the dimension of the space, and indeed this is the asymptotic behaviour of the integral for functions in general position.

However, if the integral depends on additional parameters, then for certain "caustic" parameter values there arise non-Morse critical points and the integral decreases slowly. The investigation of the asymptotic behaviour of the integral of an oscillating function in caustic cases can be regarded as a generalization of the theory of Airy functions; it is closely connected with Artin's braid theory, and the answer in the case of few parameters is expressed in terms of the Coxeter numbers of the Weyl groups of the series A, D, E, and in the case of many parameters it is expressed in terms of generalizations of them.

Contents

§1. Introduction 61
§2. Terminology 63
§3. Classification theorems 72
§4. Singularity indices 83
References 88

§1. Introduction

1.1. Trigonometric integrals. Let S: $\mathbf{R}^n \to \mathbf{R}$ be a smooth function. A *trigonometric integral with the function S* is an integral of the form

(1) $$I(h) = \int_{\mathbf{R}^n} e^{iS(x)/h} \varphi(x)\, dx,$$

where h is a small real parameter and ϕ an infinitely differentiable function of compact support.

By the stationary phase principle, the principal contribution to the asymptotic behaviour of I as $h \to 0$ is given by neighbourhoods of the critical points of S. By Morse's lemma, in a neighbourhood of a non-degenerate critical point S can be bought by a diffeomorphism to the normal form $S = \pm x_1^2 \pm \ldots \pm x_n^2 + \text{const}$. A trigonometric integral with such a function S and with $\phi \equiv 1$ can be computed explicitly: it is proportional to $h^{n/2}$.

If the support of ϕ is concentrated in a neighbourhood in which S reduces to the normal form, then clearly $|I(h)| \leqslant 0(h^{n/2})$ as $h \to 0$.

The asymptotic behaviour of a trigonometric integral with a function S for the worst ϕ is determined by the differentiable type of S (in other words, it is independent of the coordinate system): under a diffeomorphism $x = f(y)$ the trigonometric integral goes over to an integral of the same kind, but with ϕ multiplied by the Jacobian of f.

The local case when S is the germ of a smooth function at 0 can be treated in the same way. Thus, the determination of the asymptotic behaviour of trigonometric integrals is closely connected with the classification of critical points up to a diffeomorphism.

A function "in general position" has only non-degenerate critical points; hence a trigonometric integral with such a function decreases like $h^{n/2}$.

1.2. Trigonometric integrals depending on a parameter. In applications, however, we meet the situation when the trigonometric integral under discussion depends on some parameters; for example, we often come across integrals of the form

(2) $$I(h, p) = \int_{\mathbf{R}^n} e^{i[S(x) - \langle p, x\rangle]/h} \varphi(x)\, dx$$

(the so-called phase integrals, depending on an n-dimensional parameter $p \in \mathbf{R}^{n'}$), or the more general integrals

(3) $$I(h, \lambda) = \int_{\mathbf{R}^n} e^{iF(x, \lambda)/h} \varphi(x, \lambda)\, dx,$$

depending on an arbitrary number of parameters ($\lambda \in R^l$).

1.3. Estimates for trigonometric integrals. The asymptotic behaviour of such integrals depending on a small number of parameters was considered in [1] on the request of Maslov, who needed it for his investigation of focussing energy in crystals [2].

This clarified the astonishing connection [3] between the simplest degenerate critical points of the functions and the Weyl groups of the series A, D, E and their braid groups, which were introduced a little earlier by Brieskorn [4]. The asymptotic behaviour for these simplest degenerate critical points of the series A, D, E were expressed in terms of a certain integer N according to the formula

$$I(h) = O(h^{\frac{n}{2}-\beta}), \quad \text{where} \quad \beta = \frac{1}{2} - \frac{1}{N},$$

and

$$N(A_k) = k+1, \quad N(D_k) = 2k-2, \quad N(E_6) = 12, \quad N(E_7) = 18, \quad N(E_8) = 30.$$

I am grateful to B. Kostant for pointing out that N is the Coxeter number corresponding to the group generated by reflections.

At present there exist tables of normal forms to which functions reduce in the neighbourhood of a critical point such that all the cases not covered by these tables form a set of codimension 11. Thus, to complete the investigation of the asymptotic behaviour of the integrals (2), (3) with functions S and F in general position and depending on not more than 10 parameters, it remains simply to carry out the computations for reducing the functions to normal form. The answer looks like this:

1) *For every* $\varepsilon > 0$ *and for almost every*[1] *l-parameter family* $F(x, \lambda)$ *of trigonometric integrals* (3), *we have* $|I(h)| \leqslant O(h^{\frac{n}{2}-\beta_l-\varepsilon})$ *as* $h \to 0$ *for every value of* λ, *provided that* $n \geqslant 3$.

l	1	2	3	4	5	6	7	8	9	10, $n \geqslant 4$	$k(k+1)/2$, $n \geqslant k$
β_l	1/6	1/4	1/3	3/8	5/12	1/2	1/2	13/24	9/16	2/3	$k/6$?

2) *This estimate cannot be improved, in the sense that there exist l-parameter families such that the integral* (3) *decreases no faster than* $O(h^{\frac{n}{2}-\beta_l})$ *as* $h \to 0$ *for certain values of* λ: *this property of the family is preserved under small displacements.*

A similar answer is obtained for the integrals (2) with functions S in general position. See [3] for more details on the families of functions $S(x) - \langle p, x\rangle$ and their connection with Lagrange singularities.

The problem of the exact determination of the numbers β_l seems to be very difficult.

§2. Terminology

We need a whole series of concepts for our investigation of critical points of functions and the asymptotic behaviour of trigonometric integrals. This section is devoted to the definitions of these concepts.

---

[1] Excluding a thin set (that is, the union of countably many closed nowhere dense subsets) in the space of families.

2.1. The singularity index.

2.1.1. DEFINITION. *The singularity index* β *of the germ* s *of a function* S: $(\mathbf{R}^n, 0) \to \mathbf{R}$ *at the critical point* 0 is the infimum of the set of numbers γ for which the following estimate for the trigonometric integral (1) holds:

$$I(h) = O(h^{\frac{n}{2}-\gamma}) \text{ as } \quad h \to 0$$

for every representative S of the germ s and for all smooth ϕ with sufficiently small support (chosen in accordance with the representative s of S).

2.1.2. EXAMPLE. The singularity index of the germ of any function at a non-degenerate critical point is 0 (this case was discussed already by Fresnel).

2.1.3. The singularity index of the germ of the function x^3 of one variable at zero is 1/6 (this case was considered by Airy).

The function $x^3 + y^2$ of two variables has the same singularity index. In general, the definition of singularity index has been designed so that the addition of the squares of new variables does not alter it. Larger indices β correspond to "more complicated" singularities.

2.1.4. PROPOSITION. *The singularity index is an invariant of the differentiable type of the germ.*

We say that two germs s_1 and s_2 of function $\mathbf{R}^n \to \mathbf{R}$ are *differentiably equivalent* (or *have the same differentiable type*) if there exists a germ g of a diffeomorphism $(\mathbf{R}^n, 0) \to (\mathbf{R}^n, 0)$ such that $s_1 = s_2 \circ g$.

Thus, Proposition 2.1.4. means that the singularity indices of any two germs carried into each other by a smooth change of coordinates are the same.

The proof of 2.1.4. is obvious (see the Introduction).

2.1.5. PROPOSITION. *The singularity index of the germ of a polynomial at any point is a rational number.*

This proposition is non-trivial; a proof follows from recent papers of Bernshtein and Gel'fand [5] and Atiyah [6] and is based on Hironaka's theorem on resolution of singularities.

The results of these papers also give more exact information on the asymptotic behaviour of trigonometric integrals in the case where S is a polynomial. Such an integral has an asymptotic expansion with terms of the form $h^{\upsilon} \ln^{\kappa} h$; the indices υ lie in finitely many rational arithmetic progressions. On the other hand, for general l-parameter families of smooth functions the germ of each function at every point is equivalent to that of a polynomial. To formulate this assertion more precisely, we use the following concept.

2.2. Multiplicity and the local ring of a critical point of a function.

2.2.1. DEFINITION. The *local ring* of a smooth function $S = S(x_1, \ldots, x_n)$ at the critical point $0 \in \mathbf{R}^n$ is the factor ring of the ring of germs of smooth functions of $x_1, \ldots, x_n$ at 0 by the ideal generated by the partial derivatives of S:

$$Q = C^\infty((x_1, \ldots, x_n))/(\partial S/\partial x_1, \ldots, \partial S/\partial x_n).$$

It is easy to see that Q does not depend on the choice of the coordinate system $(x_1, \ldots, x_n)$.

2.2.2. DEFINITION. The *multiplicity* μ of S at the critical point 0 is the dimension of Q as a linear space over the real field **R**: $\mu = \dim_{\mathbf{R}} Q$.

The critical point 0 of S is said to be of *finite multiplicity* if μ is finite.

2.2.3. EXAMPLE. The function $S = x^3$ of one variable x has a critical point of multiplicity 2 at the origin; under small displacements of S this critical point splits into two of multiplicity 1.

In exactly the same way, the function x^n has a critical point of multiplicity $\mu = n - 1$ at the origin.

2.2.4. REMARK. The multiplicity μ at the critical point 0 of a function S should not be confused with the multiplicity of the singular point 0 on the hyperplane $S = 0$, as defined in algebraic geometry.

2.2.5. DEFINITION. *The multiplicity at the critical point* 0 *of a holomorphic function* S: $(\mathbf{C}^n, 0) \to \mathbf{C}$ is defined as the dimension of the complex local ring Q: $\mu = \dim_{\mathbf{C}} Q$, $Q = H/(\partial S/\partial x_1, \ldots, \partial S/\partial x_n)$, where H is the ring of germs of functions of $x_1, \ldots, x_n$ holomorphic at 0.

2.2.6. REMARK. For a critical point of finite multiplicity, the rings of germs of smooth functions, C^∞ and H, in Definitions 2.2.1 and 2.2.5, can be replaced by the rings of formal power series or even the ring of truncated polynomials (factored by the ideal consisting of polynomials having a zero of sufficiently high order at 0).

All of this follows from the preparation theorem of Weierstrass and Malgrange (see [7]–[9]).

2.2.7. REMARK. If it is finite, the multiplicity μ may be defined also as the degree of the gradient mapping at the critical point, or as the degree of the following mapping of a small sphere with centre at the critical point onto itself: $z \to \|z\| \, (\partial S/\partial z)/\|\partial S/\partial z\|$. Concerning this see [10], [11].

2.3. Sufficient jets. We recall that the k-jet of a function at a point is defined, in terms of a fixed coordinate system, as the section of the Taylor series of the function at the point up to the terms of degree k inclusive.

2.3.1. DEFINITION. The k-jet of a function at a point is said to be *sufficient* if all functions having the given k-jet are differentiably equivalent to each other in the neighbourhood of the point.

It is clear that sufficiency is a property of the jet and not of the function.

2.3.2. EXAMPLE. The 2-jet of the function $\pm x_1^2 \pm \ldots \pm x_n^2$ of n variables at 0 is sufficient (Morse's Lemma).

2.3.3. EXAMPLE. The 3-jet of the function x^2y of two variables at zero is not sufficient, since the germs of x^2y and $x^2y + y^4$ are not equivalent.

2.3.4. PROPOSITION. *Let* 0 *be a critical point of finite multiplicity of a smooth (real or complex) function. Then for sufficiently large k, the k-jet of this function at 0 is sufficient.*

In other words, *in a neighbourhood of a critical point the function is smoothly equivalent to its Taylor polynomial.*

The proof of Proposition 2.3.4 is by no means trivial. Apparently it was first proved by Tougeron [9]. There are other proofs (in the complex and real cases) in [8], [12], [13].

2.3.5. PROPOSITION. *The set of all ∞-jets of functions at critical points of infinite multiplicity has infinite codimension in the space of all jets of functions.*

In other words, the Taylor coefficients of a function at a critical point of infinite multiplicity are subject to an infinite number of independent (polynomial) conditions.

The proof of Proposition 2.3.5 is not difficult. Propositions 2.3.4, 2.3.5 and the transversality theorem imply that *every finite-parameter family of functions can be disturbed by arbitrarily small displacements in such a way that all functions in the disturbed family have critical points of finite multiplicity only.*

Thus, the investigation of critical points of functions that are not removable in finite-parameter families reduces to that of critical points of polynomials of finite multiplicity.

2.3.6. COROLLARY. *For almost every*[1] *finite-parameter family of functions S the singularity index at every point and for every value of the parameter is a rational number.*

The proof follows immediately from Proposition 2.1.5, by virtue of Propositions 2.3.4 and 2.3.5.

2.4. Stabilization.

2.4.1. DEFINITION. The germs at zero of two smooth functions of m and n variables are said to be *stably equivalent* if there exist non-degenerate quadratic forms which yield on direct addition differentiably equivalent germs of smooth functions in the same number of variables.

2.4.2. EXAMPLE. The germ at zero of the function x^3 of one variable is stably equivalent to the germ at zero of the function $x^3 + y^2 + z^2$ of three variables.

2.4.3. PROPOSITION. *If germs of functions of the same number of variables are stably equivalent, they are equivalent* (see [8], [14]).

Thus, transition to classes of stably equivalent functions allows us to assume that all functions have "one and the same number of variables", and nothing is changed for the classification of functions of a fixed number of variables.

2.4.4. PROPOSITION. *Stably equivalent germs have the same singularity*

[1] Excluding the union of finitely or countably many nowhere dense sets in the space of families of functions.

indices β, isomorphic local rings Q, and the same multiplicities μ of critical points.

The proofs are obvious.

2.5. Versal deformations. Versal deformations are the normal forms to which a family of functions can be brought in the neighbourhood of a parameter value corresponding to a degenerate critical point.

2.5.1. DEFINITION. A *deformation* of the germ of a smooth mapping f: $(\mathbf{R}^n, 0) \to \mathbf{R}^k$ is the germ of a smooth mapping F: $(\mathbf{R}^n \times \mathbf{R}^l, 0 \times 0) \to \mathbf{R}^k$ for which $F(., 0) = f$.

2.5.2. DEFINITION. By an *equivalence H of deformations F, G*: $(\mathbf{R}^n \times \mathbf{R}^l, 0 \times 0) \to \mathbf{R}$ of a function f we mean a deformation H: $(\mathbf{R}^n \times \mathbf{R}^l, 0 \times 0) \to \mathbf{R}^n$ of the identity mapping of $\mathbf{R}^n$ onto itself such that $F(x, \lambda) \equiv G(H(x, \lambda), \lambda)$.

2.5.3. DEFINITION. A deformation K: $(\mathbf{R}^n \times \mathbf{R}^k, 0 \times 0) \to \mathbf{R}$ of the germ of a function f at zero is said to be *induced from the deformation* F: $(\mathbf{R}^n \times \mathbf{R}^l, 0 \times 0) \to \mathbf{R}$ *under a change of parameter* ϕ: $(\mathbf{R}^k, 0) \to (\mathbf{R}^l, 0)$ if $K(x, \lambda) \equiv F(x, \phi(\lambda))$.

2.5.4. DEFINITION. A deformation F of the germ of a function f at 0 is said to be *versal* if every other deformation of f is equivalent to the deformation induced from F.

In other words, every deformation G can be expressed in terms of F,

$$G(x, \lambda) \equiv F(H(x, \lambda), \phi(\lambda)),$$

where H is a diffeomorphism depending smoothly on the parameter λ, ϕ is a smooth change of parameter, $\phi(0) = 0$, and $H(x, 0) = x$.

Differentiation of the preceding formula leads to the following concept.

2.5.5. DEFINITION. An l-parameter deformation F of the germ of a function f of n variables at 0 is said to be *infinitesimally versal* if every germ α of a smooth function of n variables has a decomposition

$$\alpha = \sum_{i=1}^{n} \frac{\partial f}{\partial x_i} h_i + \sum_{j=1}^{l} c_j g_j,$$

where the h_i are germs of smooth functions of n variables, the c_j are numbers, and the g_j are germs of the partial derivatives of F with respect to the parameters, taken at the value zero of these parameters.

2.5.6. PROPOSITION. *Every infinitesimally versal deformation of the germ of a function is versal.*

The proof of this "versality theorem" is by no means trivial: see [3], [15]–[19].

2.5.7. COROLLARY. *The germ of a function at a critical point of finite multiplicity μ has a μ-parameter versal deformation.*

It is sufficient to take this deformation of f:

$$F(x, \lambda) = f(x) + \Sigma\ \lambda_j g_j(x),$$

where $\{g_j\}$ is a set of μ functions giving a basis of the linear space Q of the local ring of f at zero after factorization.

2.5.8. EXAMPLE. The deformations $x^2 + \lambda_1$, $x^3 + \lambda_1 x + \lambda_2$, $x^4 + \lambda_1 x^2 + \lambda_2 x + \lambda_3$ are versal deformations of the functions x^2, x^3 and x^4, respectively, at zero.

The corresponding trigonometric integrals have been known in physics for a long time (see [20], [21], [22]).

2.6. Modality. Let G be a Lie group, X a smooth manifold, and $G \times X \to X$ an action of the group on the manifold. We consider any point x of X. The orbits of the action of G on X may form a finite stratification or a continuous family in the neighbourhood of x. We define the modality of a point x as the number of parameters necessary for number numbering the orbits in the neighbourhood of x.

2.6.1. DEFINITION. *The modality of a point* $x \in X$ *under the action of* G *on* X is the smallest number m such that a sufficiently small neighbourhood of x in X intersects at most finitely many at most m-parameter families of orbits of G.

2.6.2. EXAMPLE. A point is of modality 0 if and only if a sufficiently small neighbourhood of it intersects only finitely many orbits.

2.6.3. DEFINITION. The germ of a function f: $(\mathbf{R}^n, 0) \to (\mathbf{R}, 0)$ is said to be *m-modal* if some (and hence every) sufficient k-jet of the function at 0 is modal under the action of the group of k-jets of diffeomorphisms $(\mathbf{R}^n, 0) \to (\mathbf{R}^n, 0)$ on the manifold of k-jets of functions at the critical point 0 and with critical value 0.

2.6.4. EXAMPLE. The germ at zero of the function x^k of one variable is 0-modal since a sufficiently small neighbourhood of its k-jet intersects only finitely many orbits (namely the orbits of germs of the functions $\pm x^l$, where $l \leqslant k$).

So far all the 0-modal and 1-modal germs have been calculated. It turns out that there are not too many of them, and indeed an analysis of the tables of 0-modal and 1-modal germs enables us to obtain the asymptotic behaviour of trigonometric integrals of general form with not too many parameters.

2.6.5. REMARK. *The modalities of stably equivalent germs are equal.* (The proof is obvious.)

In studying the modality of a germ of multiplicity μ we may replace the function space of germs by a μ-dimensional space, namely that with the versal deformations as basis, and examine how many parameters are needed to number the orbits in a neighbourhood of the parameter value zero. Another finite-dimensional space, and one that is even more suitable for this end, is a local transversal to the orbit in the space of k-jets of functions with the critical point 0 and critical value 0. Such a transversal is of dimension $\mu - 1$ and can be obtained by taking the vectors in a basis of the linear space $\mathfrak{m}^2\, \mathfrak{m}(\partial f/\partial x)$ as tangent vectors to the transversal surface, where $\mathfrak{m}$ is a maximal ideal in the ring of germs of functions of x.

2.6.6. EXAMPLE. A convenient transversal to an orbit of the germ of x^4 is the system of germs $x^4 + \lambda_1 x^3 + \lambda_2 x^2$.

2.7. The quadratic form of a singularity. Some further topological invariants are connected with a critical point of a complex function. One of these is a manifold with boundary, the so-called local level manifold of the function. It can be defined as follows (see [23]).

Let f: $(\mathbf{C}^n, 0) \to (\mathbf{C}, 0)$ be a smooth function in a neighbourhood of 0, where 0 is an isolated critical point with the critical value 0. We fix a Hermitian metric in $\mathbf{C}^n$. We choose a number $\rho > 0$ so small that all spheres of radius $r \leqslant \rho$ centred at 0 intersect the singular set of the level set $f^{-1}(0)$ transversally. Fixing ρ, we choose a number $\varepsilon > 0$ so small that the sphere of radius ρ intersects transversally also all non-singular sets of the level sets $f^{-1}(z)$ for $|z| \leqslant \varepsilon$.

2.7.1. DEFINITION. *The local manifold at the level $z \neq 0$ of a function* f at the critical point 0 is the set

$$V_z = \{x \in \mathbf{C}^n \colon f(x) = z, \ |x| \leqslant \rho\},$$

that is, the intersection of the non-singular manifold of the level set $f^{-1}(z)$ of f with the ball of radius ρ and centre at the origin of coordinates.

It is easy to see that V_z is a smooth manifold with boundary, and that V_z has the real dimension $2n - 2$. Further, it is easy to check that the differentiable type of V_z does not depend on the choice of ρ and ε, provided only that these are small enough ($\rho < \rho_0(f)$, $\varepsilon < \varepsilon_0((\rho, f))$.

2.7.2. EXAMPLE. If $f(x) = x_1^2 + \ldots + x_n^2$, then V_z is diffeomorphic to the space of vectors of length $\leqslant 1$ in the tangent bundle of the sphere S^{n-1}. For example, for $n = 2$ V_z is an ordinary cylinder.

2.7.3. PROPOSITION (see 23). *The local level manifold V_z is homotopy equivalent to the union of as many spheres of dimension $n - 1$ as the multiplicity μ of the critical point*:

$$V_z \sim \underbrace{S^{n-1} \vee \ldots \vee S^{n-1}}_{\mu}.$$

Thus, the $(n - 1)$-dimensional homology group of the local level manifold is the μ-dimensional lattice $\mathbf{Z}^\mu$.

2.7.4. DEFINITION. By the *quadratic form* of the germ of a function at a critical point we mean the intersection index in the homology of mean dimension of the local level manifold of a function of $n \equiv 3 \bmod 4$ variables and stably equivalent to the given function:

$$H_{n-1}(V_z, \mathbf{Z}) \times H_{n-1}(V_z, \mathbf{Z}) \to \mathbf{Z}.$$

2.7.5. REMARK. The intersection index of $(n - 1)$-dimensional cycles gives a symmetric form for odd n and a skew-symmetric form for even n. For $n \equiv 3 \bmod 4$ the square of the generator in the homology group $H_{n-1}(V_z, \mathbf{Z}) \sim \mathbf{Z}$ is -2 for the function $f = x_1^2 + \ldots + x_n^2$.

The quadratic module (H_{n-1} together with the symmetric integral form) is determined by the stable equivalence class of f.

2.8. The monodromy group. The local level manifold of the germ of a function at a critical point serves as the fibre in certain fibrations. Here is the construction.

We fix some versal deformation F of the germ of f, say a μ-parameter one for definiteness. Let $\mathbf{C}^\mu$ be the space of parameters; we fix a Hermitian metric in it.

Take $\lambda \in \mathbf{C}^\mu$. Consider the set of zero level of $F(\cdot, \lambda)$:

$$V_\lambda = \{x \in \mathbf{C}^n \colon F(x, \lambda) = 0, \ |x| \leqslant \rho\}.$$

The set V_λ intersects the sphere S_ρ transversally if $|\lambda| \leqslant \varepsilon$, where ε is sufficiently small. However, it may have singularities inside this sphere.

2.8.1. DEFINITION. By the *bifurcation germ* Σ of the germ of f at $0 \in \mathbf{C}^n$ we mean the germ at $0 \in \mathbf{C}^\mu$ of the subset of a versal deformation consisting of those points λ of the parameter space for which 0 is a critical value of $F(\cdot, \lambda)$ in the ball $|x| \leqslant \rho$.

The intersection of a representative of the bifurcation germ with the ball $|\lambda| \leqslant \varepsilon$ for sufficiently small ε is called a *bifurcation diagram* Σ_ε. This complex hypersurface is not, in general, a smooth manifold. The topological type of the pair $(\Sigma_\varepsilon, D_\varepsilon)$ (where D_ε is the ball $|\lambda| \leqslant \varepsilon$) does not depend on ε, as long as ε is sufficiently small; neither does it depend on the special choice of the μ-parameter versal deformation. In particular, the complement $D_\varepsilon - \Sigma_\varepsilon$ is a base space for a fibration with fibre V_λ and projection $(x, \lambda) \to \lambda$.

2.8.2. EXAMPLE. For the germ at zero of the function x^{k+1} of a single variable we can take as a versal deformation the polynomial.

$$F(x, \lambda) = x^{k+1} + \lambda_1 x^{k-1} + \lambda_2 x^{k-2} + \ldots + \lambda_k \qquad (\mu = k).$$

A bifurcation diagram is obtained by choosing the coefficients in such a way that the polynomial $F(\cdot, \lambda)$ has multiple roots. The equation of the bifurcation diagram then has the form $\Delta(\lambda_1, \ldots \lambda_k) = 0$, where Δ is the discriminant. This hypersurface is called a *swallow-tail*, and its complement is the space $K(\pi, 1)$ for the Artin braid group with $k + 1$ threads.

2.8.3. DEFINITION. *The braid group* of the germ of a function f at a critical point is the fundamental group $\pi_1(D_\varepsilon - \Sigma_\varepsilon)$ of the complement of the bifurcation diagram.

Since the fundamental group of the base space of a fibration acts on the homology of the fibre, we get a representation

$$\pi_1(D_\varepsilon - \Sigma_\varepsilon) \to \operatorname{Aut} H_{n-1}(V_\lambda, \mathbf{Z})$$

of the braid group in the group of automorphisms of the quadratic module.

This representation is called the *monodromy representation* and its image the *monodromy group of the germ.*

Similar actions arise for the homology with closed supports and other groups connected with the fibre. In defining monodromy it is useful to bear in mind that every element of the fundamental group of the base space gives rise to a diffeomorphism of the manifold V_λ with boundary, *which is the identity in some neighbourhood of the boundary.*

2.9. Classical monodromy. The monodromy group of the germ of a function has a significant subgroup, closely connected with the asymptotic behaviour of trigonometric integrals.

We incorporate a function f: $(\mathbf{C}^n, 0) \to)\mathbf{C}, 0)$ with isolated critical point 0 in a one-parameter system of functions $F(x, \lambda) = f(x) - \lambda$.

2.9.1. DEFINITION. The *classical monodromy* of the germ of f is the automorphism

$$h: H_{n-1}(V_\lambda, \mathbf{Z}) \to H_{n-1}(V_\lambda, \mathbf{Z})$$

of the homology of the local level manifold induced by a small rotation of the parameter λ around 0 ($\lambda(t) = e^{it}\lambda(0)$, $0 \leqslant t \leqslant 2\pi$).

The one-parameter deformation F of f indicated above is induced from a versal deformation (as is every deformation of f). The rotation of λ about 0 carries it into a certain remarkable element of the braid group of the germ.

This element can also be defined in terms of a special system of generators of the braid group, which is constructed as follows.

We fix a sphere of sufficiently small radius in the space $\mathbf{C}^\mu$ of parameters of the versal deformation. Almost every line sufficiently near to 0 intersects the bifurcation diagram transversally within the sphere in exactly μ points near 0.

2.9.2. PROPOSITION (see [24], [25], [26]). *The fundamental group of the complement of the bifurcation diagram in the sphere is generated by μ paths lying in a line L as above and encircling the μ points of intersection with the bifurcation diagram (generators of $\pi_1(L - \Sigma)$).*

The classical monodromy is generated by the product of these generators, that is, by a path on the complex line encircling each of the μ points of the bifurcation diagram once.

On the other hand, having fixed the initial value of λ on a chosen line in general position, we can *approach* each of the μ bifurcation values of the parameter instead of encircling them. The local level set corresponding to the bifurcation value of the parameter passes through a non-degenerate critical point. For values of the parameter near to a bifurcation value, the part of V_λ near such a critical point is diffeomorphic to the tangent bundle of a sphere (see Example 2.7.2). In this way a differentiable sphere (zero section of this bundle) is defined in V_λ, a so-called *vanishing cycle*, or *cycle vanishing at the given critical point.*

On the arc along which we proceeded from the initial point λ to the neighbourhood of a bifurcation value of the parameter, the sets V_λ form a smooth fibration. Therefore we can define a vanishing cycle in every fibre, among them the initial one.

The sphere S^{n-1} obtained in the initial fibre V_λ is determined up to a smooth isotopy leaving the boundary of V_λ fixed, and depends on the homotopy class of paths joining the initial point to the bifurcation value.

The standard basis of the fundamental group of the complex line with μ points removed thus determines a set of vanishing cycles in the homology group group of mean dimension of the local manifold V_λ (unique apart from the sign).

2.9.3. PROPOSITION. (Tyurina, Lamotke). *The μ vanishing cycles constructed above generate $H_{n-1}(V_\lambda, \mathbf{Z})$; moreover, they can be realized by embeddings of spheres $S^{n-1} \subset V_\lambda$ such that V_λ is homotopy equivalent to the union of these spheres.*

The basis of vanishing cycles just indicated is very convenient for calculation with vanishing cycles.

2.9.4. PROPOSITION (Picard, Lefschetz). *Consider any path joining the initial parameter value λ to a bifurcation value as described above, let $e \in H_{n-1}(V_\lambda, \mathbf{Z})$ be the corresponding vanishing cycle, and let γ be an element of the braid group of the singularity corresponding to the circuit around the bifurcation value along the indicated path, encircling the bifurcation value and returning along the old path. Then the action γ_* of γ on the homology of the local level manifold is described by the formula*

$$\gamma_* c = c + (c, e)e,$$

where

$$c \in H_{n-1}(V_\lambda, \mathbf{Z}), \qquad n \equiv 3 (\mathrm{mod}\ 4).$$

This formula is valid also when c is a cycle with closed support.

2.9.5. PROPOSITION. *The classical monodromy of a singularity is the action of the product of the standard generators of the braid group of the singularity (if the standard non-intersecting generators of $\pi_1(L - \Sigma)$ go around μ bifurcation values "anticlockwise", then their product is also "anticlockwise", in the order of the arguments of the starting directions of the generators).*

The proof is obvious, because this is the way in which the path is expressed in terms of μ generators encircles all the bifurcation points once each.

§3. Classification theorems

In this section we list the normal forms to which functions can be brought in the neighbourhood of not too complicated critical points. In particular, we list

1) all 0-modal germs,
2) all 1-modal germs,
3) all elliptical germs (with a definite quadratic form),
4) all parabolic germs (with a semidefinite quadratic form),
5) all hyperbolic germs (with a quadratic form of positive index of inertia 1).

The more than 1-modal germs form a set whose codimension (in the space of germs of functions of n variables at the critical point 0 with zero critical value) is equal to

10 for $n \geqslant 4$, 11 for $n = 3$, 12 for $n = 2$, ∞ for $n = 1$.

We also list all singularities with a critical point of multiplicity $\mu < 14$ (*the critical values are everywhere assumed to be zero*).

These classification theorems reduce the determination of the asymptotic behaviour of trigonometric integrals with functions depending, in general, on not too many parameters to the investigation of a finite number of normal forms.

3.1. 0-modal singularities. We consider the following functions:

$$\begin{aligned} A_k&: \quad f(x) = \pm x^{k+1}, \quad k \geqslant 1,\\ D_k&: \quad f(x, y) = x^2y \pm y^{k-1}, \quad k \geqslant 4,\\ E_6&: \quad f(x, y) = x^3 \pm y^4,\\ E_7&: \quad f(x, y) = x^3 + xy^3,\\ E_8&: \quad f(x, y) = x^3 + y^5. \end{aligned}$$

3.1.1. THEOREM ([13]). *Every 0-modal germ is stably equivalent to one of the germs of A, D, or E at zero; these germs are themselves 0-modal.*

3.1.2. THEOREM ([13]). *The braid group of a 0-modal germ is a generalized Artin braid group (as defined by Brieskorn* [14], [17]) *of the corresponding Weyl groups of A_k, D_k and E_k.*

Brieskorn's definition goes like this. We consider a finite group W acting on $\mathbf{R}^n$ and generated by reflections in planes passing through the origin of hyperplanes (mirrors). We complexify $\mathbf{R}^n$ and the mirrors; then W acts also on the complex space $\mathbf{C}^n$ so obtained.

3.1.3. DEFINITION. *The complex fundamental domain* of W is the complement in $\mathbf{C}^n$ of the complex mirrors factored by the action of W.

3.1.4. DEFINITION. The *Artin group of W* is the fundamental group of the complex fundamental domain.

3.1.5. THEOREM ([3], [28]). *The complement to the bifurcation diagram of a germ of type A, D or E is homeomorphic to the complex fundamental domain of the corresponding Coxeter group.*

The Coxeter groups of the series A, D, E can be defined as follows. We consider graphs with k points (the so-called Dynkin diagrams):

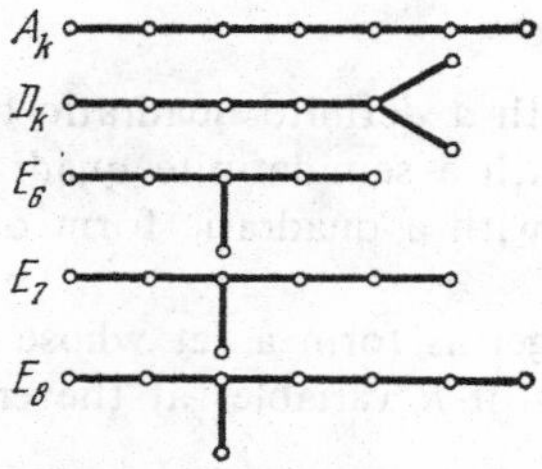

To each graph we assign the lattice $\mathbf{Z}^k$ whose generators $e_1, \ldots, e_k$ correspond to the vertices of the graph. We define an integral bilinear form on the lattice as follows:

1) the square (e_i, e_i) of each generator is -2.

2) the product (e_i, e_j) of two generators is 1 or 0 according as the corresponding points of the graph are joined by an edge or not.

3.1.6. DEFINITION. The *Coxeter group* $A_k(D_k, E_k)$ is a group of linear transformations of the linear space spanned by the lattice that preserve the relevant form, and is generated by k reflections acting according to the formula

$$c \mapsto c + (c, e)e, \; e = e_1, \ldots, e_k.$$

It is easy to check that this formula gives the reflection in the mirror orthogonal to the vector e, since $(e, e) = -2$. The quadratic forms corresponding to these graphs are negative definite.

3.1.7. THEOREM ([3], [28], [29]). *The quadratic forms of the singularities of type A, D, and E are equivalent to the forms constructed above, and a system of vanishing cycles serves as the basis* e_k. *The monodromy of each of the singularities of type A, D and E is a natural homomorphism of the Artin group into the Coxeter group.*

3.1.8. THEOREM ([3]). *The complement to the bifurcation diagram of singularities of type A, D and E is the space* $K(\pi, 1)$ *for the Artin group of the corresponding Coxeter group.*

The proof is fairly simple in the cases A and D, while for the case E we have to use a theorem of Deligne [30].

Consider now a product of the generating reflections of the Coxeter group, a so-called *Coxeter element.* Coxeter elements for different orders of the product of the generators are conjugate, and the order of a Coxeter element is called the *Coxeter number.* This invariant can be defined in many other ways (for instance, as the highest degree of a generator in the ring of polynomial invariants of the action of the group). It is not hard to check that the Coxeter numbers of the groups A_k, D_k, E_k are $k + 1$, $2k - 2$, 12, 18, 30.

3.1.9. THEOREM ([1]). *The singularity index of functions of type* A_k, D_k *or* E_k *is given by the formula* $\beta = \frac{1}{2} - \frac{1}{N}$, *where* N *is the corresponding*

Coxeter number.

The classical monodromy of singularities of types A_k, D_k or E_k is of finite order equal to the corresponding Cexeter number. The connection between N and β is not surprising, since after an N-fold circuit h around the origin the trigonometric integral returns to its original value. However, the only proofs of Theorem **3.1.9** known to me are not based on this fact, but on a direct calculation for each individual case.

3.1.10. THE CLASSIFICATION THEOREM ([3]). *Every germ of a function at a critical point of finite multiplicity is stably equivalent either to one of the germs of type A_k, D_k, E_k, or else to a germ in one of the following three sets:*

P: germs of functions of three or more variables with a 2-jet zero,

X: germs of functions of two variables with a 3-jet zero,

J: germs of functions of two variables with the 5-jet $x^3 + axy^4$.

The codimensions of the sets of germs of the types A_k, D_k, E_k (in the manifold of jets of functions at the critical point 0 with the critical value 0) is $k - 1$, and the codimensions of P, X and J are 6, 7 and 8, respectively.

Theorems **3.1.9**, **3.1.10** and Thom's transversality theorem [31] immediately give the next result:

3.1.11. COROLLARY ([1]). *For $l < 6$ every l-parameter family of functions can be brought by a small displacement to general position in such a way that the germ of every function of the family at every critical point is stably equivalent to one of the germs of A_k, D_k, E_k* (+ const), $k \leqslant 6$.

The singularities of these types are not removable by small displacements.

In particular, the only irremovable singularities in one-parameter families are of type A_2 (that is, x^3), in two-parameter families only A_3, etc.:

Number of parameters	0	1	2	3	4	5
Singularities	A_1	A_2	A_3	A_4, D_4	A_5, D_5	A_6, D_6, E_6

Thus, all singularities of codimension <6 are 0-modal.

Combining Corollary **3.1.11** and Theorem **3.1.9** we obtain:

3.1.12. COROLLARY ([1]). *The following singularities have the largest singularity index β among singularities of codimension l:*

l	1	2	3	4	5
Singularity	A_2	A_3	D_4	D_5	E_6
β_l	1/6	1/4	1/3	3/8	5/12

3.1.13. DEFINITION. We say that a set B in the space of jets *borders* a set C, and we write $B \leftarrow C$, if every neighbourhood of every point of C contains a point of B.

3.1.14. THEOREM ([3]). *All borders of the orbits of 0-modal germs are contained in the following diagram*:

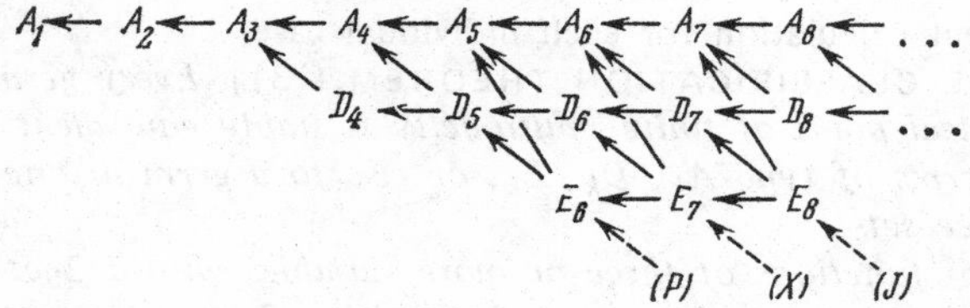

3.1.15. DEFINITION. The germ of a function at an isolated critical point is said to be *elliptic* if the quadratic form (given by the intersection index in the homology of mean dimension of the local level manifold of a function of $n \equiv 3 (\mathrm{mod}\ 4)$ variables stably equivalent to the given function) is negative definite.

3.1.16. THEOREM. *All elliptic singularities occur among the* A_k, $k \geqslant 1$, D_k, $k \geqslant 4$, E_6, E_7 and E_8.

This theorem was proved by Tyurina in [32] on the basis of arguments about the semicontinuity of indices of inertia.

Thus, the 0-modal singularities are exactly the elliptic singularities. Unfortunately, there is no direct proof ot this fact: the result is obtained by matching lists of singularities of various types and obtained in different ways.

3.2. Parabolic and hyperbolic singularities. To decrease the number of variants we restrict ourselves from now on to a complex classification.

3.2.1. DEFINITION. A germ of a function at an isolated critical point is said to be *parabolic* if the corresponding quadratic form is semidefinite. From the classification of unimodal singularities given below we obtain the next result:

3.2.2. THEOREM ([33]). *Every parabolic germ is stably equivalent to one of the germs of the following three one-parameter families*:

$$P_8\colon\ f(x, y, z) = x^2z + y^3 + \varepsilon y^2 z + az^3, \quad a(4\varepsilon + 27a) \neq 0, \quad \varepsilon^3 = \varepsilon,$$
$$X_9\colon\ f(x, y) = x^4 + \varepsilon x^2y^2 + ay^4, \quad a(4a - \varepsilon^2) \neq 0, \quad \varepsilon^3 = \varepsilon,$$
$$J_{10}\colon\ f(x, y) = x^3 + \varepsilon x^2y^2 + ay^6, \quad a(4\varepsilon + 27a) \neq 0, \quad \varepsilon^3 = \varepsilon.$$

The germs of these families are parabolic.

This was stated by Milnor as a conjecture based on the paper [34] of Wagreich, in which these germs arose in another connection. The Corollaries 3.2.3, 3.2.4 and 3.2.6 below were also stated by Milnor.

3.2.3. COROLLARY. *The quadratic forms of the parabolic germs* P_8, X_9, J_{10} *are obtained from the quadratic forms of* E_6, E_7, E_8 *respectively, by adding a zero quadratic form in two variables. In particular, the dimensions of the zero spaces of the quadratic form of a parabolic germ is equal to* 2.

3.2.4. COROLLARY. *The monodromy group of a parabolic germ has an abelian subgroup* $\mathbf{Z}^{12}$ *(or* $\mathbf{Z}^{14}$, $\mathbf{Z}^{16}$, *respectively) whose factor group is isomorphic to the Coxeter group for* E_6, E_7, *or* E_8.

Both corollaries can easily be obtained from a consideration of the Dynkin diagram of the vanishing cycles found by Gabrielov.

3.2.5. THEOREM ([35]). *There is a basis of vanishing cycles of a parabolic germ such that the monodromy group is generated by reflections in the orthogonal complements to the generators, and the Dynkin diagram has the elevant form*

P_8 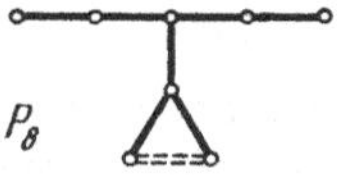 X_9 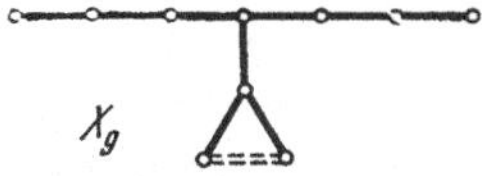 J_{10} 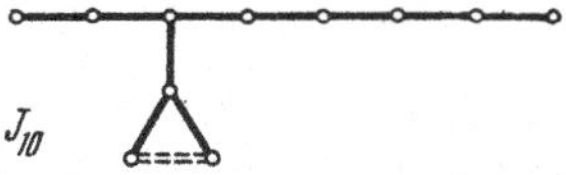

where the double dotted line indicates an intersection index -2.

3.2.6. COROLLARY ([33]). *Parabolic germs are defined by the following property: in a sufficiently small neighbourhood of a sufficient k-jet there are only finitely many orbits of dimension higher than that of the given jet (and infinitely many of the same dimension).*

3.2.7. COROLLARY. *The singularity index* β *of all parabolic germs* P_8, X_9, J_{10} *is* 1/2.

3.2.8. DEFINITION. The germ of a function at an isolated critical point is said to be *hyperbolic* if the positive index of inertia of the corresponding quadratic form (in the homology H_{n-1}, $n \equiv 3 \bmod 4$) is 1.

The classification of unimodal singularities to be given later yields:

3.2.9. THEOREM ([33]). *Every hyperbolic germ is stably equivalent to one of the germs in the following one-parameter families*:

$$T_{k,l,m}\colon\ f(x, y, z) = axyz + x^k + y^l + z^m, \qquad a \neq 0,$$

where

$$\frac{1}{k} + \frac{1}{l} + \frac{1}{m} < 1.$$

The germs in these families are hyperbolic,

$$\mu = k + l + m - 1.$$

The quadratic forms and the monodromy groups of these germs were found by Gabrielov when he proved the following theorem.

3.2.10. THEOREM. *For every hyperbolic germ there exists a basis of vanishing cycles such that the monodromy group is generated by reflections in the orthogonal complements to the generators, and the Dynkin diagram*

has the form

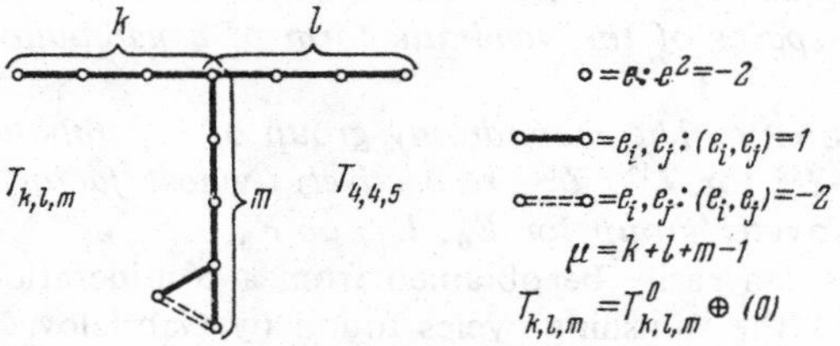

3.2.11. COROLLARY. *The quadratic form of a hyperbolic germ is the direct sum of a zero form in one variable and the form $T^0_{k,l,m}$ with indices of inertia $\mu_+ = 1$, $\mu_- = \mu - 2$ given by a T-shaped Dynkin diagram (the left lower vertex is omitted from the preceding diagram):*

$$T_{k,\,l,\,m} = T^0_{k,\,l,\,m} \oplus (0).$$

3.2.12. REMARK. The sum $\mu_+ + \mu_0$ of the number of positive and zero squares in the canonical form of the quadratic form turns out to be even in all examples known to me.

3.3. **Unimodal germs.** We call germs of modality 1 unimodal (see Definition 2.6.3).

3.3.1. THEOREM ([33]). *Every unimodal germ is either parabolic, or hyperbolic, or else stably equivalent to a germ in one of the following 14 exceptional one-parameter families (throughout, the parameter a runs over* **C**):

$$\begin{aligned}
Q_{10}&: \; x^2z + y^3 + ayz^3 + z^4,\\
Q_{11}&: \; x^2z + y^3 + yz^3 + az^5,\\
Q_{12}&: \; x^2z + y^3 + ayz^4 + z^5,\\
S_{11}&: \; x^2z + yz^2 + y^4 + ay^3z,\\
S_{12}&: \; x^2z + yz^2 + xy^3 + ay^5,\\
U_{12}&: \; x^3 + y^3 + z^4 + 3axyz^2,\\
Z_{11}&: \; x^3y + 3axy^4 + y^5,\\
Z_{12}&: \; x^3y + 3xy^4 + ay^6,\\
Z_{13}&: \; x^3y + 3axy^5 + y^6,\\
W_{12}&: \; x^4 + y^5 + 2ax^2y^3,\\
W_{13}&: \; x^4 + 4xy^4 + ay^6,\\
K_{12}&: \; x^3 + axy^5 + y^7,\\
K_{13}&: \; x^3 + axy^5 + y^8,\\
K_{14}&: \; x^3 + xy^6 + ay^8.
\end{aligned}$$

All germs of the types listed (parabolic, hyperbolic and exceptional) are 1-modal.

The quadratic forms and monodromy groups of the exceptional germs were calculated by Gabrielov.[1)]

3.3.2. THEOREM. *For every exceptional germ there exists a basis of vanishing cycles such that the Dynkin diagram takes the form*

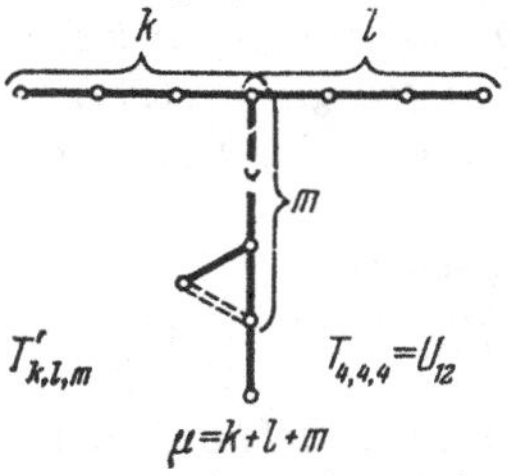

$Q_{10} \sim (3, 3, 4)$, $Q_{11} \sim (3, 3, 5)$, $Q_{12} \sim (3, 3, 6)$,
$Z_{11} \sim (2, 4, 5)$, $Z_{12} \sim (2, 4, 6)$, $Z_{13} \sim (2, 4, 7)$,
$S_{11} \sim (3, 4, 4)$, $S_{12} \sim (3, 4, 5)$, $U_{12} \sim (4, 4, 4)$,
$W_{12} \sim (2, 5, 5)$, $W_{13} \sim (2, 5, 6)$, $K_{12} \sim (2, 3, 7)$,
$K_{13} \sim (2, 3, 8)$, $K_{14} \sim (2, 3, 9)$.

It follows, in particular, that the quadratic form of an exceptional germ is non-degenerate and has the positive index of inertia 2:

$$T'_{k,\,l,\,m} = T^0_{k,\,l,\,m} \oplus \begin{pmatrix} 0 & 1 \\ 1 & 0 \end{pmatrix}.$$

The proofs of Theorems 3.2.2, 3.2.9 and 3.3.1 are based on a fairly cumbersome calculation of the orbits of the action of the group of diffeomorphisms on the space of jets of functions. To get the stated theorems it proved necessary to carry the calculations a little further so as to include the simplest non-unimodal germs.

According to the classification theorem 3.1.10 we have to deal with three cases:

P: germs of functions of three or more variables with a zero 2-jet,
X: germs of functions of two variables with a zero 3-jet,
J: germs of functions of two variables with a 5-jet $x^3 + axy^4$.

3.3.3. THEOREM. *Every germ of a function of two variables with a 5-jet $x^3 + axy^4$ lies in one of the following classes*:

$$J^8_{10} \leftarrow J^9_{11} \leftarrow J^{10}_{12} \leftarrow \dots$$

$$\nwarrow$$

$$K^{10}_{12} \leftarrow K^{11}_{13} \leftarrow K^{12}_{14} \leftarrow K^{13}_{16} \leftarrow K^{14}$$

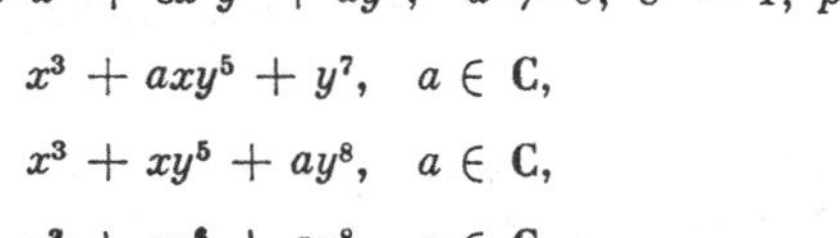

Here the letters denote classes of germs equivalent to germs in the following list:

J^8_{10}: $x^3 + \varepsilon x^2y^2 + ay^6$, $a(4\varepsilon + 27a) \neq 0$, $\varepsilon^3 = \varepsilon$,

J^{p+2}_{p+4}: $x^3 + \varepsilon x^2y^2 + ay^p$, $a \neq 0$, $\varepsilon^2 = 1$, $p > 6$,

K^{10}_{12}: $x^3 + axy^5 + y^7$, $a \in \mathbf{C}$,

K^{11}_{13}: $x^3 + xy^5 + ay^8$, $a \in \mathbf{C}$,

K^{12}_{14}: $x^3 + xy^6 + ay^8$, $a \in \mathbf{C}$,

K^{13}_{16}: $x^3 + xy^6 + ay^9 + by^{10}$, $27a^2 + 4 \neq 0$.

---

1) A special technique was developed for the computation of the quadratic forms. A sketch of the level lines of the real form of the Morse function obtained by a disturbance of a degenerate critical point turned out to be very useful for this. Using this method, S. M. Husein–Zade has obtained the Dynkin diagrams of many singularities of the form $x^2 + f(y, z)$.

The letters a and b denote parameters. In the neighbourhood of a general point of the line (plane) of parameters the same values of the parameter may give rise to non-equivalent germs. Further, $\mathbf{K}^{14}$ *denotes a certain algebraic submanifold of codimension* 14 *(in the manifold of jets of functions at the critical point* 0 *with the critical value* 0). *In general, the upper index refers to the codimension of the set formed by orbits of the given type in the relevant manifold of jets. The lower index denotes the multiplicity* μ *of the critical point.*

The polynomials listed give sufficient k-jets. The singularity of J_{10}^8 *is parabolic, while that of* J_{p+4} *for* $p > 6$ *is hyperbolic (and is equivalent to* $T_{2,3,p}$ *in Theorem* 3.2.9).

The arrows in the diagram above denote borderings (see Definition 3.1.13). Not all of these are indicated, but only the most important ones. In particular, the theorem asserts that *every germ of type J not occurring in the above list, that is, a germ in* $\mathbf{K}^{14}$, *can be transformed by an arbitrarily small disturbance of a finite jet to a germ equivalent to one of the germs* $\mathbf{K}_{16}^{13}$. For any such germ $\mu \geqslant 16$, and the modality is $\geqslant 2$. By a comparison with Theorem 3.1.10 we came to the next result:

3.3.4. COROLLARY. *The germs in the above list are unimodal, apart from* $\mathbf{K}_{16}^{13}$. *The germs* $\mathbf{K}_{16}^{13}$ *are bimodal. All the remaining germs with a 5-jet* $x^3 + axy^4$ *(that is, germs in* $\mathbf{K}^{14}$*) have modality at least* 2.

We consider now functions of two variables of type X (with a zero 3-jet).

3.3.5. THEOREM. *Every germ of a function of two variables with zero 3-jet lies in one of the following classes*:

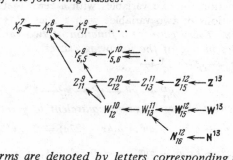

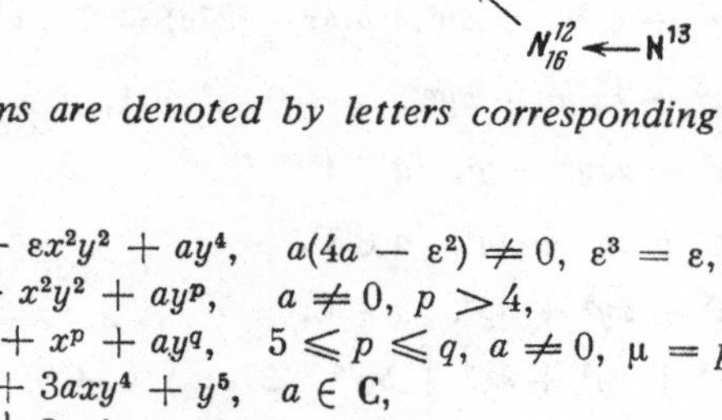

Here classes of germs are denoted by letters corresponding to the germs in the following list:

X_9^7: $\quad x^4 + \varepsilon x^2y^2 + ay^4, \quad a(4a - \varepsilon^2) \neq 0, \ \varepsilon^3 = \varepsilon,$

X_{p+5}^{p+3}: $\quad x^4 + x^2y^2 + ay^p, \quad a \neq 0, \ p > 4,$

$Y_{p,q}^{p+q-1}$: $\quad x^2y^2 + x^p + ay^q, \quad 5 \leqslant p \leqslant q, \ a \neq 0, \ \mu = p + q + 1,$

Z_{11}^9: $\quad x^3y + 3axy^4 + y^5, \quad a \in \mathbf{C},$

Z_{12}^{10}: $\quad x^3y + 3xy^4 + ay^6, \quad a \in \mathbf{C},$

Z_{13}^{11}: $\quad x^3y + 3axy^5 + y^6, \quad a \in \mathbf{C},$

Z_{15}^{12}: $\quad x^3y + 3xy^5 + ay^7 + by^8, \quad a^2 + 4 \neq 0,$

W_{12}^{10}: $x^4 + y^5 + 2ax^2y^3$, $a \in \mathbf{C}$,
W_{13}^{11}: $x^4 + 4xy^4 + ay^6$, $a \in \mathbf{C}$,
W_{15}^{12}: $x^4 + 2x^2y^3 + ay^6 + by^7$, $a^2 \neq a$,
N_{16}^{12}: $x^4y + ax^3y^2 + bx^2y^3 + xy^4 + cx^3y^3$, $\Delta(a,b) \neq 0$.

The arrows denote borderings (again, not all of them are indicated). In particular, arrows from $\mathbf{Z}^{13}$, $\mathbf{W}^{13}$ and $\mathbf{N}^{13}$ mean that 1) *all germs of functions of two variables with a zero 2-jet and not equivalent to germs in the list form a subset of codimension* 13 (*in the space of jets of functions with the critical value* 0). 2) *in every neighbourhood of a jet in this subset there is a jet of type* $\mathbf{Z}_{15}^{12}$, $\mathbf{W}_{15}^{12}$ *or* $\mathbf{N}_{16}^{12}$. *The above polynomials give sufficient jets. The singularity* X_9^7 *is parabolic*, X_{p+5}^{p+3} $(p > 4)$ *and* $Y_{p,q}$, $5 \leqslant p$ $5 \leqslant p \leqslant q$, *are hyperbolic* (*they are equivalent to* $T_{2,4,p}$ *and* $T_{2,p,q}$, *respectively, of Theorem* 3.2.9).

We note that small displacements of a germ of class X can produce only germs of classes A, D, E and J in Theorem 3.1.10.

3.3.6. COROLLARY. *The germs of the above list are unimodal, apart from* $\mathbf{Z}_{15}^{12}$, $\mathbf{W}_{15}^{12}$ *and* $\mathbf{N}_{16}^{12}$. *The germs* $\mathbf{Z}_{15}^{12}$ *and* $\mathbf{W}_{15}^{12}$ *are bimodal, and* N_{16}^{12} *are trimodal. The remaining germs, which are stably equivalent to germs of two variables with a zero 2-jet* (*that is, to germs in* $\mathbf{Z}^{13}$, $\mathbf{W}^{13}$ *or* $\mathbf{N}^{13}$), *have modality not less than* 2.

We turn finally to functions of type P (the corank of their second differential is 3 or more).

3.3.7. THEOREM. *Every germ of a function at a critical point at which the second differential has corank 3 or more is stably equivalent to a germ in one of the following classes*:

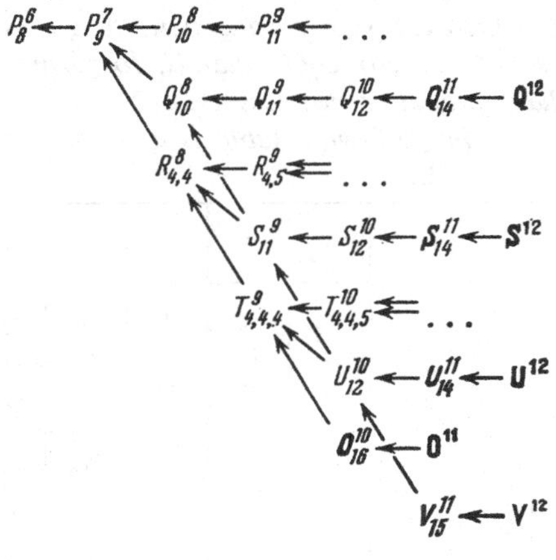

The letters denote germs in the following list:

P_8^6: $x^2z + y^3 + \varepsilon y^2z + az^3$, $a(4\varepsilon + 27a) \neq 0$, $\varepsilon^3 = \varepsilon$,
P_{p+5}^{p+3}: $x^2z + y^3 + y^2z + az^p$, $a \neq 0$, $p > 3$,
Q_{10}^8: $x^2z + y^3 + ayz^3 + z^4$, $a \in \mathbf{C}$,
Q_{11}^9: $x^2z + y^3 + yz^3 + az^5$, $a \in \mathbf{C}$,
Q_{12}^{10}: $x^2z + y^3 + ayz^4 + z^5$, $a \in \mathbf{C}$,
Q_{14}^{11}: $x^2z + y^3 + yz^4 + az^6 + bz^7$, $27a^2 + 4 \neq 0$,
$R_{p,q}^{p+q}$: $x^3 + 3xyz + y^p + az^q$, $4 \leqslant p \leqslant q$, $\mu = p + q + 2$, $a \neq 0$,
S_{11}^9: $x^2z + yz^2 + y^4 + ay^3z$, $a \in \mathbf{C}$,
S_{12}^{10}: $x^2z + yz^2 + xy^3 + ay^5$, $a \in \mathbf{C}$,
S_{14}^{11}: $x^2z + yz^2 + y^3z + ay^5 + by^6$, $a \neq 0$, $a \neq 1/4$,
$T_{p,q,r}^{p+q+r-3}$: $axyz + x^p + y^q + z^r$, $4 \leqslant p \leqslant q \leqslant r$, $\mu = p + q + r - 1$, $a \neq 0$,
U_{12}^{10}: $x^3 + y^3 + z^4 + 3axyz^2$, $a \in \mathbf{C}$,
U_{14}^{11}: $x^3 + y^3 + 3xz^3 + 3ayz^3 + 9byz^4$, $a^3 \neq 1$,
V_{15}^{11}: $x^2y + z^4 + y^2z^2 + ay^3z + by^4 + cy^4z$, $a(12b + 1)\Delta(a, b) \neq 0$,
O_{16}^{10}: $x^3 + y^3 + z^3 + u^3 + (ax + by + cz + du)^3 + exyzu$, $\Delta(a, b, c, d) \neq 0$,

where Δ is the discriminant.

The notation is as in the preceding theorem. In particular, the theorem asserts that *germs not equivalent to one in the list form the set* $\mathbf{Q}^{12} + \mathbf{S}^{12} + \mathbf{U}^{12} + \mathbf{V}^{12} + \mathbf{O}^{11}$ *of codimension* 11 *and can be brought by sufficiently small displacements to germs stably equivalent to germs of type* Q_{14}^{11}, S_{14}^{11}, U_{15}^{11}, V_{15}^{11} *or* O_{16}^{10}.

The singularity of P_8^6 is parabolic, those of P, R, and T are hyperbolic ($P_{p+5} \sim T_{3,3,p}$, $R_{p,q} \sim T_{3,p,q}$).

A comparison with Theorems 3.3.3, 3.3.5 and 3.3.7 gives the next result:

3.3.8. COROLLARY. *The germs in the above list are unimodal, apart from Q_{14}^{11}, S_{14}^{11}, U_{14}^{11} (which are bimodal), V_{15}^{11} (trimodal) and O_{16}^{10} (5-modal). The remaining germs, which are stably equivalent to germs of functions of more than 2 variables with a zero 2-jet (that is, to germs from* $\mathbf{Q}^{12}$, $\mathbf{S}^{12}$, $\mathbf{U}^{12}$, $\mathbf{V}^{12}$ *and* $\mathbf{O}^{11}$), *have modality not less than* 2.

3.3.9. COROLLARY. *The following table is a classification of critical points of codimension $l < 10$:*

l	0	1	2	3	4	5	6	7	8	9	10
	A_1	A_2	A_3	A_4	A_5	A_6	A_7	A_8	A_9	A_{10}	A_{11}
				D_4	D_5	D_6	D_7	D_8	D_9	D_{10}	D_{11}
						E_6	E_7	E_8	J_{10}	J_{11}	J_{12}
											K_{12}
								X_9	X_{10}	X_{11}	X_{12}
										$Y_{5,5}$	$Y_{5,6}$
										Z_{11}	Z_{12}
											W_{12}
							P_8	P_9	P_{10}	P_{11}	P_{12}
									Q_{10}	Q_{11}	Q_{12}

l	0	1	2	3	4	5	6	7	8	9	10
									$R_{4,4}$	$R_{4,5}$	$R_{5,5}$
											$R_{5,6}$
										S_{11}	S_{12}
										$T_{4,4,4}$	$T_{4,4,5}$
											U_{12}
											O_{16}

in the sense that the set of all germs not stably equivalent to germs in this list has codimension 11 *(in the set of jets of functions with the critical point* 0 *and the critical value* 0*). For $s < 11$ every s-parameter family of functions can be brought by a sufficiently small displacement to "general position" in such a way that the germs of all functions of the family at all critical points are stably equivalent to germs in the columns in the preceding table for which $l \leqslant s$ (up to additive constants).*

From these theorems we can extract a list of all singularities with $\mu < 14$. Concerning these theorems, we note that we have not considered the question whether different singularities in our table may not be topologically equivalent to each other. In the majority of cases, we can easily deduce that they are non-equivalent merely from the values of these invariants. Doubt remains only for certain members of the three-index series $T_{k,l,m}$ (it is not clear whether k, l, m can be recovered from the quadratic forms). We note too that the parameter a in the normal form of non-parabolic unimodal singularities can be replaced by 0 for 1, if we are not classifying functions but hypersurfaces (or functions up to a change of coordinates in preimage and image). Notwithstanding the disappearance of the "Modulus" a, the singularities remain unimodal also under the new equivalence relations (because they are close to parabolic singularities).

§4. Singularity indices

The classification theorems in the preceding sections reduce the investigation of the asymptotic behaviour of trigonometric integrals with functions in general position and depending, in general, on not too many parameters to a discussion of explicitly given polynomial cases. The calculation of the singularity index in each of these special cases reduces in principle to a finite number of arithmetical steps.

The calculations can be carried out without difficulty in the cases where the polynomial that determines the normal form is homogeneous or weighted homogeneous. From the results of the calculation for the weighted homogeneous case we can deduce the values of the singularity indices for all singularities of modality $\leqslant 1$, on the basis of a few propositions about the semicontinuity of the singularity index. The results so obtained can also be checked independently of the unproved assumptions. The author is grateful to A. Ya. Povzner for doing the checking.

4.1. Conjectures concerning the semicontinuity of the singularity index. Let S: $\mathbf{R}^n \to \mathbf{R}$ be a polynomial with the critical point 0, and let β be the singularity index of S at zero (see Definition 2.1.1). Consider any polynomial deformation F: $\mathbf{R}^n \times \mathbf{R}^l \to \mathbf{R}$ of the germ of S at zero.

4.1.1. PROPOSITION ([1]). *For every* $\varepsilon > 0$ *the trigonometric integral with a function* $F(\cdot, \lambda)$ *satisfies the inequality*

$$\left| \int\limits_{\mathbf{R}^n} e^{iF(x,\lambda)/h} \varphi(x, \lambda)\, dx \right| \leqslant C(\varphi, \varepsilon)\, h^{\frac{n}{2}-\beta-\varepsilon}$$

for every C^∞*-function* ϕ *concentrated in a sufficiently small neighbourhood of the origin in* $\mathbf{R}^n \times \mathbf{R}^l$.

If the critical point is of finite multiplicity, then it is enough to consider just one versal μ-parameter deformation F.

4.1.2. EXAMPLE. *For singularities of the series* A_k, *Proposition* 4.1.1 *takes the form of an estimate uniform in* $|\lambda| \leqslant \delta$, $|h| \leqslant h_0$:

$$\left| \int\limits_{-\infty}^{\infty} e^{\frac{i}{h}(x^{k+1}+\lambda_1 x^{k-1}+\ldots+\lambda_k)} \cdot \varphi\, dx \right| \leqslant C(\varphi, \delta)\, h^{\frac{1}{k+1}-\varepsilon}.$$

As A. A. Karatsuba has informed me, an estimate of this form (even for $\varepsilon = 0$) has been established by I. M. Vinogradov [36] (and with different methods, by Hua Lo Ken [37]). In the general form, 4.1.1 remains unproved.

4.1.3. PROPOSITION. *The singularity index* β *is semicontinuous in the sense that the singularity indices of all jets of functions at the critical point lying in a sufficiently small neighbourhood of a sufficient jet* s *does not exceed the singularity index* $\beta(s)$ *of* s.

In particular, when the critical point is split up into simpler ones, the singularity index of each one of the new points is not greater than that of the original point. Proposition 4.1.3 follows from 4.1.1; it has not been proved, in general.

4.1.4. PROPOSITION. *Suppose that we are given a family of functions depending smoothly on some parameters and having* 0 *as a critical point for all values of the parameters. Assume further that the multiplicity of the critical point* 0 *does not change under a change of parameters. Then the singularity indices at* 0 *of all functions of the family are the same.*

This proposition also remains unproved, it has been proved only that the quadratic form of the singularity does not change along the family (Tyurina [32]), nor (for $n \neq 3$) does the complex topological type of the germ (Lê Dũng Tráng and Ramanujam [38]). In this connection it is natural to conjecture that the singularity index is a topological invariant of the complex germ. It would follow, in particular, that *the singularity indices of all real forms with one and the same set of complex germs are identical.*

4.2. The rationality of the numbers β_l. The symbol β_l denotes the singularity index of the "strongest" singularity not removable in l-parameter families.

4.2.1. DEFINITION ([1]). Let $l \geqslant 0$ be an integer. Then β_l denotes the greatest lower bound of the numbers γ having the property that every γ-parameter family of functions can be brought by a small displacement to a family for which the estimate

$$\left| \int\limits_{\mathbf{R}^n} e^{\frac{i}{h} F(x,\,\lambda)} \varphi(x,\,\lambda)\, dx \right| \leqslant C(\varphi)\, h^{\frac{n}{2}-\gamma}$$

holds uniformly in the parameter $\lambda \in \mathbf{R}^l$, for all sufficiently small $h > 0$ and all ϕ with compact support.

4.2.2. EXAMPLE. By Morses's lemma, $\beta_0 = 0$. Further, $\beta_1 = 1/6$, as can be seen by considering Airy integrals (in which $F = x^3 + \lambda_1 x + \lambda_2$).

4.2.3. THEOREM. *It follows from* 4.1.1 *and* 4.1.- *that all the β_l are rational.*

PROOF. It is not hard to extract a semi-algebraic set of codimension more than l from the space of jets of functions with the critical point 0 and the critical value 0 such that the complement consists of jets that are sufficient everywhere and splits into finitely many smooth manifolds along each of which the multiplicity μ is constant. Along each of these manifolds the singularity index β is also constant (Proposition **4.1.4**). By **2.1.5** this index is rational. Since β_l is the maximum of these numbers, it is itself rational.

4.3. Singularity indices of quasi-homogeneous singularities. A polynomial $S(x_1, \ldots, x_n)$ is said to be *quasi-homogeneous with weights* $\alpha_1, \ldots, \alpha_n$ if $S(\lambda^{\alpha_1} x_1, \ldots, \lambda^{\alpha_n} x_n) \equiv \lambda S(x_1, \ldots, x_n)$, $0 < \alpha_k \leqslant 1/2$.

For example, $x^a + y^b + z^c$ has weights $1/a$, $1/b$, $1/c$, and a homogeneous polynomial of degree a has weight $1/a$ in each argument. See Saito [39] for quasi-homogeneous functions, where it is proved, in particular, that the germ of a function at a critical point is contained in the ideal generated by its partial derivatives if and only if the germ is equivalent to a quasi-homogeneous germ.

4.3.1. PROPOSITION. *A quasi-homogeneous polynomial with weights $\alpha_1, \ldots, \alpha_n$ having 0 as an isolated critical point has the singularity index*

$$\beta = \frac{n}{2} - \sum_{k=1}^{n} \alpha_k = \sum_{k=1}^{n} \left(\frac{1}{2} - \alpha_k\right) \quad \textit{there.}$$

We remark here that the normal forms of all 0-modal (that is, elliptic) singularities and all parabolic singularities are quasi-homogeneous.

Computing β from the preceding formula, we get in the elliptic case of Theorem 3.1.9: $\beta = \frac{1}{2} - \frac{1}{N}$, where N is the Coxeter number, and in the parabolic case $\beta = 1/2$, so that here we must count the Coxeter number as being ∞.

Hyperbolic singularities are not homogeneous, and we consider their singularity indices β separately. As regards the remaining families of singularities listed in §3, many of them have quasi-homogeneous representatives, and this together with Proposition 4.1.4 enables us to compute β for every family. In particular, this can be done for all the unimodal exceptional singularities listed in §3. In these cases the singularity index exceeds 1/2 by the reciprocal of a natural number: $\beta = \frac{1}{2} - \frac{1}{N}$, $N \in \mathbf{Z}$, $N < 0$, that is, the "Coxeter number" defined by the same formula as in the elliptic case (and, in general, rational) turns out to be a *negative integer* for these singularities.

4.3.2. PROPOSITION. *The weights and singularity indices of the singularities listed in* §3 *are given by the following table*:

Type	Q_{10}	Q_{11}	Q_{12}	Q_{14}	Z_{11}	Z_{12}	Z_{13}	Z_{15}	S_{11}	S_{12}
$\lvert N \rvert \alpha_k$	9,8,6	7,6,4	6,5,3	5,4,2	15,8,6	11,6,4	9,5,3	7,4,2	6,5,4	5,4,3
$\lvert N \rvert$	24	18	15	12	30	22	18	14	16	13
β	$\frac{13}{24}$	$\frac{5}{9}$	$\frac{17}{30}$	$\frac{7}{12}$	$\frac{8}{15}$	$\frac{6}{11}$	$\frac{5}{9}$	$\frac{4}{7}$	$\frac{9}{16}$	$\frac{15}{26}$

U_{12}	S_{14}	U_{14}	V_{15}	W_{12}	W_{13}	W_{15}	N_{16}	K_{12}	K_{13}	K_{14}	K_{16}
4,4,3	4,3,2	3,3,2	3,2,2	10,5,4	8,4,3	6,3,2	5,2,2	21,14,6	15,10,4	12,8,3	9,6,2
12	10	9	8	20	16	12	10	42	30	24	18
$\frac{7}{12}$	$\frac{3}{5}$	$\frac{11}{18}$	$\frac{5}{8}$	$\frac{11}{20}$	$\frac{9}{16}$	$\frac{7}{12}$	$\frac{3}{5}$	$\frac{11}{21}$	$\frac{8}{15}$	$\frac{13}{24}$	$\frac{5}{9}$

The correctness of β can be checked by using finite computations that do not use the unproved general propositions.

4.3.3. REMARK. The singularity in $K_{12}(x^3 + y^7 + z^2)$ is connected with a regular 56-hedron of genus 3 with triangular faces, having 84 edges meeting 7 at a time in each of 24 vertices (and with congruence subgroup mod 7) in much the same way as the singularity $E_8(x^3 + y^5 + z^2)$ is connected with the icosahedron with congruence subgroup mod 5 (see [40]).

4.4. Singularity indices of hyperbolic germs. Hyperbolic germs are stably equivalent to the germs $T_{k,l,m}$: $axyz + x^k + y^l + z^m$, $a \neq 0$, in the complex domain (see §3). We denote the function xyz of three variables by

T_∞. The semi-continuity hypothesis in **4.1.1** gives

4.4.1. PROPOSITION. *For every hyperbolic germ,* $\beta(T_{k,l,m}) \leqslant \beta(T_\infty)$.

This proposition can be checked using calculations not running into the indicated general hypothesis. On the other hand, the following fact can be verified directly:

4.4.2. PROPOSITION. $\beta(T_\infty) = 1/2$.

Finally, from the semi-continuity hypothesis of **4.1.1** and the fact that $\beta = 1/2$ for parabolic singularities (see §3) we deduce:

4.4.3. PROPOSITION. *The singularity index of every non-elliptic singularity is at least* $1/2$.

Combining the last three propositions we obtain the following one (which can likewise be verified independently of the conjecture **4.1.1**):

4.4.4. COROLLARY. *The singularity index of every hyperbolic germ is* $1/2$.

4.5. The strongest singularities. Combining the values of the singularity index with the list of the singularities in Corollary **3.3.9**, we can give for every $l \leqslant 10$ all singularity indices for singularities in l-parameter families in general position on submanifolds of the space of parameters. Selecting for each l the singularity in **3.3.9** with the largest singularity index, we came to the following values for β_l:

l	0	1	2	3	4	5	6	7	8	9	10, $n=3$	11, $n=3$	10, $n\geqslant 4$
Type	A_1	A_2	A_3	D_4	D_5	E_6	P_8	X_9, P_9	Q_{10}	S_{11}	U_{12}	$\mathbf{V}_{15}$	$\mathbf{O}_{16}$
β_l	0	$1/6$	$1/4$	$1/3$	$3/8$	$5/12$	$1/2$	$1/2$	$13/24$	$9/16$	$7/12$	$5/8$	$2/3$
N	$+2$	$+3$	$+4$	$+6$	$+8$	$+12$	∞	∞	-24	-16	-12	-8	-6

This reduced table contains the maximal singularity indices in general l-parameter families of functions of $n \geqslant 4$ variables for $l \leqslant 10$ and functions of 3 variables for $l \leqslant 11$. For functions of two variables the table can be extended to the case $l = 12$:

l	0	1	2	3	4	5	6	7	8	9	10	11	12
Type	A_1	A_2	A_3	D_4	D_5	E_6	E_7	X_9	X_{10}, J_{10}	Z_{11}	W_{12}	W_{13}	$\mathbf{N}_{16}$
β_l	0	$1/6$	$1/4$	$1/3$	$3/8$	$5/12$	$4/9$	$1/2$	$1/2$	$8/15$	$11/20$	$9/16$	$3/5$
N	$+2$	$+3$	$+4$	$+6$	$+8$	$+12$	$+18$	∞	∞	-30	-20	-16	-10

The calculation of the exact value of β_l for all l would be extremely difficult. On the basis of experimental evidence, I believe that one can guess the asymptotic behaviour for large l.

CONJECTURE. *A non-degenerate cubic form in k variables is the singularity with maximum singularity index β for its codimension (that is, for* $l = k(k+1)/2$). *In other words,* $\beta_{k(k+1)/2} = k/6$.

We remark that the estimate $\beta_{k(k+1)/2} \geqslant k/6$ already follows from the semi-continuity conjecture. Note also that a sum of cubes appears to be the "strongest" among singularities with the same codimension not only in the sense of the singularity index β, but also of the other parameters (μ and m).

Singularities of the type of a sum of k cubes (for $k = 1$ this is A_2, for $k = 2$ it is D_4, for $k = 3$ it is P_8, and for $k = 4$ it is O_{16}) have multiplicity $\mu = 2^k$, modality $m = 2^k - \dfrac{k(k+1)}{2} - 1$, and they form a manifold of codimension $l = k(k+1)/2$ (in the space of jets of functions with the critical point 0 and the critical value 0). Thus, we can guess that the maximum multiplicity μ_l of singularities occurring in general l-parameter families and the maximum modality m_l both grow like $2^{\sqrt{(2l)}}$ as $l \to \infty$, while β_l grows like $\sqrt{(2l)}/6$.

References

[1] V. I. Arnol'd, Integrals of quickly oscillating functions and singularities of projections of Lagrange manifolds, Funktsional. Anal. i Prilozhen. **6**:3 (1972), 61–62. MR

[2] V. P. Maslov, Focussing energy in a crystal lattice, Uspekhi Mat. Nauk **27**:6 (1972), 224.

[3] V. I. Arnol'd, Normal forms of functions near degenerate critical points, the Weyl groups A_k, D_k, E_k, and Lagrange singularities, Funktsional. Anal. i Prilozhen. **6**:4 (1972), 3–25. MR

[4] E. Brieskorn, Groupes des tresses, Séminaire Bourbaki **24**:401, November 1971.

[5] I. N. Bernshtein and S. I. Gel'fand, Meromorphy of the function P^λ, Funktsional. Anal. i Prilozhen. **3**:1 (1969), 84–86. MR **40** # 723.

[6] M. F. Atiyah, Resolution of singularities and division of distributions, Comm. Pure. Appl. Math. **23** (1970), 145–150. MR **41** # 815.

[7] B. Malgrange, Ideals of differentiable functions, Oxford Univ. Press, London 1967. MR **35** # 3446.
Translation: *Idealy differentsiruemykh funktsii*, Izdat. "Mir", Moscow 1968.

[8] V. I. Arnol'd, Singularities of smooth mappings, Uspekhi. Mat. Nauk. **23**:1 (1968), 3–44. MR **37** # 2243.
= Russian Math. Surveys **23** (1968), 1–43.

[9] J. C. Tougeron, Idéaux de fonctions differentiables. I, Ann. Inst. Fourier (Grenoble) **18** (1968), fasc.1, 177–240. MR **39** # 2171.

[10] V. P. Palamodov, On the multiplicity of a holomorphic transformation, Funktsional. Anal. i. Prilozhen. **13** (1967), 54–65. MR **38** # 4720.

[11] V. P. Palamodov, Remarks on differentiable maps of finite multiplicity, Funktsional. Anal. i Prilozhen. **6**:2 (1972), 52–61. MR **45** # 7732.

[12] A. M. Samoilenko, The equivalence of a smooth function to a Taylor polynomial in the vicinity of a critical point of finite type, Funktsional. Anal. i Prilozhen. **2**:4 (1968), 63–69. MR **42** # 5132.

[13] M. Artin, On the solution of analytic equations, Invent. Math. **5** (1968), 277–291. MR **38** # 344.

[14] A. Weinstein, Singularities of families of functions, Differential-geometrie im Grossen, Oberwolfach **4** (1971).

[15] John Mather, Structural stability of mappings, Singularities of differentiable maps, 216–267. Izdat "Mir", Moscow 1968 (in Russian). MR **40** # 4966.

[16] F. Latour, Stabilité des champs d'applications différentiables: Généralisation d'un théorème de J. Mather, C.R. Acad. Sci. (Paris) Sér. A **268** (1969), 1331–1334. MR **39** # 7617.

[17] J. Guckenheimer, Catastrophes and partial derivative equations, Princeton 1972, 1–29.

[18] V. M. Zakalyukin, A versality theorem, Funktsional. Anal. i Prilozhen. **7**:2 (1973), 28–32. MR

[19] G. N. Tyurina, Locally semi-universal flat deformations of isolated singularities in complex spaces, Izv. Akad. Nauk. SSSR Ser. Mat. **33** (1969), 1026–1058. MR **40** # 5903.
= Math. USSR–Izv.

[20] A. Fresnel, Mémoire sur la diffraction de la lumière, Mém. de l'Acad. des sciences **5** (1818), 339–353.

[21] G. B. Airy, Intensity of light in a neighbourhood of a caustic, Trans. Cambridge Philos. Soc. **6** (1838), 379–403.

[22] T. Pearcey, The structure of an electromagnetic field in the neighbourhood of a cusp of a caustic, Philos. Mag. **37** (1946), 311–315.

[23] J. Milnor, Singular points of complex hypersurfaces, Ann. of Maths. Studies No. 61, Princeton University Press, Princeton, New Jersey 1968. MR **39** # 969.
Translation: *Osobye tochki kompleksnykh giperpoverkhnostei*, Izdat. "Mir", Moscow 1971.

[24] O. Zariski, On the Poincaré group of a projective hypersurface, Ann. of Math. (2) **38** (1937), 131–141.

[25] A. N. Varchenko, Theorems on the topological equisingularity of systems of algebraic varieties and systems of polynomial mappings, Izv. Akad. Nauk. SSSR Ser. Mat. **36** (1972), 957–1019.
= Maths. USSR–Izv.

[26] H. Hamm and Lê Dũng Tráng, Un théorème du type de Lefschetz, C.R. Acad. Sci. (Paris) Sér. A **272** (1971), 946–949. MR **43** # 8094.

[27] E. Brieskorn, Die Fundamentalgruppe des Raumes der regulären Orbits einer endlichen komplexen Spiegelungsgruppe, Invent. Math. **12** (1971), 57–61.

[28] E. Brieskorn, Singular elements of semi-simple algebraic groups, Proc. Internat. Congress Mathematicians, Nice 1970.

[29] G. N. Tyurina, Resolution of singularities of flat deformations of double rational points, Funktsional. Anal. i Prilozhen. **4**:1 (1970), 77–83. MR **42** # 2031.

[30] P. Deligne, Les immeubles des groupes des tresses généralisés, Invent. Math. **17** (1972), 273–302.

[31] René Thom and Harold I. Levine, Singularities of differentiable mappings, Singularities of differentiable maps, 9–101, Izdat. "Mir", Moscow 1968 (in Russian). MR **40** # 896.

[32] G. N. Tyurina, The topological properties of isolated singularities of complex spaces of codimension one, Izv. Akad. Nauk. SSSR Ser. Mat **32** (1968), 605–620. MR **37** # 3053.

[33] V. I. Arnol'd, Classification of unimodal critical points of functions, Funktsional. Anal. i Prilozhen. 7:3 (1973), 75–76.

[34] P. Wagreich, Singularities of complex surfaces with solvable fundamental group, Topology **11** (1972), 51–72.

[35] A. M. Gabrielov, Intersection matrices for certain singularities, Funktsional. Anal. i Prilozhen. 7:3 (1973), 18–32.

[36] I. M. Vinogradov, *Metod trigonometricheskikh summ v teorii chisel* (The method of trigonometric sums in number theory), Izdat. "Nauka", Moscow 1971.

[37] L. K. Hua, On the number of solutions of Tarry's problem, Acta Scientica Sinica **1** (1953), 1–76.

[38] Lê Dũng Tráng and C. P. Ramanujan, The invariance of Milnor's number implies the invariance of the topological type, École Polytechnique, Paris 1973.

[39] K. Saito, Quasihomogene isolierte Singularitäten von Hyperflächen, Invent. Math. **14** (1971), 123–142.

[40] F. Klein, Vorlesungen über die Entwicklung der Mathematik im 19. Jahrhundert, Springer–Verlag, Berlin 1926.
Translation: *Lektsii o razvitii matematiki v xix stoletii*, ONTI, Moscow 1937.

Received 6 June 1973

Translated by J. Wiegold

NORMAL FORMS OF FUNCTIONS IN NEIGHBOURHOODS OF DEGENERATE CRITICAL POINTS

V. I. Arnol'd

An analysis of the normal forms to which functions can be reduced in neighbourhoods of degenerate critical points shows that many of them are quasihomogeneous or semiquasihomogeneous.

A semiquasihomogeneous function is a sum of a quasihomogeneous (or weighted homogeneous) polynomial with an isolated critical point and summands of a higher degree of quasihomogeneity. The normal form to which a semiquasihomogeneous function can be reduced is described in terms of the local ring of the gradient mapping given by the quasihomogeneous part of the function. The number of parameters in this normal form is called the inner modality of the quasihomogeneous part.

A classification is given of all quasihomogeneous critical points of inner modality 1: up to stable equivalence they are exhausted by three one-parameter families of parabolic singularities and 14 exceptional polynomials, 8 of which are functions of two variables, and 6 functions of three variables.

Contents

§1. Introduction 91
§2. Quasihomogeneous functions and filtrations 96
§3. Multiplicity and generators of the local ring of a semiquasihomogeneous function 97
§4. Quasihomogeneous mappings 99
§5. Quasihomogeneous diffeomorphisms and quasijets 103
§6. Quasihomogeneous vector fields 105
§7. The normal form of a semiquasihomogeneous function 108
§8. The normal form of a quasihomogeneous function 109
§9. Piecewise filtrations 111
§10. Semiquasihomogeneous functions of two variables 115
§11. Quasihomogeneous functions of three variables 121
References 131

§1. Introduction

1.1. Degenerate critical points. A critical point of a function is said to be *non-degenerate* if the second differential of the function at the point is a non-degenerate quadratic form. The behaviour of a function in the neighbourhood of a non-degenerate critical point is described by Morse's lemma: there is a system of coordinates in the neighbourhood of such a point in which the function takes a simple "normal form", namely

$$f(x_1, \ldots, x_n) = \pm x_1^2 \pm \ldots \pm x_n^2 + \text{const}.$$

A function "in general position" has only non-degenerate critical points. Every function can be brought to general position by arbitrarily small displacements, so that all its critical points become non-degenerate.

Degenerate critical points appear naturally in cases when the functions depend on parameters. For instance, in the one-parameter family

$$f(x, t) = x^3 + tx$$

of functions of a single variable x, a degenerate critical point arises for an exceptional value of the parameter ($t = 0$). This point cannot be removed by small displacement of the family: whatever the displacement, a degenerate singularity always occurs for a nearby value of the parameter.

Thus, in the investigation of critical points of functions depending on parameters we have to consider degenerate points as well as non-degenerate. The larger the number of parameters, the more complicated the critical points that can occur.

1.2. Terminology. This article is the fifth in a series ([1]–[4]) on the classification of critical points of functions. To make this paper independent of its predecessors, we give here certain definitions and results from [1]–[4].

At first glance, the problem of classifying critical points of l-parameter families of functions of n variables that are irremovable under small displacements seems hopeless for large l and n. Nonetheless, when the first piece of the classification was executed, it turned out to be not as complicated as all that, and the singularities arising were closely connected with objects that seem very remote from them at first sight: namely with Lie groups and the Weyl series A_k, D_k, E_k, Coxeter groups generated by reflections, the braid groups of Artin and Brieskorn, and finally, with the classification of regular polygons in ordinary three-space.

At first the classification of critical points is discrete: if the number l of parameters is less than six, then we can find a finite number of normal forms such that every l-parameter family of functions can be brought to "general position" by a small displacement under which every function in the family reduces to one of the given normal forms in the neighbourhood of every critical point, after a smooth change of variables.

Beginning with $l = 6$ (and the number of variables $n \geqslant 3$), "moduli" emerge, and the normal forms inevitably involve parameters.

The number of moduli (that is, the number of parameters in the normal forms) remains finite for finite l. The fact is that a smooth function is equivalent in a neighbourhood of a critical point to a fairly long segment of its Taylor series, except in the case of critical points of infinite multiplicity (which can be removed under a small displacement of an l-parameter system, for any finite l) (see [5]–[7]).

The class of functions whose Taylor series at 0 coincide up to the terms of degree k with the Taylor series of a given function is called the *k-jet* of the function at 0. A k-jet is said to be *sufficient* if all functions with this k-jet are mutually equivalent (that is, can be transformed into one another under a smooth change of variables).

In these terms we can state that *every function with a critical point of finite multiplicity has a sufficient jet.*

Finiteness of the multiplicity of a critical point is defined easily in the complex case: a function f: $(\mathbf{C}^n, 0) \to (\mathbf{C}, 0)$ has 0 as a critical point of finite multiplicity if 0 is an isolated critical point. The *multiplicity* μ of an isolated critical point of a complex function can be defined as the number of non-degenerate critical points into which it splits under a small displacement, or as the degree of the gradient function mapping a small sphere S^{2n-1} around 0 in $\mathbf{C}^n$ onto itself.

In the real case the multiplicity μ of a smooth mapping F: $(\mathbf{R}^n, 0) \to (\mathbf{R}^n, 0)$ at 0 is defined as the $\mathbf{R}$-dimension of the local ring

$$Q(F) = \mathbf{R}[[x_1, \ldots, x_n]]/(F_1, \ldots, F_n),$$

where $(F_1, \ldots, F_n)$ is the ideal spanned by the Taylor series of the functions specifying F. Here $\mathbf{R}[[\ldots]]$ is the ring of formal power series in the coordinates $(x_1, \ldots, x_n)$ in the spatial inverse image (instead of formal power series we can use convergent series, truncated polynomials, or smooth functions, provided that $\mu < \infty$).

The multiplicity μ of the critical point 0 of a function f: $(\mathbf{R}^n, 0) \to (\mathbf{R}^n, 0)$ is defined as the multiplicity of the gradient mapping: $\mu = \dim_{\mathbf{R}} Q(F)$, where $F = \operatorname{grad} f$. We denote the local ring of the gradient mapping by Q_f, and call it the *local ring of the function* at the critical point. Of course, an analogous ring is defined in the complex case, and then $\mu = \dim_{\mathbf{C}} Q_f$ [8].

In investigating critical points of finite multiplicity, the function may be replaced by a sufficient jet or a polynomial.

Thus, the question of classifying critical points reduces to an algebraic question about the orbits of the action of a finite-dimensional Lie group on a finite-dimensional manifold.

The number of moduli (or the modality) of a function in the neighbourhood of a critical point is defined as follows.

Let G be a Lie group acting on a (finite-dimensional) manifold X. The orbits of G can generate discrete stratifications or continuous families in the neighbourhood of a given point $x \in X$. We say that a point x *has modality* m (under the given action) if a sufficiently small neighbourhood of x in X can be covered by finitely many families of orbits, depending on not more than m parameters (and an arbitrarily small neighbourhood of x intersects some m-parameter family of orbits).

By the modality of a function at a critical point we understand the modality of the action of the group of changes of independent variables on the space of functionals of the function.

More accurately, we consider the finite-dimensional manifold of k-jets at 0 of the function f: $(\mathbf{C}^n, 0) \to (\mathbf{C}, 0)$ with the critical point 0 and critical value 0. The group of k-jets at 0 of diffeomorphisms $(\mathbf{C}^n, 0) \to (\mathbf{C}^n, 0)$ fixing 0 acts on this manifold. By the *modality* of f at the critical point 0 we mean the modality of its k-jet under the action of the group of k-jets of diffeomorphisms, where k is sufficiently large.

1.3. 0-modal critical points. The simplest degenerate critical points are the 0-modal ones. The 0-modal critical points can be classified as follows.

Two functions with critical point 0 and critical value 0 are said to be *equivalent* if one is transformed into the other under a diffeomorphism fixing 0 of a sufficiently small neighbourhood of 0. Two functions are *stably equivalent* if they become equivalent under direct addition of non-degenerate quadratic forms. For example, the functions $f(x) = x^3$ and $g(x, y) = x^3 + y^2$ are stably equivalent.

In a neighbourhood of a 0-modal critical point 0, every function f: $(\mathbf{C}^n, 0) \to (\mathbf{C}^n, 0)$ is equivalent to one of the following functions [2]:

$$\begin{aligned} A_k &: f(x) = x^{k+1}, \\ D_k &: f(x, y) = x^2 y + y^{k-1}, \\ E_6 &: f(x, y) = x^3 + y^4, \\ E_7 &: f(x, y) = x^3 + xy^3, \\ E_8 &: f(x, y) = x^3 + y^5. \end{aligned}$$

The classification of 0-modal critical points has given rise to the hope that the natural classes of critical points allowing a simple description are those of small modality (and not the classes with small μ, or the classes with small codimension l, which are needed in analysis and topology).

1.4. Unimodal critical points. The classification of critical points of modality 1 turned out to be not too complicated [3]. Indeed, there is one three-index series of one-parameter families of unimodal singularities,

$$T_{p,q,r}: f(x, y, z) = axyz + x^p + y^q + z^r,$$

where $\frac{1}{p} + \frac{1}{q} + \frac{1}{r} \leqslant 1$, and 14 "exceptional" one-parameter families. Every function f: $(\mathbf{C}^n, 0) \to (\mathbf{C}^n, 0)$ is equivalent in the neighbourhood of a critical point of modality 1 to one of those listed.

There are three special cases of the series T for which $\frac{1}{p} + \frac{1}{q} + \frac{1}{r} = 1$, namely (3, 3, 3), (2, 4, 4) and (2, 3, 6). These singularities are said to be *parabolic*, and the remaining singularities in the series T are *hyperbolic* (corresponding to the signature of the quadratic form given by the intersection index in the homology of average dimension of the local manifold of non-singular level of the function – see [4]; *elliptic* singularities turn up only as the 0-modal singularities A, D, E).

The 14 exceptional families of unimodal singularities are also given by simple formulae:

$$\begin{aligned}
K_{12} &= x^3 + y^7 + axy^5, & K_{13} &= x^3 + xy^5 + ay^8,\\
K_{14} &= x^3 + y^8 + axy^6, & Z_{11} &= x^3y + y^5 + axy^4,\\
Z_{12} &= x^3y + xy^4 + ax^2y^3, & Z_{13} &= x^3y + y^6 + axy^5,\\
W_{12} &= x^4 + y^5 + ax^2y^3, & W_{13} &= x^4 + xy^4 + ay^6,\\
Q_{10} &= x^3 + y^4 + yz^2 + axy^3, & Q_{11} &= x^3 + y^2z + xz^3 + az^5,\\
Q_{12} &= x^3 + y^5 + yz^2 + axy^4, & S_{11} &= x^4 + y^2z + xz^2 + ax^3z,\\
S_{12} &= x^2y + y^2z + xz^3 + az^5, & U_{12} &= x^3 + y^3 + z^4 + axyz^2.
\end{aligned}$$

An analysis of these formulae shows that in each such family there is exactly one quasihomogeneous (weighted homogeneous) polynomial, and that the whole family can be considered as a one-parameter deformation of this polynomial having the special property of "semiquasihomogeneity".

The quasihomogeneous polynomials of these 14 types are connected with the 14 triangles in the Lobachevskii plane, as was shown by I. V. Dolgachev (and for K_{12} already by Klein).

1.5. Semiquasihomogeneous critical points. The aim of this paper is the investigation of quasihomogeneous functions and their semiquasihomogeneous deformations. We indicate the normal forms to which functions having non-degenerate quasihomogeneous "principal part" can be reduced. We call the number of moduli in these normal forms the *inner modality* of the quasi-homogeneous function (the exact definition of the inner modality μ_0 is in §8.6; μ_0 does not depend on the choice of the normal form).

Quasihomogeneous functions of inner modality 1 can be listed independently of the classification of all unimodal functions. The main result of our article is an algebraic classification of quasihomogeneous functions of inner modality 1.

It turns out that (up to stable equivalence) there are just three one-parameter families of parabolic singularities and 14 polynomials generating exceptional families.

The technique developed here for working with semiquasihomogeneous singularities enables us to shorten significantly the calculations necessary for classifying unimodal singularities. The technique is based on a discussion of various filtrations in the ring of functions (or power series). With every such filtration we can associate its space of jets of functions, its filtered group of jets of diffeomorphisms, and the Lie algebra of vector fields. In the case of a "quasihomogeneous" filtration there arises also a Lie group of quasihomogeneous diffeomorphisms, which plays a rôle in this theory like that of the full linear group in the case of ordinary jets (which corresponds to the filtration given by the powers of the maximal ideal).

The author is grateful to A. G. Kushirenko for numerous discussions that have put on a solid foundation calculations that had previously seemed a matter of pure luck. The author is also grateful to A. M. Gabrielov and D. B. Fuks for helpful discussions.

§2. Quasihomogeneous functions and filtrations

In this section we give the definitions of the principal objects that are studied in the sequel: quasihomogeneous and semiquasihomogeneous functions.

2.1. DEFINITION. We consider the arithmetical space $\mathbf{C}^n$ with fixed coordinates $x_1, \ldots, x_n$. *A function* f: $(\mathbf{C}^n, 0) \to (\mathbf{C}, 0)$ *is said to be quasihomogeneous of degree* d *with exponents* $\alpha_1, \ldots, \alpha_n$ if $f(\lambda^{\alpha_1} x_1, \ldots, \lambda^{\alpha_n} x_n) = \lambda^d f(x_1, \ldots, x_n)$ *for all* λ.

In terms of the Taylor series $f = \Sigma f_{\mathbf{k}} \mathbf{x}^{\mathbf{k}}$, quasihomogeneity of degree 1 means that all the exponents of non-zero terms lie on the hyperplane

$$\Gamma = \{\mathbf{k} : \alpha_1 k_1 + \ldots + \alpha_n k_n = 1\}.$$

In what follows, we consider quasihomogeneous functions of degree 1 with rational exponents, $0 < \alpha_s \leqslant 1/2$. Such functions are automatically polynomials. We call the hyperplane Γ the *diagonal.* The diagonal cuts the coordinate axes in segments of length $a_s = 1/\alpha_s$.

2.2. DEFINITION. A quasihomogeneous function f is said to be *non-degenerate* if 0 is an isolated critical point (that is, if the multiplicity μ of the critical point 0 is finite). The degenerate quasihomogeneous functions form an algebraic hypersurface in the linear space of all quasihomogeneous polynomials with fixed quasihomogeneous exponents.

With every type of quasihomogeneity (that is, with every set $\boldsymbol{\alpha}$ of quasihomogeneous exponents) there is associated a filtration in the ring of power series (functions, germs etc), defined as follows:

2.3. DEFINITION. We say that a monomial $\mathbf{x}^{\mathbf{k}} = x_1^{k_1} \ldots x_n^{k_n}$ has (generalized) *degree* d if $\langle \boldsymbol{\alpha}, \mathbf{k} \rangle = \alpha_1 k_1 + \ldots + \alpha_n k_n = d$.

The degree of a monomial is a rational number. The exponents of all monomials of degree d (of given type) lie on a hyperplane parallel to the diagonal Γ. We fix the quasihomogeneous type, that is, we fix the set $\boldsymbol{\alpha}$ of exponents.

2.4. DEFINITION. A polynomial (power series, germ, function) *has filtration* d if all its monomials are of degree d or higher: when the (generalized) degree of all monomials is d, we call d the (generalized) *degree* of the polynomial; the degree of 0 is $+\infty$.

The polynomials (series, germs) of filtration d form a linear space E_d; $E_{d'} \subseteq E_d$ if $d < d'$; the filtration of a product is the sum of the filtrations of the factors, so that E_d is an ideal in the ring of polynomials (series, germs). Denoting this ring by A, we call the factor-ring A/E_d the *ring of d-jets*, and its elements *d-jets*.

By the filtration $\varphi(f)$ of a polynomial (series, germ) f we understand, as a rule, the largest d such that $f \in E_d$. The filtrations of all polynomials (series, germs) lie in a rational arithmetical progression: $\varphi(f) \in \mathbf{Z}_+ d_0$, where d_0 is the greatest common divisor of the numbers α_s (an initial segment of the progression may not be completely filled out by the values of φ).

2.5. DEFINITION. A polynomial (power series, germ) is said to be *semiquasihomogeneous of degree d with exponents* $\alpha_1, \ldots, \alpha_s$ if it is of the form $f = f_0 + f'$, where f_0 is a non-degenerate quasihomogeneous polynomial of degree d with exponents α, and f' is a polynomial (series, germ) of filtration strictly greater than d.

In other words, a semiquasihomogeneous function is obtained from a non-degenerate quasihomogeneous function by adding monomials whose exponents lie *above the diagonal*. Note that a quasihomogeneous function is not semiquasihomogeneous if it is degenerate.

§3. Multiplicity and generators of the local ring of a semiquasihomogeneous function

In this section we show that a monomial basis for the local ring of a quasihomogeneous non-degenerate function is a basis for all semiquasihomogeneous functions with the given quasihomogeneous part.

3.1. THEOREM. *The multiplicity of the critical point* 0 *of a semiquasihomogeneous function f is that of its quasihomogeneous part*: $\mu(f) = \mu(f_0)$.

PROOF. We consider the family of topological spheres

$$S_t = \{\mathbf{x} \in \mathbf{C}^n : |x_1|^{a_1} + \ldots + |x_n|^{a_n} = t\}, \qquad a_s = 1/\alpha_s, \quad d = 1.$$

The number $\mu(f)$ is the degree of the mapping $\mathbf{x} \to (\partial f/\partial \mathbf{x})/\|\partial f/\partial \mathbf{x}\|$, $\mathbf{x} \in S_t$, for small values of t. For every point $\mathbf{x} \in S_1$, at least one of the derivatives $\partial f_0/\partial x_s$ is different from 0 (since f_0 is non-degenerate). Thus, there is a constant c such that $\max\limits_s |\partial f_0/\partial x_s| \geqslant c > 0$ on S_1.

Note that $S_t = T_t S_1$, where $T_t(x_1, \ldots, x_n) = (t^{\alpha_1} x_1, \ldots, t^{\alpha_n} x_n)$. Further, the partial derivative $\partial f_0/\partial x_s$ is quasihomogeneous of degree $1 - \alpha_s$ and of type α. Therefore, at every point of S_t, $|\partial f_0/\partial x_s| \geqslant ct^{1-\alpha_s}$ for at least one s.

On the other hand, the function f' has filtration not less than $1 + d_0$. Therefore there exists a constant C such that $|\partial f'/\partial x_s| \leqslant Ct^{1+d_0-\alpha_s}$ on S_t for all s.

Comparing with the preceding inequality, we see that for small enough t there are no critical points of the functions $f_0 + \theta f'$, $0 \leqslant \theta \leqslant 1$, on S_t. Thus, the degree of the mappings of the sphere onto itself given by the gradients of f_0 and $f_0 + f'$ coincide, as required.

REMARK. It can be shown in a similar way that *all quasihomogeneous functions sufficiently near f_0 and having the same degree of quasihomogeneity* *have the same multiplicity* μ. Further, since the set of non-degenerate quasihomogeneous functions of given degree is connected, *the multiplicity* μ *is the same for all non-degenerate quasihomogeneous functions of given degree (and therefore for all semiquasihomogeneous functions of given degree and given type).*

The multiplicity μ of the critical point 0 of a function f may also be defined as the dimension of the local ring

$$Q_f = \mathbf{C}[[x_1, \ldots, x_n]]/(\partial f/\partial x_1, \ldots, \partial f/\partial x_n).$$

Allowing freedom of speech, we call a set of μ series (polynomials, germs) *a basis of the local ring of* f if they become a basis for Q_f over $\mathbf{C}$ after factorization by this ideal.

From the theorem just proved we deduce:

3.3. COROLLARY. *Assume that the system of monomials* $e_1, \ldots, e_\mu$ *is a basis of the local ring of the quasihomogeneous part* f_0 *of a semiquasihomogeneous function* f. *Then this same system of monomials is a basis of the local ring of* f.

The proof is based on the following general lemma.

3.4. LEMMA. *Assume that a system of smooth functions* f *depending continuously on a finite number of parameters has* 0 *as a critical point of constant multiplicity* μ *for all values of the parameters. Then every basis of the local ring of the function corresponding to the value* 0 *of the parameter remains a basis for nearby values of the parameter.*

PROOF OF THE LEMMA. The lemma follows from this fact: given a subspace of a finite-dimensional vector space depending smoothly on some parameters, and a system of vectors forming a basis of the transversal space, then the system remains a basis of the transversal space for nearby values of the parameters.

To make the space finite-dimensional, it is enough to factor the ring $\mathbf{C}[[x_1, \ldots, x_n]]$ by a sufficiently high power of the maximal ideal (the power depending only on μ, see [5]).

PROOF OF COROLLARY 3.3. We consider a semiquasihomogeneous function $f = f_0 + f'$. We claim that the transition from f_0 to f may be regarded as a *small* deformation. We construct a one-parameter family of functions, $f_t(x) = t^{-1}f(T_t x)$, where $T_t x = (t^{\alpha_1}x_1, \ldots, t^{\alpha_n}x_n)$, $d = 1$. We have $f_t(x) = f_0 + t^{-1}f'(T_t x)$, where all the coefficients in the second summand depend continuously on t, since the filtration of f' is greater than 1. By the lemma, a basis of the local ring for f_0 is a basis for f_t for sufficiently small t. A basis of the local ring for f_t goes over to a basis of that for f under the action of the diffeomorphism T_t connecting the functions f and f_t. But under T_t, every monomial goes over to a monomial proportional to it. Thus, a *monomial* basis of Q_{f_0} is not only a basis of Q_{f_t} for small t, but also of Q_f; this is what we wanted to prove.

3.5. REMARK. *The number of basis monomials of the local ring of a* *quasihomogeneous or semiquasihomogeneous function f having given generalized degree d does not depend on the choice of the basis in the local ring.*

PROOF. We consider the factor-space

$$E_\delta/(E_{>\delta} + E_\delta \cap I), \text{ where } I = (\partial f/\partial x_1, \ldots, \partial f/\partial x_n),$$

E_δ is the space of series of filtration δ in $\mathbf{C}[[x_1, \ldots, x_n]]$,
$E_{>\delta}$ is the space of series of filtration greater than δ.
The number of basis monomials of degree δ is equal to the dimension of this factor-space, therefore, does not depend on the basis.

3.6. COROLLARY. *The number of basis monomials of the local ring of a function having given generalized degree δ (for given type $\boldsymbol{\alpha}$) is the same for all semiquasihomogeneous functions f of type $\boldsymbol{\alpha}$ and degree d.*

PROOF. It is enough to consider non-degenerate quasihomogeneous functions (for a semiquasihomogeneous function the basis is the same, by Corollary 3.3). The manifold of non-degenerate quasihomogeneous functions of given degree d (and type $\boldsymbol{\alpha}$) is arcwise connected (it is the complement of a hypersurface in a linear space). The number of basis monomials of degree δ of the local ring is locally constant along a curve joining two points of this manifold, because the same basis works for neighbouring functions (Lemma 3.4). Thus, it is constant, as required.

§4. Quasihomogeneous mappings

In this section we compute various numerical invariants of quasihomogeneous mappings, in particular, the multiplicity μ and the generating polynomial χ.

We fix a quasihomogeneous type $\boldsymbol{\alpha} = (\alpha_1, \ldots, \alpha_n)$ in $\mathbf{C}^n$ with a fixed coordinate system. We consider a mapping F: $(\mathbf{C}^n, 0) \to (\mathbf{C}^n, 0)$ and use the older notation $F(x_1, \ldots, x_n) = (F_1(x), \ldots, F_n(x))$; Let $\mathbf{d} = (d_1, \ldots, d_n)$ be a vector with non-negative components.

4.1. DEFINITION. The mapping F is said to be *quasihomogeneous of degree* $\mathbf{d}$ (and type $\boldsymbol{\alpha}$) if every component F_s is a quasihomogeneous function of degree d_s of one and the same type $\boldsymbol{\alpha}$.

The local ring of F is the factor ring

$$Q(F) = \mathbf{C}[[x_1, \ldots, x_n]]/(F_1, \ldots, F_n).$$

F is said to be *non-degenerate* if its multiplicity at 0 is finite, that is, if the local ring $Q(F)$ has finite dimension over $\mathbf{C}$; this dimension $\mu = \dim_{\mathbf{C}} Q(F)$ is called the *multiplicity of F at* 0.

F is said to be *semiquasihomogeneous* if $F = F_0 + F'$, where F_0 is a non-degenerate quasihomogeneous mapping, and each component F_s' has filtration greater than the degree of the corresponding component F_{0s}.

If a function f is semiquasihomogeneous of degree d (and type $\boldsymbol{\alpha}$), then the mapping $x \mapsto \mathrm{grad}(f(x))$ is semiquasihomogeneous of degree $d_s = d - \alpha_s$.

4.2. PROPOSITION. *The assertions of Theorem* 3.1, *Remark* 3.2, *Corollary* 3.3, *Lemma* 3.4, *Remark* 3.5 *and Corollary* 3.6 *hold not only for the gradient mappings* $x \mapsto \mathrm{grad}(f(x))$ *as formulated above, but also for any quasihomogeneous or semiquasihomogeneous mappings.*

For example, Remark 3.2 runs as follows: *all semiquasihomogeneous mappings of the same degree* $\mathbf{d}$ *(and type* α*) have the same multiplicity* μ.

Kushirenko has proved analogues of these assertions over an arbitrary algebraically closed field. For $\mathbf{C}$ the proof proceeds as in §3.

The value of the class of quasihomogeneous mappings for our problem lies in the facts that in them homotopies can be carried out more freely, and changes of variables are easier than for gradients. In particular, the following proposition (see [10]) makes it possible to go by a simple substitution from a quasihomogeneous to a genuine homogeneous mapping.

4.3. PROPOSITION. *Suppose that* $F\colon \mathbf{C}^n \to \mathbf{C}^n$ *is a quasihomogeneous mapping whose type and degree have the common denominator* N: $\alpha_s = A_s/N$, $d_s = D_s/N$, *where now* A_s, D_s, N *are integers. Consider the mapping* $T\colon \mathbf{C}^n \to \mathbf{C}^n$ *given by the formula* $T(y_1, \ldots, y_n) = (y_1^{A_1}, \ldots, y_n^{A_n})$. *Then*

1) *the mapping* $F \circ T\colon \mathbf{C}^n \to \mathbf{C}^n$ *has as its components homogeneous functions in the usual sense, of degrees* $D_1, \ldots, D_n$.

2) $$\mu(F \circ T) = \mu(F) \prod A_s.$$

3) *If* $e_1, \ldots, e_\mu$ *is a monomial basis of the local ring of* F, *then the following functions form a monomial basis of the local ring of* $F \circ T$:

$$e'_{i,u} = (T^* e_i) y_1^{u_1} \ldots y_n^{u_n}, \text{ where } 1 \leqslant i \leqslant \mu,\ 0 \leqslant u_s < A_s.$$

PROOF. 1) The monomial $\mathbf{x}^{\mathbf{k}}$ determines in the s-th component of $F \circ T$ the monomial $\prod y_s^{k_s A_s}$ of degree $\sum k_s A_s = N(\mathbf{k}, \alpha) = N d_s = D_s$. 2) The formula for the multiplicity is obtained from considering the system of equations $(F \circ T)(y) = \varepsilon$ in y. 3) The functions $e'_{i,u}$ generate the whole local ring. For every function of y can be written in the form $\varphi = \sum_u y^u T^* \varphi_u$, and every function of x in the form $\varphi_u = \sum c_{i,u} e_i + \sum F_s h_{s,u}$. Thus,

$$\varphi = \sum_{u,i} c_{i,u} y^u T^* e_i + \sum_{u,s} y^u h_{s,u} T^* F_s,$$

that is, the $e'_{i,u}$ generate $Q(F \circ T)$. Since the number of functions $e'_{i,u}$ is $\mu(F \circ T)$, they form a basis. This proves Proposition 4.3.

4.4. DEFINITION. The *generating function* of type α of a semiquasihomogeneous mapping F (where $\alpha_s = A_s/N$, A_s and N integers) is the polynomial $\chi_F(z) = \sum \mu_i z^i$, where μ_i is the number of basis monomials in the local ring of F having generalized degree i/N.

We remark that even for fixed quasihomogeneous type χ depends on the integer N. However, the possible values of N are multiples of one of them (the least common denominator of the fractions α_s), and $\chi_{F;\,kN,\,\alpha}(z) = \chi_{F;\,N,\,\alpha}(z^k)$.

The dimension of the local ring is given by the formula $\mu = \chi_F(1)$. Further, the degree of χ_F is the greatest of the (generalized) degrees of the monomial generators in a basis of the local ring.

4.5. THEOREM. (see [9], [14]). *The generating function of a (semi)-quasihomogeneous mapping F of degree d and type α for which $\alpha_s = A_s/N$, $d_s = D_s/N$, where A_s, D_s, N are integers, is given by the formula*

$$\chi_F(z) = \prod_{s=1}^{n} \frac{z^{D_s}-1}{z^{A_s}-1}.$$

EXAMPLE. If $F = \operatorname{grad} f$, where f is a (semi)quasihomogeneous function of type α (and degree 1), then

$$\chi_F(z) = \prod_{s=1}^{n} \frac{z^{N-A_s}-1}{z^{A_s}-1}.$$

Several useful formulae follow immediately from this theorem.

4.6. COROLLARY (see [10], [14]). *The dimension of the local ring of a semiquasihomogeneous mapping is given by the "generalized Bezout formula":*

$$\mu = \prod_{s=1}^{n} \frac{d_s}{\alpha_s}.$$

4.7. COROLLARY (see [11]). *The local ring of a semiquasihomogeneous mapping F has exactly one basis monomial of degree*

$$d_{\max} = \sum_{s=1}^{n} (d_s - \alpha_s);$$

all monomials of higher filtration lie in the ideal generated by the components $(F_1, \ldots, F_s)$.

Let us look, in particular, at the local ring of a semiquasihomogeneous function f of type $(\alpha_1, \ldots, \alpha_n)$ of degree 1. In this case $\mathbf{d}_s = 1 - \alpha_s$, and we obtain the next result.

4.8. COROLLARY (see [10], [14]). *The dimension of the local ring of a semiquasihomogeneous function f of type $(\alpha_1, \ldots, \alpha_n)$ and degree 1 is given by the formula*

$$\mu = \prod \left(\frac{1}{\alpha_s} - 1\right).$$

4.9. COROLLARY ([11]). *A monomial basis of the local ring of a semi-quasihomogeneous function f of type $(\alpha_1, \ldots, \alpha_n)$ and degree 1 has exactly* *one generator of (generalized) degree* $d_{\max} = \Sigma(1 - 2\alpha_s)$; *all monomials of higher degree lie in the ideal* $(\partial f/\partial x_1, \ldots, \partial f/\partial x_s)$.

Here are some more immediate corollaries of Theorem 4.5.

4.10. COROLLARY. *The generating polynomial of a semiquasihomogeneous mapping is always recurrent*:

$$\mu_i = \mu_{k-i} \text{ where } k = \Sigma D_s - \Sigma A_s.$$

4.11. COROLLARY. *The generalized degree of the penultimate monomial (according to filtration) in a monomial basis of the local ring of a quasihomogeneous function of type* $(\alpha_1, \ldots, \alpha_n)$ *and degree* 1 *is* $d_{\max} - \alpha_{\min}$, *where* $\alpha_{\min} = \min(\alpha_1, \ldots, \alpha_n)$.

4.12. COROLLARY. *A non-degenerate quasihomogeneous mapping of type* $\boldsymbol{\alpha} = \mathbf{A}/N$ *and degree* $\mathbf{d} = \mathbf{D}/N$ *can exist only when the polynomial* $\prod(z^{D_s} - 1)$ *is divisible by* $\prod(z^{A_s} - 1)$.

4.13. COROLLARY. *A non-degenerate quasihomogeneous mapping of type* $\boldsymbol{\alpha}$ $(\alpha_s = A_s/N)$ *can exist only when* $\prod(z^{N-A_s} - 1)/\prod(z^{A_s} - 1)$ *is a polynomial.*

REMARK. In the case of functions of two or three variables, the fact that the fraction $\Pi(z^{N-A_s} - 1)/\Pi(z^{A_s} - 1)$ can be cancelled is sufficient as well as necessary for the existence of a non-degenerate quasihomogeneous function with exponents A_s/N (see §§ 10, 11). This is not true for four variables, as can be seen from the following example, which V. M. Izlev has shown to me:

$$N = 265, A_1 = 1, A_2 = 24, A_3 = 33, A_4 = 58.$$

In this example the quotient is a polynomial with non-negative coefficients, while all quasihomogeneous functions with exponents A_s/N are degenerate.

The results of this section have been rediscovered over and over again. At the time when this paper is going to press, Gabrielov has informed me that Theorem 4.5 and Corollaries 4.6 and 4.8 can be found in papers of Milnor, Orlik and Wagreich. Saito and Hironaka have given other proofs of Corollaries 4.7 and 4.9. Kushirenko has told me that all the results, are, in fact, in Bourbaki [14] (see Proposition 2 in Ch. 5, § 5.1, in the section "Poincaré series of a graded algebra").

PROOF OF THEOREM 4.5. It is enough to consider the case of a non-degenerate quasihomogeneous mapping F (see Corollary 3.6 and Proposition 4.2). We change T as in Proposition 4.3. It follows from the form of the generators of the local ring of $F \circ T$ that

$$\chi_{F\circ T;\,1,1}(z) = \chi_{F;\,N,\alpha}(z)\chi_{T;\,1,1}(z), \text{ where } \mathbf{1} = (1, \ldots, 1).$$

The generating polynomials of T and $F \circ T$ occurring in this formula are homogeneous in the usual sense, and can be calculated explicitly. Indeed, for the mapping $x = y^A$ we have $\chi(z) = \dfrac{z^A - 1}{z - 1}$. It follows that

$$\chi_T(z) = \prod_{s=1}^{n} \frac{z^{A_s} - 1}{z - 1}$$

(here and later the pair $(N, \alpha) = (1, 1)$ is omitted from the notation for χ).

On the other hand, $F \circ T$ is a non-degenerate mapping whose components are homogeneous functions of degree D_s. Thus (by Proposition 4.2 and Corollary 3.6), it has the same generating polynomial as every other non-degenerate homogeneous mapping with these degrees. We can take, for example, the mapping T' given by the formula

$$T'(y_1, \ldots, y_n) = (y_1^{D_1}, \ldots, y_n^{D_n}).$$

Thus,

$$\chi_{F \circ T}(z) = \chi_{T'}(z) = \prod_{s=1}^{n} \frac{z^{D_s} - 1}{z - 1}.$$

The formula for χ_F is obtained by dividing the formulae for $\chi_{F \circ T}$ and for χ_T. This proves Theorem 4.5.

§5. Quasihomogeneous diffeomorphisms and quasijets

Several Lie groups and Lie algebras are connected with the filtrations defined by a quasihomogeneous type $\boldsymbol{\alpha}$. In the case of ordinary homogeneity, these are the full linear group, the group of k-jets of diffeomorphisms, the subgroup of k-jets with identity $(k-1)$-jets, and their factor groups. The analogues for quasihomogeneous filtrations are defined as follows.

We consider the space $\mathbf{C}^n$, with a fixed coordinate system $(x_1, \ldots, x_n)$. The ring of formal power series[1] in these coordinates is denoted by $A = \mathbf{C}[[x_1, \ldots, x_n]]$. We assume that a type of quasihomogeneity $\alpha = (\alpha_1, \ldots, \alpha_n)$ is given. We denote by E_d the ideal of A generated by series of filtration d. Further, let $E_{>d}$ stand for the ideal of E_d consisting of series of filtration strictly greater than d.

A *formal diffeomorphism* g: $(\mathbf{C}^n, 0) \to (\mathbf{C}^n, 0)$ is given by a collection of n power series without free terms and gives a ring isomorphism g^*: $A \to A$ by the formula $g^*f = f \circ g$, where $\circ$ denotes the substitution of a series in a series.

5.1. DEFINITION. A diffeomorphism g has *filtration* d if, for all λ,

$$(g^* - 1)E_\lambda \subset E_{\lambda+d}.$$

5.2. PROPOSITION. *Let $d \geqslant 0$. Then the set $G_d = G_d(\boldsymbol{\alpha})$ of all diffeomorphisms of filtration d is a group under $\circ$.*

PROOF. We remark, first of all, that $g^*E_\lambda = E_\lambda$ for all λ when $d \geqslant 0$ ($g^*E_\lambda \supset E_\lambda$, since $d \geqslant 0$, and $g^{*-1}E_\lambda \supset E_\lambda$, since the factor space A/E_λ is finite-dimensional). Thus, for a, b in G_d we have

---

[1] The greater part of what follows carries over at once to the case when A is the ring of convergent series over $\mathbf{C}$ or $\mathbf{R}$, or the ring of germs of smooth functions.

$$[(a \circ b)^* - 1]E_\lambda = [b^*(a^* - 1) + (b^* - 1)]E_\lambda \subset E_{\lambda+d},$$
$$(a^{-1*} - 1)E_\lambda = a^{-1*}(1 - a^*)E_\lambda \subset E_{\lambda+d},$$

as required.

5.3. PROPOSITION. *For $q > p \geqslant 0$, G_q is a normal subgroup of G_p.*

PROOF. The definition of G_q uses only the filtration $\{E_\lambda\}$. This filtration is invariant under G_0 and a fortiori under G_p. Thus, a subgroup defined in terms of this filtration is normal, as required.

The group G_0 is especially important, because it plays for the quasihomogeneous case the rôle played by the full group of jets of diffeomorphisms in the homogeneous case. It must be stressed that in the quasihomogeneous case certain diffeomorphisms have negative filtrations and do not lie in G_0.

5.4. DEFINITION. *The group of d-jets of type α* is the factor group of the group of diffeomorphisms by the subgroup consisting of the diffeomorphisms of filtration greater than d,

$$J_d = J_d(\alpha) = G_0/G_{>d}.$$

It is clear that J_d is a finite-dimensional Lie group. There are natural factorizations $\pi_{p,q}\colon J_p \to J_q$ $(p > q \geqslant 0)$.

Attention should be drawn to the fact that in the ordinary homogeneous case our numbering differs by 1 from the standard one: our J_0 is called the group of 1-jets, etc.

5.5. PROPOSITION. *The group J_p is obtained from J_0 by a chain of extensions with commutative factors. More accurately, let E_p be the term of the filtration immediately following E_q. Then the kernel K of the homomorphism $\pi_{p,q}$ is commutative.*

PROOF. Let $A, B \in K$; we consider any representatives $a, b \in G_0$. Then

$$(ab)^* - 1 = (a^* - 1) + (b^* - 1) + (b^* - 1)(a^* - 1).$$

Further, for every λ,

$$(a^* - 1)E_\lambda \subset E_{\lambda+p}, \quad (b^* - 1)E_\lambda \subset E_{\lambda+p},$$

since the q-jets a and b are trivial. Thus, $[(ab)^* - (ba)^*]E_\lambda \subset E_{\lambda+2p}$. Therefore, ab and ba determine the same element of J_p, which is what we wanted to prove.

The group J_0 has especial value, because it is the quasihomogeneous generalization of the full linear group.

5.6. DEFINITION. A diffeomorphism $g \in G_0$ is said to be *quasihomogeneous of type α* if every space of quasihomogeneous functions of degree d (and type α) is mapped into itself by g.

The set of all quasihomogeneous diffeomorphisms (of fixed type) forms a group. We denote it by $H(= H(\alpha))$ and call it the *group of quasihomogeneous diffeomorphisms.*

We consider the natural embedding $i\colon H \to G_0$ and factorization $\pi\colon G_0 \to J_0$.

5.7. PROPOSITION. *The group J_0 is naturally isomorphic to the group H of quasihomogeneous diffeomorphisms; in fact, the compound map $\pi i\colon H \to J_0$ is an isomorphism of Lie groups.*

PROOF. a) $\operatorname{Ker} \pi i = e$. For $\operatorname{Ker} \pi i = H \cap G_{>0}$. Hence for $h \in \operatorname{Ker} \pi i$ and for every monomial f of degree d, $(h^* - 1)f$ lies in the space of homogeneous functions of degree d and also has filtration greater than d. Thus, $(h^* - 1)f = 0$ for every monomial f, and so $h = e$.

b) $\operatorname{Im} \pi i = J_0$. For this proof we construct the inverse mapping $J_0 \to H$ explicitly. Let $x_1, \ldots, x_n$ be coordinates in $\mathbf{C}^n$, and let a diffeomorphism $g \in G_0$ be a representative of the jet $j \in J_0$. We consider the series $g^*x_i \in E_{\alpha_i}$. We select in it the homogeneous component y_i of degree α_i, so that $g^*x_i = y_i + z_i$, $z_i \in E_{>\alpha_i}$. We define a polynomial mapping $h^{-1}\colon \mathbf{C}^n \to \mathbf{C}^n$ by the relation $h^*x_i = y_i$. To check that h is a diffeomorphism, we calculate the Jacobian:

$$\det\left|\frac{\partial\,(y_i+z_i)}{\partial x_j}\right| = \det\left|\frac{\partial y_i}{\partial x_j}\right| + R.$$

The term R containing the derivatives with respect to z is 0 at the origin. For every summand of the determinant y is homogeneous of degree 0 in x. All other summands containing z have positive filtration since $z_i \in E_{>d_i}$. Therefore, $R \in E_{>0}$ and $R(0) = 0$. Thus, the Jacobians of g and h at 0 are the same, so that the Jacobian of h at zero is different from 0, which means that h is a diffeomorphism. The ring automorphism h^* preserves degrees of all monomials, because it preserves the degrees of the coordinates x_i. Thus, $h \in H$. It is clear that $\pi i h = j$, and this proves the assertion.

5.8. PROPOSITION. *Suppose that $d \geqslant 0$. Then the group J_d of d-jets of diffeomorphisms acts as a group of linear transformations on the space $A/E_{>d}$ of d-jets of functions.*

PROOF. Let $g \in G_{>d}$. Then application of g does not change the α-jet of a function f, since $f \circ g - f \in E_{>d}$. Thus, $(h, f) \mapsto f \circ h$ gives a mapping $J_d \times (A/E_{>d}) \to A/E_{>d}$, as required.

5.9. REMARK. In the case of ordinary homogeneity, even the group of $(d - 1)$-jets of diffeomorphisms acts on the space of d-jets (all this in our notation). This fact has an analogue in the quasihomogeneous case in the action of the group of $(d - \min \alpha_s)$-jets.

§6. Quasihomogeneous vector fields

The infinitesimal analogues of the concepts we have introduced run as follows.

6.1. DEFINITION. A formal vector field $\mathbf{v} = \Sigma v_i \partial/\partial x_i$ *has filtration d* if the directional derivative of $\mathbf{v}$ raises the filtration by not less than

d: $L_{\mathbf{v}}E_\lambda \subset E_{\lambda+d}$. We denote the set of all vector fields of filtration d by $\mathfrak{g}_d$. Our filtration in the module of vector fields (that is, derivation of A) is compatible with the filtration of the ring:

$$a \in E_d,\ \mathbf{v} \in \mathfrak{g}_\delta \Rightarrow a\mathbf{v} \in \mathfrak{g}_{d+\delta},\ L_{\mathbf{v}}a \in E_{d+\delta}.$$

6.2. PROPOSITION. *Suppose that* $d \geqslant 0$. *Then* 1) *the Poisson bracket of vector fields defines on* $\mathfrak{g}_d$ *a Lie algebra structure*; 2) *the Poisson bracket of elements* $\mathfrak{g}_{d_1}$ *and* $\mathfrak{g}_{d_2}$ *lies in* $\mathfrak{g}_{d_1+d_2}$ *so that each* $\mathfrak{g}_d$ *is an ideal in the Lie algebra* $\mathfrak{g}_0$.

PROOF. If $f \in E_\lambda$, $\mathbf{v}_1 \in \mathfrak{g}_{d_1}$, $\mathbf{v}_2 \in \mathfrak{g}_{d_2}$ then $(L_{\mathbf{v}_1}L_{\mathbf{v}_2} - L_{\mathbf{v}_2}L_{\mathbf{v}_1})f \in E_{\lambda+d_1+d_2}$, as required.

The filtration of a vector field is connected with the filtrations of its components in the following manner.

6.3. PROPOSITION. *The field* $\mathbf{v} = \Sigma v_i \partial/\partial x_i$ *has filtration* d *(and type* α*) if and only if each component* v_i *is a function (series) of filtration* $d + \alpha_i$.

For the proof we introduce the following notation. A vector field of the form $\mathbf{x}^{\mathbf{k}}\partial/\partial x_i$ is called a *vector monomial.* The *degree* of such a vector monomial (for given quasihomogeneous type) is the (possibly negative) rational number $\langle \mathbf{k}, \alpha\rangle - \alpha_i = \langle \mathbf{k} - \mathbf{1}_i, \alpha\rangle$ in the arithmetical progression containing the degrees of the ordinary monomials. A vector field is said to be *homogeneous of degree d* if all vector monomials occurring in it with non-zero coefficients are of degree d.

PROOF OF PROPOSITION 6.3. If $\mathbf{v} \in \mathfrak{g}_d$, then $v_i = L_{\mathbf{v}}x_i \in E_{d+\alpha_i}$, since $x_i \in E_{\alpha_i}$. Let $v_i = \Sigma\ v_{i,\mathbf{k}}\mathbf{x}^{\mathbf{k}}$. For every monomial $f = \mathbf{x}^{\mathbf{l}}$ we have

$$L_{\mathbf{v}}f = \sum v_i \frac{\partial f}{\partial x_i} = \sum l_i v_{i,\mathbf{k}}\mathbf{x}^{\mathbf{l}+\mathbf{k}-\mathbf{1}_i}.$$

Here, $\langle \mathbf{k}, \alpha\rangle \geqslant d + \alpha_i$ if $v_i \in E_{d+\alpha_i}$. Thus, $\langle \mathbf{l} + \mathbf{k} - \mathbf{1}_i, \alpha\rangle \geqslant \langle \mathbf{l}, \alpha\rangle + d$, that is, $L_{\mathbf{v}}E_\lambda \subset E_{\lambda+d}$, as required.

We obtain immediately from Proposition 5.5:

6.4. COROLLARY. *The Lie algebra* $\mathfrak{j}_d$ *of the group* J_d *of d-jets of diffeomorphisms is the factor algebra* $\mathfrak{j}_d = \mathfrak{g}_0/\mathfrak{g}_{-d}$. *The mapping* $\pi_{p,q}$: $J_p \to J_q$ *induces a homomorphism* $\pi_{p,q*}$: $\mathfrak{j}_p \to \mathfrak{j}_q$ *of Lie algebras. The kernel of the mapping of* $\mathfrak{j}_p$ *into the algebra of jets that immediately precedes it in filtration is commutative.*.

Finally, from Proposition 5.7:

6.5. COROLLARY. *The quasihomogeneous vector fields of degree* 0 *form a finite-dimensional Lie subalgebra* $\mathfrak{h}$ *of the Lie algebra of all vector fields. The Lie algebra* $\mathfrak{h}$ *is naturally isomorphic to the Lie algebra of the group* j_0 *of* 0*-jets of diffeomorphisms.*

In what follows we sometimes identify the vector field $\mathbf{v}$ with the set of n functions (or series) v_i. The next two propositions are used later in the reduction of semiquasihomogeneous functions to normal form.

6.6. LEMMA. *Let F be a power series of filtration d, and let $\mathbf{v}$ be a formal vector field of positive filtration δ. Then the Taylor series*

$$F(\mathbf{x}+\mathbf{v}(\mathbf{x}))=F(\mathbf{x})+\frac{\partial F}{\partial \mathbf{x}}\mathbf{v}+R$$

has remainder term R of filtration strictly greater than $d+\delta$.

PROOF. In view of the linearity of R relative to F, it is enough to show this for the case when F is a monomial. Let $F=\mathbf{x}^{\mathbf{k}}$, $\mathbf{v}=\Sigma v_i\partial/\partial x_i$. We consider the term of the Taylor series containing $\partial^{|m|}F/\partial\mathbf{x}^{\mathbf{m}}$ ($\mathbf{m}=(m_1,\ldots,m_n)$). The monomials occurring in this term have exponents $\mathbf{p}=\mathbf{k}-\mathbf{m}+\sum_{i=1}^{n}\mathbf{l}_i$, where $\mathbf{l}_i$ is the exponent of one of the monomial functions $v_i^{m_i}$. Thus, $\mathbf{l}_i=\sum_{j=1}^{m_i}\mathbf{l}_{i,j}$, where $\mathbf{l}_{i,j}$ is one of the exponents $\mathbf{l}$ in the decomposition $v_i=\Sigma v_{i,\mathbf{l}}\mathbf{x}^{\mathbf{l}}$. So

$$\mathbf{p}=\mathbf{k}+\sum_{i=1}^{n}\sum_{j=1}^{m_i}(\mathbf{l}_{i,j}-\mathbf{1}_i).$$

But $\langle\mathbf{k},\alpha\rangle\geqslant d$, $\langle\mathbf{l}_{ij}-\mathbf{1}_i,\alpha\rangle\geqslant\delta>0$, by assumption. Therefore, $\langle\mathbf{p},\alpha\rangle\geqslant d+|\mathbf{m}|\delta$, where $|\mathbf{m}|=m_1+\ldots+m_n$. Hence all monomials occurring in terms of degree higher than 1 in the Taylor series relative to v have filtration not less than $d+2\delta$, as required.

6.7. COROLLARY. *Suppose that $F=F_0+F_1+F_2$, where $F_0\in F_d$, $F_1\in E_{>d}$, $F_2\in E_{>d+\delta}$: $\mathbf{v}=\mathbf{v}_0+\mathbf{v}_1$, where $v_0\in\mathfrak{g}_\delta$, $v_1\in\mathfrak{g}_{>\delta}$, $\delta>0$. Then*

$$F(\mathbf{x}+\mathbf{v}(\mathbf{x}))=F_0(\mathbf{x})+\left[F_1(\mathbf{x})+\frac{\partial F_0}{\partial\mathbf{x}}\mathbf{v}_0\right]+R',\quad R'\in E_{>d+\delta}.$$

PROOF. We set $F_0+F_1=F'$. We have $R'=R_1+R_2+R_3+R_4$, where

$$R_1=F'(\mathbf{x}+\mathbf{v}(\mathbf{x}))-F'(\mathbf{x})-\frac{\partial F'}{\partial\mathbf{x}}\mathbf{v}\in E_{>d+\delta}$$

(Lemma 6.6),

$$R_2=\frac{\partial F_0}{\partial\mathbf{x}}\mathbf{v}_1=L_{\mathbf{v}_1}F_0\in E_{>d+\delta},$$

$$R_3=\frac{\partial F_1}{\partial\mathbf{x}}\mathbf{v}=L_{\mathbf{v}}F_1\in E_{>d+\delta},$$

$$R_4=F_2(\mathbf{x}+\mathbf{v}(\mathbf{x}))\in E_{>d+\delta},$$

as required.

§7. The normal form of a semiquasihomogeneous function

We consider the local ring of a quasihomogeneous or semiquasihomogeneous function f of degree d. We fix a system of monomials forming a basis for this ring.

7.1. DEFINITION. A monomial is said to be *upper* or *to lie above the diagonal* (or *lower*, or *diagonal*) *if it has degree greater than* d (or less than d, or equal to d) for given quasihomogeneous exponents.

Note that the *number of upper, diagonal, and lower basis monomials does not depend on the basis of the local ring* (see Remark 3.5).

Let $e_1, \ldots, e_s$ be the system of all upper basis monomials in a fixed basis of the local ring of the function f_0.

7.2. THEOREM. *Every semiquasihomogeneous function with quasihomogeneous part f_0 is equivalent to a function of the form $f_0 + \Sigma c_k e_k$, where the c_k are constants.*

The proof of Theorem 7.2 is obtained on application of the following lemma.

7.3. LEMMA. *Let f_0 be a quasihomogeneous function of degree d and $e_1, \ldots, e_r$ the set of all basis monomials of fixed degree $d' > d$ in the local ring of f_0. Then every series of the form $f_0 + f_1$, where the filtration of f_1 is greater than d, can be brought by a formal diffeomorphism to the form $f_0 + f_1'$, where the terms in f_1' of degree less than d' are the same as in f_1, and the terms of degree d' reduce to $c_1 e_1 + \ldots + c_r e_r$.*

PROOF. Let g denote the sum of the terms of degree d' in f_1. There exists a decompostion (if convenient, as far as the terms of filtration higher than d', but certainly without them)

$$g = \sum \frac{\partial f_0}{\partial x_i} v_i(\mathbf{x}) + c_1 e_1 + \ldots + c_r e_r,$$

since $e_1, \ldots, e_r$ are basis monomials. The vector field $\mathbf{v}$ occurring in this formula can be replaced by a homogeneous one of degree $\delta = d' - d > 0$ without invalidating the formula (to prove this it is enough to decompose $\mathbf{v}$ into its homogeneous parts).

We consider now the formal substitution $\mathbf{x} = \mathbf{y} - \mathbf{v}(\mathbf{y})$, where $\mathbf{v}$ is the vector field with components v_i as defined above. We claim that this is a formal diffeomorphism. For the field $\mathbf{v}$ has positive degree δ, so that if the coordinates are numbered according to decreasing exponents α_i, the Jacobian matrix of the substitution at 0 is unitriangular. Applying Corollary 6.7, we find that

$$f(\mathbf{y} - \mathbf{v}(\mathbf{y})) = f_0(\mathbf{y}) + [f_1(\mathbf{y}) + (c_1 e_1(\mathbf{y}) + \ldots + c_r e_r(\mathbf{y})) - g(\mathbf{y})] + R'(\mathbf{y})$$

(in the old notation). Since the filtration of R' is greater than d', this proves Lemma 7.3.

PROOF OF THEOREM 7.2. Applying Lemma 7.3 to the function f_0 and the monomials next to those of highest degree d', we come to the desired form of the term of degree d'. Applying the same lemma to the series $f_0 + f_1'$ so obtained and the monomials following those of degree d', we come to the desired form of the term of degree d' without changing the terms of degree d and d'. Continuing in this way, we obtain the desired normal form up to terms of degree as high as required (and even, if convenient, *we can reduce the formal series completely to formal normal form by a formal diffeomorphism*; this follows from the fact that the degrees of fields **v** that arise at various stages grow).

Up to this moment we have not used anywhere the finiteness of the multiplicity μ, so that the formal assertion has been proved without this assumption. If μ is finite, then a fairly long section of the Taylor series (of length bounded in terms of μ; see, for example, [5]) of the function is equivalent to the function itself, so that reduction to normal form is realized by a genuine diffeomorphism.

§8. The normal form of a quasihomogeneous function

Let f_0 be a non-degenerate quasihomogeneous function. We consider the linear space E of all quasihomogeneous functions of the same type and degree of quasihomogeneity as f_0.

8.1. THEOREM. *Every quasihomogeneous function of the type and degree of f_0 and sufficiently near to f_0 is equivalent to a function of the form $f_0 + c_1 e_1 + \ldots + c_r e_r$, where $e_1, \ldots, e_r$ is the collection of all diagonal basis monomials of the local ring of f_0.*

PROOF. We consider a diffeomorphism of the form $\mathbf{x} \to \mathbf{x} + \varphi(\mathbf{x})$, where φ is a quasihomogeneous vector field of degree 0. All such diffeomorphisms form a Lie group. This group acts (linearly) on E. An orbit through f_0 and one through any nearby point have one and the same dimension (see Remark 3.2 and Lemma 3.4). They intersect the plane $f_0 + Ce_1 + \ldots + Ce_r$ transversally at the point f_0 and neighbouring points. Thus, the union of the orbits intersecting this plane near f_0 contains a neighbourhood of f_0 in the whole space E, as we wanted to show.

8.2. EXAMPLE. We show that *every non-degenerate binary form of degree $n \geqslant 4$ reduces to a normal form with $n - 3$ parameters*:

$$x^{n-1}y + c_1 x^{n-2}y^2 + \ldots + c_{n-3}x^2y^{n-2} + xy^{n-1}.$$

For the non-degeneracy of this form yields the existence of two simple linear factors, which can be taken as the coordinates x and y. Then the terms x^n and y^n are absent from the expression of the form; after this we make the coefficients of $x^{n-1}y$ and xy^{n-1} equal to 1. In this way the form reduces to the one above.

It is easy to check that the monomials $x^{n-2}y^2, \ldots, x^2y^{n-2}$ form a diagonal basis for every form of the type indicated (even in the degenerate case).

8.3. PROPOSITION. *A non-degenerate binary form of degree 4 can be reduced by a linear transformation to the (Legendre) normal form* X_9: $x^4 + ax^2y^2 + y^4$, $a^2 \neq 4$.

PROOF. By Proposition 8.2, the form reduces to $xy(x^2 + 2cxy + y^2)$. This is the product of the quadratic forms $a = xy$, $b = x^2 + 2cxy + y^2$. There are two *independent* degenerate forms among the linear combinations $pa + qb$ of a and b ($p{:}\, q$ is defined from the characteristic equation $p^2 + 2cpq + (c^2 - 1)q^2 = 0$, whose discriminant is not 0). Taking the square roots of these degenerate forms as coordinates, we reduce the original form to $c_1x^4 + c_2x^2y^2 + c_3y^4$. The condition for such a form to be non-degenerate is $c_1c_3(4c_1c_3 - c_2^2) \neq 0$. By a magnification of coordinates we reduce the form to that indicated in Proposition 8.3, which is thereby proved.

8.4. PROPOSITION. *A non-degenerate quasihomogeneous function of degree* 1 *with quasihomogeneous exponents* 1/3, 1/6 *can be reduced by a quasihomogeneous diffeomorphism to the normal form* J_{10}: $x^3 + ax^2y^2 + y^6$, *where* $4a^3 + 27 \neq 0$.

PROOF. By definition, the function has the form $c_1x^3 + c_2x^2y^2 + c_3xy^4 + c_4y^6$. From the non-degeneracy it follows that $c_1 \neq 0$. By changing x to $x + \lambda y^2$ we can achieve $c_3 = 0$. Then $c_4 \neq 0$ as a consequence of non-degeneracy, and after a magnification of coordinates the form reduces to that in Proposition 8.4, which is thereby proved.

8.5. PROPOSITION. *Every non-degenerate quasihomogeneous function of degree* 1 *with exponents* $\left(\frac{1}{3}, \frac{1}{3k}\right)$, $\left(\frac{k}{3k+1}, \frac{1}{3k+1}\right)$, $\left(\frac{1}{4}, \frac{1}{2k}\right)$, $\left(\frac{k}{4k+2}, \frac{1}{2k+1}\right)$ *can be reduced by a quasihomogeneous diffeomorphism to one of the respective normal forms*:

$$x^3 + ax^2y^k + y^{3k}, \quad x^3y + ax^2y^{k+1} + y^{3k+1} \quad (4a^3 + 27 \neq 0);$$
$$x^4 + ax^2y^k + y^{2k}, \quad x^4y + ax^2y^{k+1} + y^{2k+1} \quad (a^2 - 4 \neq 0).$$

The proof repeats the calculations in Propositions 8.3 and 8.4.

8.6. DEFINITION. The *inner modality* of a quasihomogeneous function is the total number of diagonal and superdiagonal monomials in some (and thus in any) monomial basis of the local ring.

8.7. THEOREM. *The modality of any semiquasihomogeneous function (in particular, of a non-degenerate quasihomogeneous function) is not less than the inner modality of its quasihomogeneous part.*

For suppose that $f = f_0 + f_1$ is the function in question and that $e_1, \ldots, e_r$ is the set of all diagonal and superdiagonal monomials in any basis of the local ring. Then the same monomials form the diagonal and superdiagonal part of a basis of the local ring of each of the functions $f + c_1e_1 + \ldots + c_re_r$, at least for sufficiently small c (see Corollary 3.3 and Lemma 3.4). Thus, there exists an s-dimensional plane through f in the space of functions with critical point and critical value 0 having the property that all non-zero tangent vectors to the orbits of points of the plane near f do not lie in the plane. Thus, the modality of f is not less than s.

8.8. REMARK. Gabrielov has shown that the modality is equal to the dimension of the submanifold of a base of versal deformations along which μ does not change. In all the examples of semiquasihomogeneous functions I know this dimension is the total number of diagonal and superdiagonal basis monomials.

§9. Piecewise filtrations

It often turns out to be useful to consider filtrations in which the rôle of the diagonal is played by a Newton open polygon (or, in the case of several dimensions, by a polyhedron convex towards 0). The formal definition is as follows.

Let $\boldsymbol{\alpha}_1, \ldots, \boldsymbol{\alpha}_p$ be a fixed collection of p quasihomogeneous types. We recall that the monomial $\mathbf{x}^{\mathbf{k}}$ is of degree $\langle \boldsymbol{\alpha}_i, \mathbf{k} \rangle = \varphi_i(\mathbf{k})$ in the i-th filtration. We define the *piecewise degree* of $\mathbf{x}^{\mathbf{k}}$ to be $\varphi(\mathbf{k}) = \min[\varphi_i(\mathbf{k}), \ldots, \varphi_p(\mathbf{k})]$.

9.1. DEFINITION. A power series has *piecewise filtration d* if all its monomials have piecewise degree d or higher.

Note that the equation $\varphi(\mathbf{k}) = 1$ defines a hypersurface Γ in the space of exponents $\mathbf{k}$ that is convex towards 0. We call Γ an *open polyhedron.* In these terms we can say that a monomial has (piecewise) degree d if and only if its exponent vector lies on the open polyhedron $d\Gamma$ obtained from Γ by a homothety with the coefficient d. In exactly the same way a series has (piecewise) filtration d if the exponent vectors of all its monomials lie on or outside $d\Gamma$.

The sum of the terms of lowest (piecewise) degree in a given power series is called the *principal part* of the series. A (piecewise) *homogeneous function of degree d* is a polynomial whose monomials all have (piecewise) degree d.

Analogous concepts are defined for vector fields; the degree of the monomial $\mathbf{x}^{\mathbf{l}}\partial/\partial x_i$ is defined to be

$$\varphi(\mathbf{l}-\mathbf{1}_i) = \min_{1 \leqslant j \leqslant p} \langle \boldsymbol{\alpha}_j, \mathbf{l}-\mathbf{1}_i \rangle.$$

Note that for all functions f, g and every vector field $\mathbf{v}$ we have[1]

filtration of $fg \geqslant$ filtration of f + filtration of g,

filtration of $\sum \frac{\partial f}{\partial x_i} v_i \geqslant$ filtration of f + filtration of $\mathbf{v}$.

---

[1] The filtration φ is naturally connected with the filtration arising naturally from the Koszul complex constructed from the derivative of a piecewise-homogeneous function; this was pointed out to me by Kushirenko.

The *group of diffeomorphisms of filtration d, the group of d-jets* of diffeomorphisms and the corresponding Lie algebras are defined just as in the case of quasihomogeneous filtrations. There is no analogue for piecewise filtrations, except for the group of quasihomogeneous diffeomorphisms.

9.2. DEFINITION. A piecewise-homogeneous function f_0 of degree d *satisfies condition A* if for every function g of filtration $d + \delta > d$ in the ideal spanned by the derivatives of f_0 there is a decomposition

$$g = \sum \frac{\partial f_0}{\partial x_i} v_i + g',$$

where the vector field v has filtration δ, and the function g' has filtration greater than $d + \delta$.

Note that a quasihomogeneous function always satisfies condition A.

We consider a basis of the local ring of a piecewise-homogeneous function f_0 of finite multiplicity μ.

9.3. DEFINITION. A basis $e_1, \ldots, e_\mu$ of homogeneous elements is said to be *regular* if, for each D, the elements of the basis of degree D are independent modulo the sum of the ideal $I = (\partial f/\partial x_0)$ and the space $E_{>D}$ of functions of filtrations greater than D.

9.4. PROPOSITION. *There always exists a regular basis, in fact, one consisting entirely of monomials.*

PROOF. The monomials whose exponent vectors lie on $D\Gamma$ generate $E_D \bmod E_{>D}$. Thus, their images in $E_D/(E_D \cap I) + E_{>D}$ generate this linear space, hence a basis for the factor space can be extracted from their images. The inverse images of the basis vectors so chosen are monomials in E_D, and we include these monomials in a basis of the local ring.

For large enough D we have $E_D \subset I$ (since $\mu < \infty$). Therefore, the system of monomials constructed is finite. It is clear from the construction that every vector in A is representable as a linear combination of the chosen monomials and of elements of this ideal. Finally, if the least degree of a monomial occurring in a relation $c_1 e_1 + \ldots \in I$ with non-zero coefficient is $D < \infty$, then the images of monomials e_i of degree D in the factor space $E_D/(E_D \cap I) + E_{>D}$ would be dependent, against the choice of the e_i. Thus, $\{e_i\}$ is a basis of the local ring, as required.

The number of elements in a regular monomial basis having given (piecewise) homogeneous degree does not depend on the choice of a basis of the local ring. A monomial in a regular basis is said to be *diagonal* (*superdiagonal*) if its degree is equal to (greater than) the degree of the function f_0 under discussion.

9.5. THEOREM. *If the principal part f_0 of a function f satisfies condition A and has finite multiplicity μ, then f can be reduced by a diffeomorphism to the form $f_0 + c_1 e_1 + \ldots + c_s e_s$, where $e_1, \ldots, e_s$ are the superdiagonal monomials in a regular basis.*

The proof of Theorem 9.5 repeats that of Theorem 7.2.

9.6. EXAMPLE. *We consider the function* $f_0 = x^a + \lambda x^2 y^2 + y^b$, *where*

$a \geq 4$, $b \geq 5$, $\lambda \neq 0$. *We claim that* 1) $\mu = a + b + 1$; 2) *the system* $1, x, \ldots, x^{a-1}, y, \ldots, y^b, xy$ *of monomials is a regular basis*; 3) *condition A is satisfied by the filtration given by the open polygon* Γ *with vertices* $(a, 0)$, $(2, 2)$, $(0, b)$.

For the proof of all these assertions it is useful to carry out certain geometrical constructions in the plane of exponents. These constructions reduce the analysis of the local ring to a sequence of geometrical operations reminiscent of the solution of crossword puzzles. This "crossword solution" technique is applicable not only in this example, and we explain it to a somewhat wider extent than is necessary for the analysis of our example.

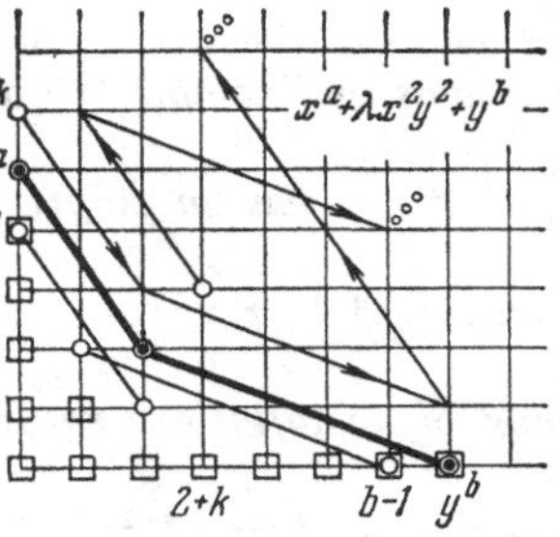

Fig. 1.

So let us assume that each partial derivative of the given function contains not more than two monomials (which is obviously satisfied by f_0). We join the monomials of the partial derivative with respect to x by a segment, which we call a *fundamental x-segment.* Similar fundamental segments are defined for the other variables. In our example there are two fundamental segments parallel to the open polygon Γ (Fig. 1).

The fundamental segments depict relationships between the images of their monomials in the local ring. We consider some consequences of these relationships. We call every translate of a fundamental segment by an integral vector with non-negative components a *permissible x-segment*. We note that two permissible segments may lie one on the other and even coincide geometrically (if one is an x-segment and the other a y-segment). But this does not happen in our example.

Two permissible segments are said to be *joined* if they have a common end. A *permissible chain* is a collection of permissible segments such that every one of the segments is joined to every other by a sequence of consecutive joined segments in the collection. A chain is said to be *maximal* if it is not part of any larger permissible chain.

A *cycle* is a finite sequence of consecutive joined permissible segments in which the last segment is joined to the first. A cycle is said to be *regular* if in its journey along the x-(y-)segments each direction is traversed the same number of times as every other. (In our example cycles are trivial, but if we consider the case $a = b = 4$, we get a non-trivial cycle $(1, 3) \to (3, 1) \to (1, 3)$ with one x-segment and one y-segment.)

We can now formulate the "crossword solution" rule for a function f with a critical point of finite multiplicity.

9.7. PROPOSITION. 1. *If the exponent of a monomial lies in an infinite permissible chain, then it lies in the ideal generated by the partial derivatives of the function.*

2. *If all cycles are regular, then the dimension μ of the local ring is equal to the number of maximal cycles. In this case we can find a basis of the local ring by taking an (arbitrary) monomial from each of the maximal chains.*

3. *Moreover, given a filtration, we obtain a regular basis by choosing a monomial of highest filtration from each of the maximal chains.*

The proof of Proposition 9.7 follows immediately from the definitions.

9.8. PROPOSITION. *The maximal permissible chains in Example 9.6 are as follows:*

1) *each of the points* $1, x, \ldots, x^{a-2}, y, \ldots, y^{b-2}, xy$ *has an empty permissible chain;*

2) *there are three finite maximal permissible chains* $x^{a-1} \to xy^2$, $x^2y \to y^{b-1}$, $x^a \to x^2y^2 \to y^b$;

3) *each of the remaining points is the beginning of an infinite permissible chain.*

It is clear from Fig. 1 that the permissible segment $x^p y^q \to y^{p-2} y^{q-1+b}$ raises the filtration for $p > q \geqslant 1$, while the segment $x^p y^q \to x^{p-1+a} y^{q-2}$ raises the filtration for $q > p \geqslant 1$. The filtrations of monomials with $p = 0$ or $q = 0$ or $p = q$ rises in two steps:

$$x^{2+k} \to x^{2+k}y^2 \to x^k y^{b+1}, \quad x^{2+k}y^{2+k} \to x^{k+a-1}y^{k+b-1}.$$

Consequently, apart from those described in 1) and 2), every maximal permissible chain of every monomial contains monomials of arbitrarily high degree and so is infinite.

Thus, the monomials figuring in 9.6 form a regular basis, and $\mu = a + b + 1$. To check condition A it is enough to compute the filtrations of the coefficients of the relations constructed above that describe the permissible chains. This calculation is not complicated and we omit it.

Therefore, condition A is satisfied, and the number of basis monomials of a regular basis above Γ turns out to be 0. Thus, from Theorem 9.5 we derive:

9.9. COROLLARY. *Every function f with principal part $f_0 = x^a + \lambda x^2 y^2 + y^b$, where $\lambda \neq 0$, $a \geqslant 4$, $b \geqslant 5$, is equivalent to its principal part.*

REMARK. The following rule for calculating the modality of functions of two variables with a non-degenerate Newton diagram is corroborated by a large collection of diverse examples, though it has not been proved in general as yet.

From the point (2, 2) of the plane of exponents, we draw horizontal and vertical rays in the direction of increasing exponents and consider the polygon bounded by the segments of these rays in the Newton open polygon. *The modality of the function is the number of integral points inside and on the boundary of the polygon.* The non-degeneracy condition can be stated here as follows. The coefficients of the monomials corresponding to the links in the Newton open polygon must be such that the polynomial obtained on adding them is the product of a monomial and a non-degenerate quasihomogeneous function.

Kushirenko has proved that the inner modality (that is, the number of monomials in a regular basis on the Newton open polygon and above) of a function of two variables with a non-degenerate Newton diagram is, in fact, equal to the number of integral points in the polygon described above.

9.10. CONDITION FOR SUFFICIENCY OF A JET. *Suppose that, in addition to the conditions of Theorem* 9.5, *all monomials of degree* $s + 1$ *lie in the ideal* $(\partial f_0/\partial x_i)$ *and are superdiagonal. Then the s-jet of f is sufficient.*

This is because the algorithm for reducing to the normal form as indicated in the proof of Theorem 7.2 does not alter the coefficients of superdiagonal monomials, no matter how we alter the terms of degree $s + 1$ and above in the series for f.

In particular, the b-jets in Example 8.6 are sufficient for $b \geqslant a$.

§ 10. Semiquasihomogeneous functions of two variables

In this section and the next we give tables in which we describe generators for local rings, normal forms of semiquasihomogeneous singularities, and sufficient jets for the simplest quasihomogeneous singularities of degree 1. The tables include all inner unimodal singularities and the first non-unimodal ones.

The non-degenerate quasihomogeneous functions of two variables fall into three (intersecting) classes. We recall that a function of two variables of corank 1 is a function whose second differential at zero is identically 0.

10.1. PROPOSITION. *Every non-degenerate quasihomogeneous function of corank* 2 *of two variables* x, y *can contain only the following monomials with non-zero coefficients:* x^a *and* y^b, *or* x^a *and* xy^b, *or* $x^a y$ *and* y^b *or* $x^a y$ *and* $y^b x$.

PROOF. Otherwise the function would be divisible by x^2 or y^2, and 0 would not be an isolated critical point.

We recall that all the exponents of the monomials of a quasihomogeneous function of degree 1 lie on a line Γ, the so-called diagonal (given by the equation $\langle \mathbf{k}, \boldsymbol{\alpha} \rangle = 1$).

10.2. THEOREM. *We assume that exactly two monomials of a non-degenerate function of two variables lie on the diagonal* Γ. *Then the following system of monomials forms a basis for the local ring:*

f	α_1, α_2	μ	Basis monomials $x^k y^l$
$x^a + y^b$	$\frac{1}{a}, \frac{1}{b}$	$(a-1)(b-1)$	$0 \leqslant k \leqslant a-2,\ 0 \leqslant l \leqslant b-2$
$x^a y + y^b$	$\frac{b-1}{ab}, \frac{1}{b}$	$(a-1)b+1$	$0 \leqslant k \leqslant a-2,\ 0 \leqslant l \leqslant b-1;\ x^{a-1}$
$x^a y + y^b x$	$\frac{b-1}{ab-1}, \frac{a-1}{ab-1}$	ab	$0 \leqslant k \leqslant a-1,\ 0 \leqslant l \leqslant b-1$

(a function f can be reduced to the form shown in the table by magnification and renumbering of coordinates).

PROOF. Consider Fig. 2. In this figure the domain of monomials contained in the ideal (f_x, f_y) is hatched-in. The thin slanting lines

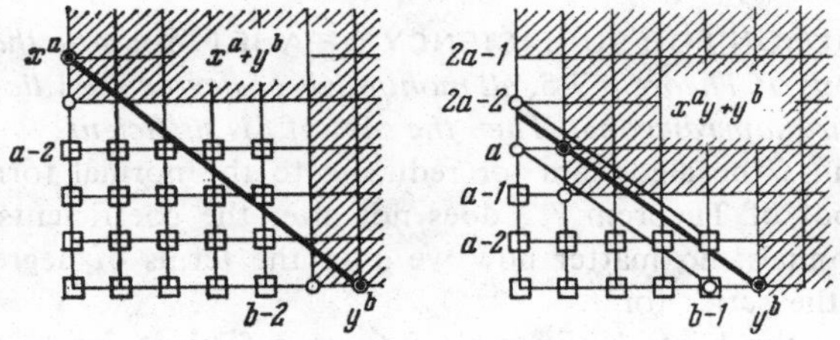

Fig. 2.

are permissible chains (see 9.6). It is clear from the figure that every monomial not contained in the ideal is joined to a monomial in the table by a permissible path. This means that the monomials in the table generate the local ring. But they are equal in number to the dimension of the local ring as calculated in 4.8. Thus they form a basis, as asserted.

By Theorem 7.2, we derive from Theorem 10.2 the normal forms for semiquasihomogeneous singularities with the quasihomogeneous parts shown in the table, and also a formula for their versal deformations. These formulae allow us to extract explicitly the characteristic values of the classical monodromy operator (see [10]).

We recall that we have called the total number μ_0 of diagonal and superdiagonal basis monomials of the local ring the *inner modality.*

10.3. THEOREM. 1) *The quasihomogeneous functions of two variables with $\mu_0 = 0$ are exhausted (up to equivalence) by the following list:*

Type	Normal form	v_1, v_2; N	Basis monomials and their weights
A_k	$x^{k+1}+y^2$	2, $k+1$; $2k+2$	1, x, ..., x^{k-1} 0, 2, ..., $2k-2$
D_k	x^2y+y^{k-1}	$k-2$, 2; $2k-2$	1, y, ..., y^{k-2}, x 0, 2, ..., $2k-4$, $k-2$
E_6	x^3+y^4	4, 3; 12	1, y, x, y^2, xy, xy^2 0, 3, 4, 6, 7, 10
E_7	x^3+xy^3	6, 4; 18	1, y, x, y^2, xy, y^3, y^4 0, 4, 6, 8, 10, 12, 16
E_8	x^3+y^5	10, 6; 30	1, y, x, y^2, xy, y^3, xy^2, xy^3 0, 6, 10, 12, 16, 18, 22, 28

All non-degenerate functions with quasihomogeneous exponents U_i/N as in the table can be reduced to the normal forms of the table.

2) *The quasihomogeneous functions of two variables with $\mu_0 = 1$ are exhausted (up to equivalence) by two one-parameter families and 8 individual functions as in the following table:*

Type	Normal form	$U_1, U_2; N$	Basis monomials and their weights
X_9	$x^4+ax^2y^2+y^4$, $a^2-4\neq 0$	1, 1; 4	$1, x, y, x^2, xy, y^2, x^2y, xy^2, x^2y^2$ 0, 1, 1, 2, 2, 2, 3, 3, 4
Z_{11}	x^3y+y^5	4, 3; 15	$1, y, x, y^2, xy, x^2, y^3, xy^2, y^4, xy^3, xy^4$ 0, 3, 4, 6, 7, 8, 9, 10, 12, 13, 16
Z_{12}	x^3y+xy^4	3, 2; 11	$1, y, x, y^2, xy, x^2, y^3, xy^2, x^2y, xy^3, x^2y^2, x^2y^3$ 0, 2, 3, 4, 5, 6, 6, 7, 8, 9, 10, 12
Z_{13}	x^3y+y^6	5, 3; 18	$1, y, x, y^2, xy, y^3, x^2, xy^2, y^4, xy^3, y^5, xy^4, xy^5$ 0, 3, 5, 6, 8, 9, 10, 11, 12, 14, 15, 17, 20
W_{12}	x^4+y^5	5, 4; 20	$1, y, x, y^2, xy, x^2, y^3, xy^2, x^2y, xy^3, x^2y^2, x^2y^3$ 0, 4, 5, 8, 9, 10, 12, 13, 14, 17, 18, 22
W_{13}	x^4+xy^4	4, 3; 16	$1, y, x, y^2, xy, x^2, y^3, xy^2, x^2y, y^4, x^2y^2, y^5, y^6$ 0, 3, 4, 6, 7, 8, 9, 10, 11, 12, 14, 15, 18
J_{10}	$x^3+ax^2y^2+y^6$, $4a^3+27\neq 0$	2, 1; 6	$1, y, x, y^2, xy, y^3, xy^2, y^4, xy^3, xy^4$ 0, 1, 2, 2, 3, 3, 4, 4, 5, 6
K_{12}	x^3+y^7	7, 3; 21	$1, y, y^2, x, y^3, xy, y^4, xy^2, y^5, xy^3, xy^4, xy^5$ 0, 3, 6, 7, 9, 10, 12, 13, 15, 16, 19, 22
K_{13}	x^3+xy^5	5, 2; 15	$1, y, \ldots, y^8; x, xy, xy^2, xy^3$ 0, 2, ..., 16; 5, 7, 9, 11
K_{14}	x^3+y^8	8, 3; 24	$1, y, \ldots, y^6; x, xy, \ldots, xy^6$ 0, 3, ..., 18; 8, 11, ..., 26

All non-degenerate functions with quasihomogeneous exponents U_i/N as in the table can be reduced to the normal forms of the table.

In these tables, the lower index in the notation for functions is the value of μ. The numbers below the monomials are their weights, that is, the numerators of the generalized degrees (the monomial x^py^q is of degree $(U_1p + U_2q)/N$).

PROOF OF THE THEOREM. 1) If the second differential is not identically 0, then the function is equivalent to A_k (see [2], for example). If the corank of the function is 2, then the homogeneous exponents are given by the table in 10.2. By 4.9, the generalized degree of a basis monomial of highest degree is $d_{\max} = n - 2\Sigma\alpha_i = 2 - 2(\alpha_1 + \alpha_2)$. The condition

$d_{max} < 1$ is equivalent to $\mu_0 = 0$, and in each of the three cases of Proposition 10.1 it defines the domain under a hyperbola in the (a, b)-plane. Calculation of the integral points in these domains gives the series A, D, E.

What has been said becomes a little more understandable, perhaps, if we observe that the *classification of singularities with* $\mu_0 = 0$ *reduces to an enumeration of the lines in the plane passing below the point* (2, 2), *intersecting the coordinate axes at distances not less than* 2 *from* 0, *and having integral points in the positive octant with abscissae and ordinates not exceeding* 1.

This is because the condition $d_{max} < 1$ means that the diagonal Γ lies below (2, 2).

It is easy to check that such lines are exhausted by our list (apart from re-naming the axes), and that non-degenerate quasihomogeneous functions with quasihomogeneous exponents A, D or E reduce to the forms shown in the table.

The basis monomials are easy to find from the crosswords described in 9.6; the formula for μ_i from 4.5 is also useful.

2) Suppose that $\mu_0 = 1$. Then a basis monomial of highest generalized degree lies on the diagonal or above, and the last but one strictly below it. But by 4.11, the last and last but one of the (generalized) degrees

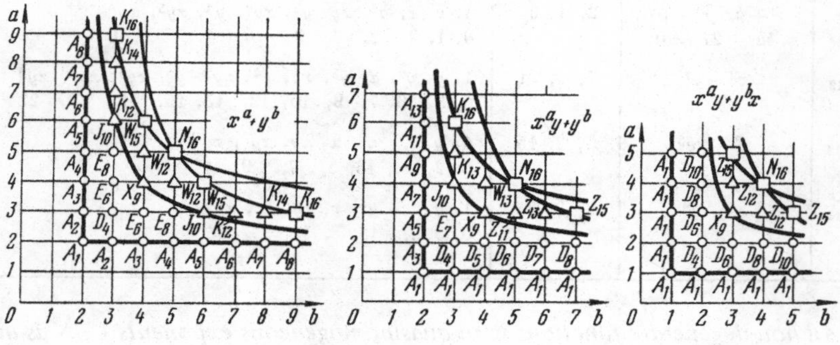

Fig. 3.

of the basis monomials are d_{max} and $d_{max} - \alpha_{min}$ (because the second degree is $\alpha_{min} = \min(\alpha_1, \alpha_2)$). Thus, the condition $\mu_0 = 1$ assumes the form

$$2\alpha_1 + 2\alpha_2 \leqslant 2, \; 3\alpha_1 + 2\alpha_2 > 2, \; 2\alpha_1 + 3\alpha_2 > 2.$$

Substituting the values of α_1 and α_2 from the table 10.2, we obtain in each of the three cases a domain between hyperbolae in the (a, b)-plane. The enumeration of the integral points in these domains (Fig. 3) gives the exponents of homogeneity shown in Table 10.3 ($X_9 - K_{14}$). For all these exponents there are exactly two integral points on the diagonal, apart from the cases X_9 and J_{10}. Magnification of coordinates reduces the function to a sum of monomials; the normal forms for X_9 and J_{10} are given in Propositions 8.3 and 8.4. The basis monomials are computed using 9.6, 4.5. This proves the theorem.

10.4. REMARK. The classification of singularities with $\mu_0 = 1$ comes to an enumeration of the lines Γ in the plane of exponents passing through the point (2, 2) or above it, but below (2, 3) and (3, 2).

We can also consider the boundary cases when Γ passes through one of the points (2, 3), (3, 2). The quasihomogeneous singularities corresponding to these boundary points are:

Type	Normal form	Non-degeneracy condition	U_1, U_2	N
Z_{15}	$x^3y+ax^2y^3+y^7$	$4a^3+27 \neq 0$	2, 1	7
W_{15}	$x^4+ax^2y^3+y^6$	$a^2-4 \neq 0$	3, 2	12
K_{16}	$x^3+ax^2y^3+y^9$	$4a^3+27 \neq 0$	3, 1	9
N_{16}	$x^4y+ax^3y^2+bx^2y^3+y^4x$	$4(a^3+b^3)-18ab-a^2b^2+27 \neq 0$	1, 1	5

The following systems of monomials are examples of basis of the local rings:

Type	Basis monomials below the diagonal	On	Above
Z_{15}	$1, y, x, y^2, xy, y^3, x^2, xy^2, y^4, xy^3, y^5, xy^4, y^6$ 0, 1, 2, 2, 3, 3, 4, 4, 4, 5, 5, 6, 6	xy^5 7	xy^6 8
W_{15}	$1, y, x, y^2, xy, x^2, y^3, xy^2, x^2y, y^4, xy^3, x^2y^2, xy^4$ 0, 2, 3, 4, 5, 6, 6, 7, 8, 8, 9, 10, 11	x^2y^3 12	x^2y^4 14
K_{16}	$1, y, y^2, y^3, x, y^4, xy, y^5, xy^2, y^6, xy^3, y^7, xy^4, xy^5$ 0, 1, 2, 3, 3, 4, 4, 5, 5, 6, 6, 7, 7, 8	xy^6 9	xy^7 10
N_{16}	$1, x, y, x^2, xy, y^2, x^3, x^2y, xy^2, y^3, x^3y, x^2y^2, xy^3$ 0, 1, 1, 2, 2, 2, 3, 3, 3, 3, 4, 4, 4	x^2y^3, x^3y^2 5, 5	x^3y^3 6

In the case of N_{16}, the system of monomials written down forms a basis only when $ab \neq 9$. If we replace the last monomial (x^3y^3) by x^4y^2, the resulting system determines a basis if $35a + 9b^2 \neq 0$. In the cases Z_{15}, W_{15}, K_{16} one and the same system works for all values of the parameters for which the function remains non-degenerate.

10.5. COROLLARY. *Every semiquasihomogeneous function of inner modality* $\mu_0 = 1$ *(that is, having as principal part one of the types* $X_9 - K_{14}$ *or with quasihomogeneous exponents as in Table* 10.3 *can be reduced by a diffeomorphism to a normal form of one of the following types:*

Type	Normal form	Non-degeneracy condition	s	Type	Normal form	Non-degeneracy condition	s
X_9	$x^4+ax^2y^2+y^4$	$a^2-4\neq 0$	4	W_{13}	$x^4+xy^4+ay^6$	$a\in C$	6
Z_{11}	$x^3y+y^5+axy^4$	$a\in C$	5	J_{10}	$x^3+ax^2y^2+y^6$	$4a^3+27\neq 0$	6
Z_{12}	$x^3y+xy^4+ax^2y^3$	$a\in C$	6	K_{12}	$x^3+y^7+axy^5$	$a\in C$	7
Z_{13}	$x^3y+y^6+axy^5$	$a\in C$	6	K_{13}	$x^3+xy^5+ay^8$	$a\in C$	8
W_{12}	$x^4+y^5+ax^2y^3$	$a\in C$	5	K_{14}	$x^3+y^8+axy^6$	$a\in C$	8

Bases for the local rings of these singularities are the same as for the quasihomogeneous singularities of Theorem 10.3. *The number s in the table denotes the (ordinary) degree of a sufficient jet.*

10.6. COROLLARY. *Semiquasihomogeneous functions with homogeneous exponents* Z_{15}, W_{15}, K_{16}, N_{16} *(see* 10.4) *can be reduced to the following forms:*

Type	Normal form	Non-degeneracy condition	s
Z_{15}	$x^3y+ax^2y^3+y^7+bxy^6$	$4a^3+27\neq 0$	8
W_{15}	$x^4+ax^2y^3+y^6+bx^2y^4$	$a^2-4\neq 0$	7
K_{16}	$x^3+ax^2y^3+y^9+bxy^7$	$4a^3+27\neq 0$	10
N_{16}	$x^4y+ax^3y^2+bx^2y^3+xy^4+cx^3y^3$	$4(a^3+b^3)+27\neq 18ab+a^2b^2$	6

In the case N_{16} it is assumed that $ab\neq 9$. If $ab = 9$, but $35a+9b^2\neq 0$, then cx^3y^3 must be replaced by cx^4y^2.

10.7. COROLLARY. *Each of the singularities in the list* 10.5 *has modality not less than* 1. *The singularities* Z_{15}, W_{15}, K_{16} *have modality not less than* 2, *and* N_{16} *not less than* 3.

For the modalities of all these singularities are exactly these values, but we do not prove this here.

§11. Quasihomogeneous functions of three variables

The non-degenerate quasihomogeneous functions of three variables fall into seven (intersecting) classes. The classification has been considered by Orlik and Wagreich [12], but they missed two classes (III and VI).

11.1. PROPOSITION. *Every non-degenerate quasihomogeneous function of three-variables of degree* 1 *and corank* 3 *contains at least one of the seven systems of monomials in the following tables with non-zero coefficients (with a suitable numbering of the variables):*

Class	Monomials	α_1,	α_2,	α_3	μ
I	x^a, y^b, z^c	$\frac{1}{a}$,	$\frac{1}{b}$,	$\frac{1}{c}$	$(a-1)(b-1)(c-1)$
II	x^a, y^b, $z^c y$	$\frac{1}{a}$,	$\frac{1}{b}$,	$\frac{b-1}{bc}$	$(a-1)(bc-b+1)$
III	x^a, $y^b x$, $z^c x$	$\frac{1}{a}$,	$\frac{a-1}{ab}$	$\frac{a-1}{ac}$	$\frac{(ab-a+1)(ac-a+1)}{a-1}$
IV	x^a, $y^b z$, $z^c y$	$\frac{1}{a}$,	$\frac{c-1}{bc-1}$,	$\frac{b-1}{bc-1}$	$(a-1)bc$
V	x^a, $y^b z$, $z^c x$	$\frac{1}{a}$,	$\frac{ac-a+1}{abc}$,	$\frac{a-1}{ac}$	$ac(b-1)+a-1$
VI	$x^a y$, $y^b x$, $z^c x$	$\frac{b-1}{ab-1}$,	$\frac{a-1}{ab-1}$,	$\frac{(a-1)b}{(ab-1)c}$	$\frac{a(abc-c-ab+b)}{a-1}$
VII	$x^a y$, $y^b z$, $z^c x$	$\frac{bc-c+1}{abc+1}$,	$\frac{ac-a+1}{abc+1}$,	$\frac{ab-b+1}{abc+1}$	abc

PROOF. We begin a classification for arbitrarily many variables $x_1, \ldots, x_n$. We fix the suffix i of the coordinate x_i. When all monomials of the form $x_i^a x_j$ are absent, the x_i-axis consists entirely of critical points.

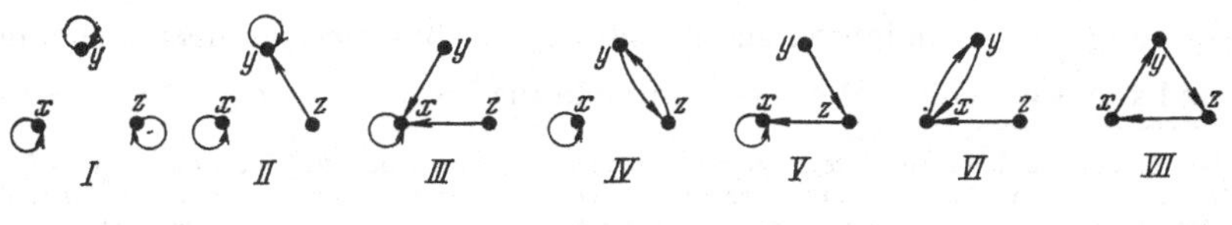

Fig. 4.

Therefore, at distance not more than 1 from each coordinate axis in the space of exponents there is the exponent of an existing monomial. Taking a monomial near to each of the axes (which is possible since the second differential is $\equiv 0$), we obtain a mapping $i \to j$ of the set of coordinate axes to itself. Thus, we have to classify mappings of a finite set to itself. For $n = 3$ this is not hard to do. A set with three elements has seven endomorphisms (apart from a re-naming of points), and this is what gives the seven classes of the table.

The values of the homogeneous exponents α_i and the multiplicity μ are calculated according to the formulae in §4.

11.2. PROPOSITION. 1) *A non-degenerate quasihomogeneous function of class* III *exists if and only if the least common multiple* $[b, c]$ *of b and c is divisible by* $a - 1$.

2) *A non-degenerate quasihomogeneous function of class* VI *exists if and only if* $(b - 1)c$ *is divisible by the product of* $a - 1$ *and the greatest common divisor* (b, c) *of b and c.*

3) *Non-degenerate quasihomogeneous functions of the remaining five classes exist for arbitrary* a, b, c.

PROOF. For the proof of 3) it is enough to add up the monomials shown in the table. To prove 1) and 2), we note that a quasihomogeneous function of class III or VI is degenerate if it contains none of the monomials $y^p z^q$ $(p \geqslant 0, q \geqslant 0)$. For the zero level set consists in that case of two components (one of which is the plane $x = 0$). This means that the critical point is not isolated (the set of critical points contains the line of intersection of the components), and the function is degenerate.

Conversely, it is easy to check that the quasihomogeneous functions

III: $x^a + xy^b + xz^c + \varepsilon y^p z^q$ and VI: $x^a y + y^b x + z^c x + \varepsilon y^p z^q$

are non-degenerate for almost all ε.

It remains to prove that a diagonal monomial $y^p z^q$ exists under exactly the above divisibility conditions.

In case III the generalized degree of $y^p z^q$ is $(pc + qb)(a - 1)/abc$. The monomial $y^p z^q$ is diagonal if and only if the degree is 1, that is, $(pc + qb)(a - 1) = (a - 1)bc + bc$. Thus, bc is divisible by $a - 1$, and the quotient (which is $pc + qb + bc$) is divisible by (b, c). In other words, bc is divisible by the product of $a - 1$ and (b, c), that is, $[b, c]$ is divisible by $a - 1$.

Conversely, suppose that $[b, c]$ is a multiple of $a - 1$. Then the number $\frac{abc}{a-1} = bc + \frac{bc}{a-1}$ is an integer and divisible by bc. But every number greater than bc and a multiple of (b, c) can be written in the form[1] $pc + qb$ $(p \geqslant 0, q \geqslant 0)$.

[1] For there are not less than 2 integral points on the line $\{p, q: pc + qb = bc\}$ in the quadrant $p \geqslant 0$, $q \geqslant 0$. The distance between consecutive integral points on every parallel line $pc + qb = m > bc$ is the same, so that there is an integral point on the segment of the line within the square (even when $m \geqslant (b - 1)(c - 1)$).

Thus, $abc/(a-1) = pc + qb$, and the monomial $y^p z^q$ is the diagonal.

In case VI the diagonality condition takes the form

$$(a-1)(pc+qb) = (a-1)bc + (b-1)c.$$

Thus, $(b-1)c$ is divisible by $(a-1)(b, c)$. Conversely, if $(b-1)c$ is divisible by $(a-1)(b, c)$, then $bc + \frac{(b-1)c}{a-1}$ can be written in the form $pc + qb$, where $p \geqslant 0$, $q \geqslant 0$, as required.

An example of a non-degenerate function of class III is $x^7 + xy^3 + xz^4 + \varepsilon x^3y^2$; and one of class VI is $x^5y + xy^3 + xz^4 + \varepsilon x^3y^2$.

We recall that the inner modality μ_0 of a quasihomogeneous function is the number of monomials in the local ring lying on the diagonal or above.

11.3. THEOREM. *The inner unimodal quasihomogeneous functions (that is, those with $\mu_0 = 1$) of corank 3 in three variables are exhausted (up to equivalence) by the following list:*

Type	Normal form	$\alpha_1, \quad \alpha_2, \quad \alpha_3$
P_8	$x^2z+y^3+ay^2z+z^3,\ 4a^3+27 \neq 0$	$\frac{1}{3}, \frac{1}{3}, \frac{1}{3}$
Q_{10}	$x^3+y^4+yz^2$	$\frac{1}{3}, \frac{1}{4}, \frac{3}{8}$
Q_{11}	$x^3+y^2z+xz^3$	$\frac{1}{3}, \frac{7}{18}, \frac{2}{9}$
Q_{12}	$x^3+y^5+yz^2$	$\frac{1}{3}, \frac{1}{5}, \frac{2}{5}$
S_{11}	$x^4+y^2z+xz^2$	$\frac{1}{4}, \frac{5}{16}, \frac{3}{8}$
S_{12}	$x^2y+y^2z+xz^3$	$\frac{4}{13}, \frac{5}{13}, \frac{3}{13}$
U_{12}	$x^3+y^3+z^4$	$\frac{1}{3}, \frac{1}{3}, \frac{1}{4}$

Every non-degenerate quasihomogeneous function with quasihomogeneous exponents shown in the table can be reduced to the indicated normal form

PROOF. By §4, the last and last but one of the generalized degrees of the basis monomials are d_{max} and $d_{max} - \alpha_{min}$. The condition $\mu_0 = 1$ can therefore be rewritten in the form $d_{max} \geqslant 1$, $d_{max} - \alpha_{min} < 1$. By §4, $d_{max} = n - 2\Sigma\alpha_s$. Thus, for $n = 3$ the condition $\mu_0 = 1$ takes the form

$$\alpha_1 + \alpha_2 + \alpha_3 \leqslant 1, \quad 2\alpha_1 + 2\alpha_2 + 2\alpha_3 > 2 - \alpha_s, \quad s = 1, 2, 3.$$

Now we must sort out the seven classes I–VII of Proposition 11.1, and choose an α in each case for which the given inequalities hold. We discard the case when monomials of degree 2 occur, since we are interested in functions of corank 3. Substituting the values of α, from Table 11.1, we get the following integral points corresponding to singularities of corank 3:

Class	(a, b, c) and notation for the function
I	$(3, 3, 3) = P_8$; $(3, 3, 4)$, $(3, 4, 3)$, $(4, 3, 3) = U_{12}$
II	$(3, 3, 2) = P_8$; $(4, 3, 2) = U_{12}$; $(3, 4, 2) = Q_{10}$; $(3, 5, 2) = Q_{12}$
III	$(3, 2, 2)$, $(2, 3, 2)$, $(2, 2, 3) = P_8$
IV	$(3, 2, 2) = P_8$; $(4, 2, 2) = U_{12}$; $(3, 3, 2)$, $(3, 2, 3) = Q_{12}$
V	$(3, 2, 2) = P_8$; $(3, 2, 3) = Q_{11}$; $(4, 2, 2) = S_{11}$
VI	$(2, 2, 2) = P_8$
VII	$(2, 2, 3)$, $(2, 3, 2)$, $(3, 2, 2) = S_{12}$; $(2, 2, 2) = P_8$

On the boundary of the domain of permissible α (that is, for $2\alpha_1 + 2\alpha_2 + 2\alpha_3 \geqslant 2 - \alpha_s$), there are some further points with integral a, b, c, namely:

Class	(a, b, c) and notation for the function
II	$(3, 3, 3) = U_{14}$; $(4, 4, 2) = V_{15}$; $(3, 6, 2) = Q_{14}$
III	$(3, 2, 3)$, $(3, 3, 2) = U_{14}$; $(4, 2, 3)$, $(4, 3, 2) = V_{15}$
V	$(3, 2, 4) = Q_{14}$; $(4, 2, 3) = V_{15}$; $(5, 2, 2) = S_{14}$; $(3, 3, 2) = U_{14}$
VI	$(2, 3, 2) = S_{14}$; $(2, 2, 3) = U_{14}$; $(3, 3, 2) = V_{15}$
I, IV, VII	—

For classes III and VI we have shown here only the integral points satisfying the divisibility relations of Proposition 11.2.

We show now that non-degenerate quasihomogeneous functions of the types listed really exist, and exhibit normal forms for them.

By Table 11.1, the homogeneous exponents α_s correspond to integral points (a, b, c). Sometimes triples (a, b, c) in different classes give rise to the same exponents. Such triples have been denoted in the preceding tables by the same letters (for instance, (4,3,3) in class I and (4, 2, 2) in class IV are denoted by U_{12}, and both give $\alpha_1 = 1/4$, $\alpha_2 = \alpha_3 = 1/3$).

When we know the homogeneous exponents, we can find all diagonal monomials. For example, for the values of α_s of type U_{12} just indicated, the equation of the diagonal Γ has the form $3p + 4q + 4r = 12$. The diagonal monomial $x^p y^q z^r$ corresponds to a non-negative solution of this equation in integers. There are five such solutions:

$$(4,0,0),\ (0,3,0),\ (0,2,1),\ (0,1,2),\ (0,0,3).$$

When we carry out the analogous computations for all the cases listed above, we come to the following table (in which the cases reducing to a permutation of the coordinate axes are omitted):

Type	$A_1,\ A_2,\ A_3;\ N$	Exponents of diagonal monomial $x^p y^q z^r$
$I_{333} = P_8$	1, 1, 1; 3	(300), (210), (201), (120), (102), (111), (021), (012), (030), (003)
$II_{342} = Q_{10}$	8, 6, 9; 24	(300), (040), (012)
$V_{323} = Q_{11}$	6, 7, 4; 18	(300), (021), (103)
$II_{352} = Q_{12}$	5, 3, 6; 15	(300), (050), (031), (012)
$V_{422} = S_{11}$	4, 5, 6; 16	(400), (021), (102)
$VII_{223} = S_{12}$	4, 5, 3; 13	(210), (021), (103)
$I_{433} = U_{12}$	3, 4, 4; 12	(400), (030), (021), (012), (003)

Here $\alpha_s = A_s/N$, so that $A_1 p + A_2 q + A_3 r = N$.

The boundary points of the domain of permissible α give four more types of quasihomogeneity:

Type	$A_1, A_2, A_3; N$	Exponents of diagonal monomials $x^p y^q z^r$
$\mathrm{II}_{362} = Q_{14}$	4, 2, 5; 12	(300), (220), (140), (060), (012)
$\mathrm{VI}_{232} = S_{14}$	4, 2, 3; 10	(050), (130), (210), (102), (022)
$\mathrm{II}_{333} = U_{14}$	3, 3, 2; 9	(300), (210), (120), (030), (103), (013)
$\mathrm{II}_{442} = V_{15}$	2, 2, 3; 8	(400), (310), (220), (130), (040), (102), (012)

We now choose normal forms for the functions with these 11 types of quasihomogeneity.

CASES Q_{10}, Q_{11}, S_{11} and S_{12}. All of these have three diagonal monomials each. The coefficients must be different from 0 (otherwise the function is degenerate). After magnification of coordinates the function reduces to a sum of diagonal monomials, that is,

$$Q_{10} = x^3 + y^4 + yz^2, \quad Q_{11} = x^3 + y^2z + xz^3,$$
$$S_{11} = x^4 + y^2z + xz^2, \quad S_{12} = x^2y + y^2z + xz^3.$$

CASE Q_{12}. The general linear combination of the diagonal monomials in the table is

$$Ax^3 + By^5 + Cy^3z + Dyz^2.$$

For a non-degenerate function, $AD \neq 0$. Replacing z by $z + \lambda y^2$ we reduce the function to a form in which $C = 0$. After this, magnification of the coordinates x, y, z converts the coefficients A, B, C to 1, and we obtain

$$Q_{12} = x^3 + y^5 + yz^2.$$

CASE U_{12}. According to the table, the general form of a quasihomogeneous function of type U_{12} is $Ax^4 + By^3 + Cy^2z + Dyz^2 + Ez^3$. For a non-degenerate function $A \neq 0$, and the cubic form in y and z is non-degenerate. A non-degenerate binary cubic form can be reduced by a linear change of variables to $y^3 + z^3$. The coefficient A can be made unity by magnifying the x-axis. So we obtain

$$U_{12} = x^4 + y^3 + z^3.$$

CASE P_8. It is well known that a non-degenerate ternary cubic form can be reduced to the Weierstrass normal form

$$x^2z - 4y^3 + g_2yz^2 + q_3z^3, \quad g_2^3 - 27g_3^2 \neq 0.$$

Replacing y by $y + \lambda z$, we reduce the form to $x^2z - 4y^3 + Ay^2z + Bz^3$. By non-degeneracy, $B \neq 0$. By a magnification of the x- and z-coordinates we can reduce the form to

$$P_8 = x^2z + y^3 + ay^2z + z^3.$$

This form is non-degenerate if $4a^3 + 27 \neq 0$.

This completes the proof of Theorem 11.3.

11.4. PROPOSITION. *Non-degenerate quasihomogeneous functions with quasihomogeneous exponents Q_{14}, S_{14}, U_{14} and V_{15} (see above) can be reduced by a quasihomogeneous diffeomorphism to the following normal forms*:

Type	Normal form	Non-degeneracy condition
Q_{14}	$yz^2+x^3+ax^2y^2+y^6$	$4a^3+27\neq 0$
S_{14}	$xz^2+x^2y+axy^3+y^5$	$a^2-4\neq 0$
U_{14}	$x^3+y^3+xz^3+ayz^3$	$a^3-1\neq 0$
V_{14}	$xz^2+ayz^2+x^4+bx^2y^2+y^4$	$(b^2-4)(a^4+a^2b+1)\neq 0$

PROOF. CASE Q_{14}. From the table of diagonal monomials we find

$$Q_{14} = A_1x^3 + A_2x^2y^2 + A_3xy^4 + A_4y^6 + A_5yz^2.$$

The non-degeneracy condition gives $A_1 \neq 0$. Replacement of x by $x + \lambda y^2$ removes A_3. In the expression so obtained, $A_4A_5 \neq 0$ (non-degeneracy). By magnifying coordinates we make $A_1 = A_4 = A_5 = 1$. When this is done, the non-degeneracy conditions takes the form $4A_2^3 + 27 \neq 0$.

CASE S_{14}. From the table of diagonal monomials,

$$S_{14} = A_1xz^2 + A_2x^2y + A_3xy^3 + A_4y^5 + A_5y^2z^2.$$

It follows from non-degeneracy that $A_1 \neq 0$. Replacement of x by $x + \lambda y^2$ removes A_5. After this, non-degeneracy gives $A_1A_2A_4 \neq 0$. We make $A_1 = A_2 = A_4 = 1$ by magnifiying coordinates. The non-degeneracy condition then takes the form $A_3^2 \neq 4$.

CASE U_{14}. From the table of diagonal monomials,

$$U_{14} = A_1x^3 + A_2x^2y + A_3xy^2 + A_4y^3 + A_5xz^3 + A_6yz^3.$$

The cubic form $A_1x^3 + \ldots + A_4y^3$ is non-degenerate, and by a linear change of the variables x, y it can be reduced to the form $x^3 + y^3$. It follows from non-degeneracy that one of the coefficients A_5, A_6 is not zero, say A_5. We make $A_5 = 1$ by magnifying the z-coordinate. The non-degeneracy condition then takes the form $A_6^3 \neq 1$.

CASE V_{15}. From the table of diagonal monomials,

$$V_{15} = A_1z^2x + A_2z^2y + A_3x^4 + A_4x^3y + A_5x^2y^2 + A_6xy^3 + A_7y^4.$$

By non-degeneracy, the binary 4-form $A_3x^4 + \ldots + A_7y^4$ is itself non-degenerate. By a linear change of x and y we can annihilate A_4 and A_6 (see 8.3). By non-degeneracy, one of A_1, A_2 is not 0, say A_1. By a magnification of coordinates we achieve that $A_1 = A_3 = A_7 = 1$. The non-degeneracy condition then assumes the form $A_5^2 \neq 4$, $A_2^4 + A_5A_2^2 + 1 \neq 0$. This proves Proposition 11.4.

11.5. PROPOSITION. *The following sets of monomials are bases for the local rings of the functions in Theorem* 11.3:

Type	Normal form	Basis monomials and their weights
P_8	$x^2z+y^3+ay^2z+z^3$ $4a^3+27\neq 0 \quad N=3$	$1,\ x,\ y,\ z,\ xy,\ yz,\ z^2,\ yz^2$ $0,\ 1,\ 1,\ 1,\ 2,\ 2,\ 2,\ 3$
Q_{10}	$x^3+y^4+yz^2$ $N=24$	$1,\ y,\ x,\ z,\ y^2,\ xy,\ yz,\ y^3,\ xy^2,\ xy^3$ $0,\ 6,\ 8,\ 9,\ 12,\ 14,\ 17,\ 18,\ 20,\ 26$
Q_{11}	$x^3+y^2z+xz^3$ $N=18$	$1,\ z,\ x,\ y,\ z^2,\ xz,\ z^3,\ xy,\ xz^2,\ z^4,\ z^5$ $0,\ 4,\ 6,\ 7,\ 8,\ 10,\ 12,\ 13,\ 14,\ 16,\ 20$
Q_{12}	$x^3+y^5+yz^2$ $N=15$	$1,\ y,\ x,\ z,\ y^2,\ xy,\ y^3,\ xz,\ xy^2,\ y^4,\ xy^3,\ xy^4$ $0,\ 3,\ 5,\ 6,\ 6,\ 8,\ 9,\ 11,\ 11,\ 12,\ 14,\ 17$
S_{11}	$x^4+y^2z+xz^2$ $N=16$	$1,\ x,\ y,\ z,\ x^2,\ xy,\ xz,\ x^3,\ x^2y,\ x^2z,\ x^3z$ $0,\ 4,\ 5,\ 6,\ 8,\ 9,\ 10,\ 12,\ 13,\ 14,\ 18$
S_{12}	$x^2y+y^2z+xz^3$ $N=13$	$1,\ z,\ x,\ y,\ z^2,\ xz,\ yz,\ xy,\ xz^2,\ yz^2,\ z^4,\ z^5$ $0,\ 3,\ 4,\ 5,\ 6,\ 7,\ 8,\ 9,\ 10,\ 11,\ 12,\ 15$
U_{12}	$x^3+y^3+z^4$ $N=12$	$1,\ z,\ x,\ y,\ z^2,\ xz,\ yz,\ xy,\ xz^2,\ yz^2,\ xyz,\ xyz^2$ $0,\ 3,\ 4,\ 4,\ 6,\ 7,\ 7,\ 8,\ 10,\ 10,\ 11,\ 14$

Here the number below the monomial denotes its weight: $x^py^qz^r$ has weight $i = A_1p + A_2q + A_3r$, where the $\alpha_s = A_s/N$ are the homogeneous exponents (the values of A_s and N are shown in pages 44/5).

The proof is based on the formula in §4 for the number μ_i of monomials of weight i. The choice of μ_i monomials out of all solutions of the equation $i = A_1p + A_2q + A_3r$ proceeds by means of the "crossword solution" as described in 9.6.

Similar calculations for the functions $Q_{14} - V_{15}$ lead to the following result:

11.6. PROPOSITION. *The following sets of monomials are bases for the local rings of the functions in Proposition* 11.4:

Type	Monomials below the diagonal	On	Above
Q_{14}	$1, y, x, y^2, z, y^3, xy, y^4, xy^2, xz, y^5, xy^3$ 0, 2, 4, 4, 5, 6, 6, 8, 8, 9, 10, 10	y^6 12	y^7 14
S_{14}	$1, y, z, x, y^2, yz, y^3, xy, y^2z, xy^2, y^4, y^3z$ 0, 2, 3, 4, 4, 5, 6, 6, 7, 8, 8, 9	xy^3 10	xy^4 12
U_{14}	$1, z, x, y, z^2, xz, yz, xy, z^3, yz^2, z^4, xyz$ 0, 2, 3, 3, 4, 5, 5, 6, 6, 7, 8, 8	yz^3 9	yz^4 11
V_{15}	$1, x, y, z, x^2, xy, y^2, yz, x^2y, xy^2, z^2, y^2z$ 0, 2, 2, 3, 4, 4, 4, 5, 6, 6, 6, 7	yz^2, x^2y^2 8, 8	y^2z^2 10

Thus, in all the cases listed we have succeeded in giving a single system of monomials furnishing a basis of the local ring for all values of the parameters for which the function is non-degenerate. The possibility of such a choice of normal forms and bases in the cases P_8, Q_{14}, S_{14}, U_{14}, V_{15} has not been clear up to now.

11.7. COROLLARY. *Every semiquasihomogeneous function with quasihomogeneous part of any of the types $P_8 - U_{12}$ (or with quasihomogeneous exponents as in Table 11.3) reduces under diffeomorphism to one of the following normal forms:*

Type	Normal form	s
P_8	$x^3z+y^3+ay^2z+z^3,\quad 4a^3+27\neq 0$	3
Q_{10}	$x^3+y^4+yz^2+axy^3$	4
Q_{11}	$x^3+y^2z+xz^3+az^5$	5
Q_{12}	$x^3+y^5+yz^2+axy^4$	5
S_{11}	$x^4+y^2z+xz^2+ax^3z$	4
S_{12}	$x^2y+y^2z+xz^3+az^5$	5
U_{12}	$x^3+y^3+z^4+axyz^2$	4

Bases for the local rings are the same as for the quasihomogeneous singularities shown in Table 11.5. The number s is the order of a sufficient s-jet.

11.8. COROLLARY. *Every semiquasihomogeneous function with quasihomogeneous exponents of the singularities Q_{14}, S_{14}, U_{14}, V_{15} can be* *reduced by a diffeomorphism to one of the following normal forms:*

Type	Normal form	Non-degeneracy condition	s
Q_{14}	$yz^2+x^3+ax^2y^2+y^6+by^7$	$4a^3+27 \neq 0$	7
S_{14}	$xz^2+x^2y+axy^3+y^5+bxy^4$	$a^2-4 \neq 0$	6
U_{14}	$x^3+y^3+xz^3+ayz^3+byz^4$	$a^3-1 \neq 0$	5
V_{15}	$xz^2+ayz^2+x^4+bx^2y^2+y^4+cy^2z^2$	$b^2-4 \neq 0$ $a^4+a^2b+1 \neq 0$	5

Proofs of these corollaries are obtained from Theorem 7.2 and Tables 11.3, 11.4, 11.5, 11.6. The number s is computed on the basis of 9.10 (the proof of the equation $s(S_{11}) = 4$ uses the fact that x^5 lies in the ideal for arbitrary a).

11.9. COROLLARY. *Each of the singularities in the list* 11.7 *has inner modality* $\mu_0 = 1$ *and modality not less than* 1. *Singularities of the types* Q_{14}, S_{14}, U_{14} *have* $\mu_0 = 2$ *and modality not less than* 2; *the type* V_{15} *has* $\mu_0 = 3$ *and modality not less than* 3.

11.10. REMARK. In fact *the modalities of the singularities listed in* §§10 *and* 11 *are the same as the corresponding inner modalities* μ_0 (see [3]); however, we do not prove this here.

11.11. THEOREM. *Every quasihomogeneous function of inner modality* $\mu_0 = 1$ *is stably equivalent to a function of one of the* 17 *types in Theorems* 10.3 (part 2) *and* 11.3.

PROOF. The finiteness of μ follows from that of μ_0. By a theorem of Saito (1.3 of [13]), every quasihomogeneous function with finite μ is stably equivalent to a quasihomogeneous function (possibly in fewer variables) whose second differential is identically 0. The inner modality of the new function is the same as that of the original one. The new function is not a function of a single variable, or else $\mu_0 = 0$. If it is a function of two variables, then by Theorem 10.3(2), the function can be reduced to one of the 10 forms shown in this theorem. If it is a function of three variables, Theorem 11.3 says that it can be reduced to one of the seven forms shown in this theorem.

For corank greater than 3 there are no functions of inner modality 1. For Saito proved that the (generalized) degree of a basis monomial of highest degree in the local ring of any quasihomogeneous function of generalized degree 1 with zero second differential is not less than that of a non-degenerate cubic form in the same number of variables (Theorem 2.11 of [11]): $d_{\max} = n - 2\sum \alpha_s \geqslant n - \frac{2}{3}n$. On the other hand, it follows from this inequality that $\Sigma \alpha_s \leqslant n/3$, which means that $\min \alpha_s$ is not more than

1/3. On the other hand, for $n \geqslant 4$ we have $d_{max} = n - 2\sum \alpha_s \geqslant 4/3$. Thus, for $n \geqslant 4$ we have not only $d_{max} \geqslant 1$, but $d_{max} - \alpha_{min} \geqslant 1$. Therefore, $\mu_0 \geqslant 2$ by Corollary 7.11. This proves the theorem.

References

[1] V. I. Arnol'd, Integrals of quickly oscillating functions and singularities of projections of Langrange manifolds, Funktsional. Anal. i Prilozhen. **6** : 3 (1972), 61–62. = Functional. Anal. Appl. **6** (1972). 222–224.

[2] V. I. Arnol'd, Normal forms of functions near degenerate critical points, the Weyl groups A_k, D_k, E_k and Lagrange singularities, Funktsional. Anal. i Prilozhen. **6** : 4 (1972), 3–25. = Functional. Anal. Appl. **6** (1972), 254–272.

[3] V. I. Arnol'd, Classification of unimodal critical points of functions, Funktsional. Anal. i Prilozhen. **7** : 3 (1973), 75–76. = Functional. Anal. Appl. **7** (1973), 230–231.

[4] V. I. Arnol'd, Remarks on the stationary phase method and Coxeter numbers, Uspekhi Mat. Nauk **28** : 5((1973), 17–44. = Russian Math. Surveys **28** : 5 (1973), 19–48.

[5] V. I. Arnol'd, Singularities of smooth mappings, Uspekhi Mat. Nauk. **23** : 1 (1968), 3–44. MR **37** #2243. = Russian Math. Surveys **23** : 1 (1968), 1–43.

[6] A. M. Samoilenko, The equivalence of a smooth function to a Taylor polynomial in the vicinity of a critical point of finite type, Funktsional. Anal. i Prilozhen, **2** : 4 (1968), 63–69. MR **42** #5132. = Functional Anal. Appl. **2** (1968), 318–323.

[7] J. C. Tougeron, Idéaux de fonctions differentiables, Ann. Inst. Fourier (Grenoble) **18** (1968), 177–240.

[8] V. P. Palamodov, On the multiplicity of a holomorphic transformation, Funktsional. Anal. i Prilozhen. **1** : 3 (1967), 54–65. MR **38** #4720. = Functional. Anal. Appl. **1** (1967), 218–226.

[9] P. Orlik and P. Wagreich, Singularities of algebraic surfaces with C^* action, Math. Ann. **93** (1971), 121--135.

[10] J. Milnor and P. Orlik, Isolated singularities defined by weighted homogeneous polynomials, Topology **9** (1970), 385–393.

[11] K. Saito, Einfach elliptische Singularitäten, Göttingen 1973, 1–46.

[12] P. Orlik and P. Wagreich, Isolated singularities of algebraic surfaces with C^* action, Ann. of Math. (2) **93** (1971), 205–228.

[13] K. Saito, Quasihomogene isolierte Singularitäten von Hyperflächen, Inventiones Math. **14** (1971), 123–142.

[14] N. Bourbaki, Groupes et algèbres de Lie, Hermann & Cie, Paris 1960. Translation: *Gruppy i algebry Li,* Mir, Moscow 1972.

Received by the Editors 28 September 1973

Translated by J. Wiegold

CRITICAL POINTS OF SMOOTH FUNCTIONS AND THEIR NORMAL FORMS

V. I. Arnol'd

This paper contains a survey of research on critical points of smooth functions and their bifurcations. We indicate applications to the theory of Lagrangian singularities (caustics), Legendre singularities (wave fronts) and the asymptotic behaviour of oscillatory integrals (the stationary phase method). We describe the connections with the theories of groups generated by reflections, automorphic forms, and degenerations of elliptic curves. We give proofs of the theorems on the classification of critical points with at most one modulus, and also a list of all singularities with at most two moduli. The proofs of the classification theorems are based on a geometric technique associated with Newton polygons, on the study of the roots of certain Lie algebras resembling the Enriques-Demazure technique of fans, and on spectral sequences that are constructed with respect to quasihomogeneous filtrations of the Koszul complex defined by the partial derivatives of a function.

Contents

Chapter I. General survey

§1. Classification of critical points 134

§2. Quotient singularities 135

§3. Quadratic forms of singularities 136

§4. A strange duality 138

§5. Versal deformation and the level bifurcation set 139

§6. Orbit transversal and the function bifurcation set 141

§7. Lagrangian singularities and the classification of caustics 142

§8. Legendre singularities and the classification of wave fronts 143

§9. Normal forms for caustics and wave fronts 145

§10. Integrals of oscillatory functions 147

§11. Semiquasihomogeneous functions and the Newton diagram 149

Chapter II. The hierarchy of singularities

§12. Normal forms 152

§13. Lists of singularities 158

§14. Determination of singularities 166

§15. Proofs of the theorems on the rotation of a ruler (Theorems 6, 18, 26, 33, 59, 67, 74, 83, 88) 175

§16. Normal forms of quasihomogeneous singularities (proofs of Theorems 3, 10, 13, 22, 29, 36, 47, 50, 52, 54, 56, 58, 63, 66, 70, 77, 82, 85, 90, 97, 98) 178
§17. Proofs of the theorems on the normal forms of semiquasi-homogeneous functions (Theorems 2, 4, 5, 7, 8, 9, 11, 14, 19, 20, 21, 23, 27, 28, 30, 34, 35, 37, 48, 51, 60, 61, 62, 64, 68, 69, 71, 75, 76, 78, 84, 86, 89, 91, 99) 192
§18. Proofs of the theorems on the normal forms of singularities of series T (Theorems 53, 55, 57) 195
References 198

CHAPTER I [1]

General Survey

A *critical point* of a smooth function is a point at which the differential of the function is equal to 0. A critical point is said to be *non-degenerate* if the second differential is a non-degenerate quadratic form. Every degenerate critical point splits into several non-degenerate critical points under a small perturbation (a so-called *morsification*).

Functions in general position have only non-degenerate critical points. Degenerate critical points show up naturally only when the functions depend on parameters. For example, the family of functions $f(x, t) = x^3 - tx$ has a degenerate critical point for the parameter value $t = 0$, and every nearby family also has a degenerate critical point for some small value of the parameter. For a large number of parameters more complicated critical points appear.

The study of families of functions depending on parameters arises everywhere in analysis (or mathematical physics). Below we shall discuss three applications: to the study of *Lagrangian singularities* (or *caustics*), *Legendre singularities* (or *wave fronts*) and *oscillatory integrals* (or the *stationary phase method*).

The classification of the simplest singularities in all these problems turns out to be related to the Lie, Coxeter, and Weyl groups of the series A, D and E, to the theory of braids of Artin and Brieskorn [22], and to the classification of regular polyhedra in Euclidean three-space.

The appearance of the diagrams A, D and E and of the Coxeter groups in such diverse situations as the theory of simple Lie groups, the classification of simple categories of linear spaces (in papers of Gabriel [*133]–[*135], Gel'fand and Ponomarev [*138]–[*140], Roiter and Nazarova [*136]–[*137], and of Dlab and Ringel [*141]–[*142], Kodaira's classification of the degenerations of elliptic curves, the theory of regular polyhedra and the

[1]* This chapter is a slightly modified version of the author's one hour address to the International Congress of Mathematicians, Vancouver 1974.

theory of simple singularities seems an astonishing series of coincidences in the results of independent classifications. As we now see, the classification of more complicated singularities also leads to astonishing coincidences, in which Lobachevskii triangles and automorphic functions are involved.

§1. Classification of critical points

Let f be the germ of a smooth function at an isolated critical point $O \in \mathbf{C}^n$. The *multiplicity* (or *Milnor number*) μ of O is the number of non-degenerate critical points into which O splits under a morsification.

Two germs of functions at O are said to be *equivalent* if one of them is taken to the other under a diffeomorphism of the domain space that leaves O fixed.

The jet (Taylor polynomial) of a function at O is *sufficient* if it determines the germ of that function at O up to equivalence.

A smooth function with an isolated critical point is always equivalent to a polynomial, namely, to a segment of its Taylor series, and it has a sufficient jet (Tougeron [79], M. Artin [14], Arnol'd [3], Samoilenko [67])

Thus, the problem of classifying isolated critical points reduces to algebraic problems concerning the action of finite-dimensional Lie groups on finite-dimensional spaces of jets (Thom [74], Mather [54], Siersma [69]).

The classification of the simplest singularities is discrete, but strongly degenerate singularities depend on parameters (moduli).

The *modality* m of a point $x \in X$ under the action of a Lie group G on a variety X is the smallest number such that a sufficiently small neighbourhood of x is covered by a finitely many m-parameter families of orbits. The point x is called *simple* if its modality is 0, that is, if a neighborhood of it is intersected by finitely many orbits.

The modality of the germ of a function at a critical point with critical value 0 is defined as the modality of its sufficient jet in the space of jets of functions with critical point O and critical value 0.

Two germs are said to be *stably equivalent* if they become equivalent after direct addition of non-degenerate quadratic forms of a suitable number of variables.

THEOREM I (see [6]). *Up to stable equivalence, the simple germs (germs with $m = 0$) are exhausted by the following list*:

$$A_k\colon f(x) = x^{k+1},\quad k \geqslant 1;\qquad D_k\colon f(x, y) = x^2y + y^{k-1},\quad k \geqslant 4;$$
$$E_6 : f(x, y) = x^3 + y^4;\quad E_7 : f(x, y) = x^3 + xy^3;\quad E_8 : f(x, y) = x^3 + y^5$$

THEOREM II (see [7]). *Up to stable equivalence, the unimodal germs (germs with $m = 1$) are exhausted by the three-index series of one-parameter families*

$$T_{p,q,r}: f(x, y, z) = axyz + x^p + y^q + z^r, \quad \frac{1}{p} + \frac{1}{q} + \frac{1}{r} < 1, \quad a \neq 0;$$

$$T_{3,3,3}: f(x, y, z) = x^3 + y^3 + z^3 + axyz, \quad a^3 + 27 \neq 0;$$
$$T_{2,4,4}: f(x, y, z) = x^4 + y^4 + z^2 + ax^2y^2, \quad a^2 \neq 4;$$
$$T_{2,3,6}: f(x, y, z) = x^3 + y^6 + z^2 + ax^2y^2, \quad 4a^3 + 27 \neq 0,$$

and 14 *more exceptional one-parameter families, listed in Table* 1 *below (the meaning of its columns will be explained later).*

Table 1

Notation	Normal form	Indices of homogeneity	Coxeter numbers	Dolgachev numbers	Gabrielov numbers	Dual class
Q_{10}	$x^2z + y^3 + z^4 + ayz^3$	8 9 6	—24	2 3 9	3 3 4	K_{14}
Q_{11}	$x^2z + y^3 + yz^3 + az^5$	7 6 4	—18	2 4 7	3 3 5	Z_{13}
Q_{12}	$x^2z + y^3 + z^5 + ayz^4$	6 5 3	—15	3 3 6	3 3 6	Q_{12}
S_{11}	$x^2z + yz^2 + y^4 + ay^3z$	6 5 4	—16	2 5 6	3 4 4	W_{13}
S_{12}	$x^2z + yz^2 + xy^3 + ay^5$	5 4 3	—13	3 4 5	3 4 5	S_{12}
U_{12}	$x^3 + y^3 + z^4 + axyz^2$	4 4 3	—12	4 4 4	4 4 4	U_{12}
Z_{11}	$x^3y + y^5 + z^2 + axy^4$	15 8 6	—30	2 3 8	2 4 5	K_{13}
Z_{12}	$x^3y + xy^4 + z^2 + ay^6$	11 6 4	—22	2 4 6	2 4 6	Z_{12}
Z_{13}	$x^3y + y^6 + z^2 + axy^5$	9 5 3	—18	3 3 5	2 4 7	Q_{11}
W_{12}	$x^4 + y^5 + z^2 + ax^2y^3$	10 5 4	—20	2 5 5	2 5 5	W_{12}
W_{13}	$x^4 + xy^4 + z^2 + ay^6$	8 4 3	—16	3 4 4	2 5 6	S_{11}
K_{12}	$x^3 + y^7 + z^2 + axy^5$	21 14 6	—42	2 3 7	2 3 7	K_{12}
K_{13}	$x^3 + xy^5 + z^2 + ay^8$	15 10 4	—30	2 4 5	2 3 8	Z_{11}
K_{14}	$x^3 + y^8 + z^2 + axy^6$	12 8 3	—24	3 3 4	2 3 9	Q_{10}

There are also classifications of all functions of two variables with non-zero 3-jet [9] or with non-zero 4-jet (§13). Similar tables exist in the real case. Furthermore, the following theorem holds.

THEOREM III (see [6], [7]). *The set of non-simple germs of functions of $n \geqslant 3$ variables has codimension* 6, *and the set of germs of modality greater than* 1 *has codimension* 10 *in the space of germs of functions with critical value* 0.

Thus, every s-parameter family of functions, where $s < 6$ (respectively, $s < 10$), can by a sufficiently small perturbation be brought into general position, so that the germs of the functions of the family at all the critical points become stably equivalent to the germs of Theorem I (respectively, to those of Theorems I and II), up to additive constants.

§2. Quotient singularities

The group $SU(2)$ acts linearly on $\mathbf{C}^2$. The discrete subgroups of $SU(2)$ are called *binary groups*: the binary polygonal group, the binary dihedral group, binary tetrahedral group, binary group of the cube (or quaternion group), and the binary icosahedral group. (After factoring $SU(2)$ by its centre $\{\pm E\}$ they go to the corresponding subgroups of the group of rotations of the sphere.)

The quotient variety $M = \mathbf{C}^2/\Gamma$ of $\mathbf{C}^2$ by the action of a binary group Γ is an algebraic surface with a singular point. The ring of Γ-invariant polynomials of two variables has 3 generators. The relation (syzygy) connecting these 3 generators is also the equation of the quotient variety M as a surface in $\mathbf{C}^3$. At the time of H. A. Schwarz the following result was already known.

THEOREM IV (see [43], [45], [57]). *The surface* $\mathbf{C}^2/\Gamma$ *has a simple singularity of one of the types* A_k *(for a polygon),* D_k *(for a dihedron),* E_6 *(for the tetrahedron),* E_7 *(for the cube),* E_8 *(for the icosahedron).*

The group $SU(1, 1)$ of (2×2)-unimodular matrices that preserve the quadratic form $|z_1|^2 - |z_2|^2$ acts on the set P of vectors in $\mathbf{C}^2$ on which the form is positive. Each discrete group Γ_0 of motions of the Lobachevskii plane (with compact fundamental domain) determines a "binary group" $\Gamma \subset SU(1, 1)$ and an algebraic surface $M = (P/\Gamma) \cup O$ with singular point O. It can be shown that the coordinate ring of M is isomorphic to the algebra of integral automorphic forms with respect to Γ_0 (I. V. Dolgachev).

Let Δ be a Lobachevskii triangle with the angles π/p, π/q, π/r. The reflections in the sides of Δ generate a discrete group. The motions form a subgroup of it of index 2; we call it the group of Δ.

In $SU(1, 1)$ the *binary triangular group* $\Gamma(\Delta)$ corresponds to it.

An analysis of the 14 exceptional singularities led Dolgachev to the following result.

THEOREM V (see [28]). *There exist exactly* 14 *Lobachevskii triangles for which the surface M is imbedded in* $\mathbf{C}^3$ *(in other words, precisely* 14 *triangles for which the algebra of integral automorphic forms has* 3 *generators. These* 14 *surfaces M have exceptional unimodal singularities at O (see Theorem* II).

The corresponding numbers (p, q, r) are indicated under the heading "Dolgachev numbers" in Table 1 on p.4.

§3. Quadratic forms of singularities

With each isolated critical point of a holomorphic function there is associated a manifold V with boundary ∂V, called the *non-singular level manifold*. More precisely, we consider a sufficiently small ball with centre at the critical point. Then the part of the level set inside the ball and sufficiently close to the critical set is a smooth manifold with boundary (Fig. 1, p. 6).

Fig. 1

(The boundary ∂V supplies standard examples in differential topology. For example, for the simple singularity E_8 in $\mathbf{C}^5$ the manifold ∂V is an exotic 7-sphere of Milnor, which is homeomorphic, but not diffeomorphic, to S^7. Contracting the boundary of V (E_8) in $\mathbf{C}^7$ to a point we obtain a non-smoothable topological 12-dimensional manifold. See [44], [19], [57], [46].)

Milnor [57] has proved that the manifold V^{2n-2} is homotopy equivalent to a bouquet of μ spheres S^{n-1} of middle dimension, so that $H_{n-1}(V, \mathbf{Z}) = \mathbf{Z}^\mu$. The intersection index defines an integral bilinear form on H_{n-1}.

The *quadratic form of a singularity* is the index of self-intersection on homology of the non-singular level manifold of a function of $n \equiv 3 \pmod 4$ variables that is stably equivalent to the given function (the behaviour of the intersection index on writing down the squares of the new variables is described by a theorem of Thom and Sebastiani [73]; by adding two squares the intersection index "changes sign" (that is, becomes symmetric)).

A singularity is said to be *elliptic* (respectively, *parabolic, hyperbolic*) if its quadratic form is negative definite (respectively, semidefinite, has positive index of inertia 1).

THEOREM VI (see [76], [6], [7]). *The elliptic singularities are precisely the simple singularities A, D and E of Theorem* I; *the parabolic* (*respectively, hyperbolic*) *singularities are the* $T_{p,q,r}$ *of Theorem* II *with* $1/p+1/q+1/r=1$ (*respectively, with* $1/p+1/q+1/r<1$).

The assertion about parabolic singularities was conjectured by Milnor, based on work of Wagreich [83].

It is convenient to describe the quadratic forms of singularities by *Dynkin diagrams.* Such a diagram is a graph whose vertices correspond to the vanishing cycles (basis vectors in H_{n-1} with square -2). Two vertices are joined by k simple (respectively, dotted) edges if the scalar product of the corresponding vectors is equal to k (respectively, $-k$). For example, the diagram •—• corresponds to the form $-2x_1^2 + 2x_1x_2 - 2x_2^2$.

Very effective methods for computing Dynkin diagrams have been worked out by Gabrielov ([34]–[36]) and Gusein-Zade ([39]–[40]). The latter's method allows us to draw the diagram of any function of two variables immediately from the picture of the level curves of a real morsification. (At the congress in Vancouver, N. A'Campo ([*131]) announced that he had independently discovered Gusein-Zade's method.)

The quadratic forms of the simple singularities A, D, and E are given by the usual Dynkin diagrams (see Hirzebruch [43]):

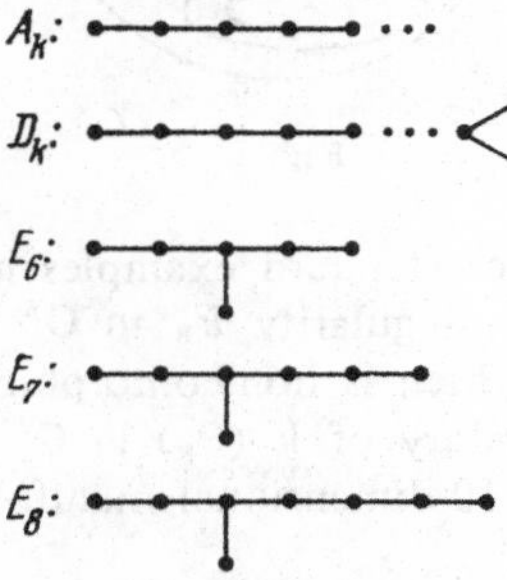

The quadratic forms of all unimodal singularities have been found by Gabrielov ([34]–[35]). We denote by $\tau_{p,q,r}$ the quadratic form with a diagram in the form of the letter T, on the three segments of which there are, respectively, p, q, r vertices (for example $E_7 = \tau_{2,3,4}$).

THEOREM VII (see [7]). *The quadratic form of each hyperbolic or parabolic singularity* $T_{p,q,r}$ *is a direct sum* $\tau_{p,q,r} \oplus 0$, *where* 0 *is the zero form of one variable. The quadratic form of each of the* 14 *exceptional singularities is the direct sum* $\tau_{p,q,r} \oplus \begin{pmatrix} 0 & 1 \\ 1 & 0 \end{pmatrix}$, *where* p, q, r *are indicated under the heading "Gabrielov number" in Table* 1.

§4. A strange duality

A comparison of the Dolgachev numbers and the Gabrielov numbers of the 14 exceptional singularities leads to the following conclusions:

THEOREM VIII. *The Gabrielov numbers of each exceptional singularity coincide with the Dolgachev numbers of some (generally speaking, another) singularity, and the Gabrielov numbers of the latter are the Dolgachev numbers of the former.*

Thus, the involution

$$Q_{10} \leftrightarrow K_{14}, \quad Q_{11} \leftrightarrow Z_{13}, \quad Z_{11} \leftrightarrow K_{13}, \quad S_{11} \leftrightarrow W_{13},$$

acts on the set of the 14 singularities, leaving fixed all six singularities with $\mu = 12$. We note that a priori no connection whatever is apparent between the singularities that are dual to each other (or their Lobachevskii triangles or their quadratic forms), nor between the Gabrielov and Dolgachev numbers.

Upon analysing the table of Dolgachev numbers, D. B. Fuks remarked that their sum with μ is always equal to 24 (this observation permitted him to correct an error in the original computations).

Thus, the sum of all six Gabrielov and Dolgachev numbers of each of the fourteen singularities is equal to 24. We also note that the singularities that are dual to each other are precisely those for which the Coxeter numbers (defined below) coincide. At the present stage the theory of singularities is an experimental science.

§5. Versal deformation and the level bifurcation set

A *deformation* of a function f is the germ at O of a smooth mapping of a finite-dimensional space (the *base* of the deformation) into a function space, taking O into f. A deformation is said to be *versal* if this mapping is transversal to the orbit of f under the action of the (pseudo-) group of diffeomorphisms of the argument space; if the dimension of the base is minimal (equal to the codimension of the orbit), then the deformation is said to be *miniversal*.

The germ of f at O with finite multiplicity μ always has a μ-parameter miniversal deformation ([77], [54], [47], [87]).

The *local ring* of the germ of f at the critical point $O \in \mathbf{C}^n$ is the factor ring of the ring of (formal or convergent) power series of the coordinates at O by the ideal spanned by the series of the partial derivatives of f. The dimension of this ring over $\mathbf{C}$ is equal to the multiplicity μ (Palamodov [61]).

As a miniversal deformation of the function f one can take the deformation $\lambda \mapsto f + \lambda_1 e_1 + \ldots + \lambda_\mu e_\mu$, where the functions e_s generate the local ring of f at O as a $\mathbf{C}$-module. Every deformation of f is equivalent to one induced from a versal deformation of the germ of f at O.

The *level bifurcation set* for f is the germ at O of the hypersurface in the base that is formed by those λ for which 0 is the critical value of the corresponding function close to O.

The complement to the level bifurcation set is the base of a fibration whose fibre is a non-singular level manifold of f. The action of the fundamental group of the base of the fibration on the homology of a fibre is called the *monodromy* of the singularity and its image is the *monodromy group*.

THEOREM IX (see [6]). *The complement to the level bifurcation set of a simple singularity has the homotopy type of a* $K(\pi, 1)$*-space, where* π *is the Artin-Brieskorn braid group (see* [21]–[23]). *This complement is the variety of regular orbits of the action of the corresponding Coxeter group on the complexification of the Euclidean space* $\mathbf{R}^\mu$. *The monodromy group of a simple singularity is the natural representation of the braid group on the*

Coxeter group.

The proof of Theorem IX in the cases E_6, E_7, E_8 is not simple: it is based on theorems of Deligne [26] and Brieskorn [20]. The topology of the complements to bifurcation sets has been little studied (see [4], [22]), but the recently discovered connections with loop spaces (Segal [68], Fuks [32]–[33]), with pseudo-isotopies (Cerf [25], Thom [75]) with algebraic *K*-theory (Volodin [82], Wagoner [84], Hatcher [41], [*41a]) seem very promising.

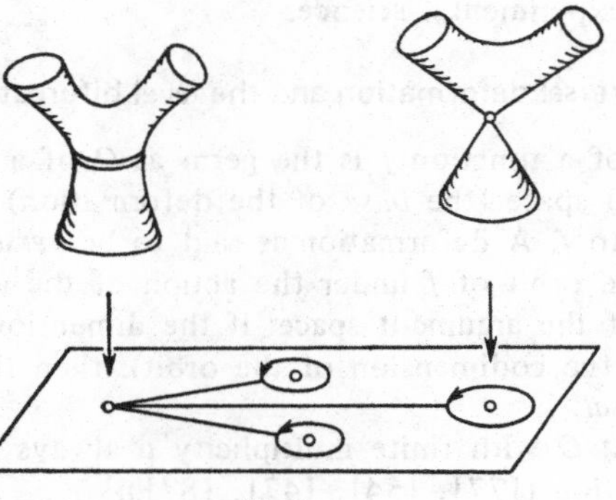

Fig. 2

We return to the level bifurcation set Σ for an arbitrary singularity. A line $\mathbf{C}^1$ in general position in the base intersects Σ at μ points close to O. We call these points *distinguished.*

On the line $\mathbf{C}^1$ we fix a base point (different from the distinguished points); this point corresponds to a non-singular fibre V. We choose μ non-intersecting *distinguished paths* in $\mathbf{C}^1$, going from the base point to the distinguished points (Fig. 2). In the fibre over a point of a distinguished path, close to the end of the path, a Picard-Lefschetz *vanishing cycle* (a middle-dimensional sphere) is defined. Returning it along the path to the fibre over the base point, we define there a *distinguished vanishing cycle.* These μ cycles form a *distinguished basis* of the homology of the fibre (Lamotke [129], Gabrielov [34]).

The fundamental group of the complement to the bifurcation set is generated by μ *distinguished loops* on $\mathbf{C}^1$; they are obtained from the distinguished paths if, without having reached a distinguished point, we go around it and return to the start (theorem of Zariski [88]; see also [80], [49]).

For simplicity we assume that the number of variables is $n \equiv 3$ (mod 4). Then the action of each distinguished generator of the fundamental group on the homology of the fibre reduces to reflection in the orthogonal complement to the distinguished vanishing cycle (Picard-Lefschetz theorem).

Thus, to compute the monodromy group of a singularity it suffices to find the Dynkin diagram of some distinguished basis. The first examples were found by Pham (see Pham [62], Brieskorn [19]), using the function $f = \Sigma x_i^{a_i}$; Pham's basis is in fact distinguished. The papers of Gabrielov [34]–[35] and Gusein-Zade [39]–[40] contain many other examples. Lazzeri proved that the Dynkin diagram of a distinguished basis is always connected, and Gusein-Zade [40] proved that for any function of two variables this diagram can be obtained from the picture of the level lines of a real morsification.

The *monodromy operator* of a singularity is the action on $H_{n-1}(V)$ of the product of the distinguished generators of the fundamental group. This operator is associated with the asymptotic behaviour of integrals in which f is contained, and many articles are devoted to computing it (Milnor and Orlik [58], Brieskorn [24], A'Campo [2]).

A'Campo has proved that the monodromy operator of any degenerate singularity is non-trivial. The calculation of the monodromy operator with respect to the Dynkin diagram of a distinguished basis reduces to multiplication of matrices. The monodromy operator of a simple singularity is the Coxeter element of the corresponding Coxeter group. Its order is called the *Coxeter number* of the simple singularity.

§6. Orbit transversal and the function bifurcation set

Consider the set $\mathfrak{m}^2$ of germs at $O \in \mathbf{C}^n$ of functions with critical point O and critical value 0. The group of O-fixing germs of diffeomorphisms of the argument space acts on $\mathfrak{m}^2$. A minimal transversal to the orbit of a function f in $\mathfrak{m}^2$ has dimension $\mu - 1$. This transversal T gives a $(\mu - 1)$-parameter deformation of f. This deformation, like any other, is induced from a miniversal deformation of f under some mapping of the base τ: $T^{\mu-1} \to \mathbf{C}^\mu$.

The level bifurcation set Σ is precisely the image of the transversal under the mapping τ. The set Σ is irreducible and has a non-singular normalization T (see [36], [71]).

Next, we consider the set $\mathfrak{m}$ of functions equal to 0 at O. Deformations in the class of such functions are called *restricted.* A restricted miniversal deformation has $\mu - 1$ parameters, and if we add an arbitrary constant to functions of a family, then we obtain an *extended* (μ-parameter) miniversal deformation in the usual sense of the term miniversal.

We fix a restricted miniversal deformation of f. A point of the base $\mathbf{C}^{\mu-1}$ of this deformation is called a *function bifurcation point* if it corresponds to a function with fewer than μ critical values close to O. The set of all such points forms a hypersurface Δ in a neighbourhood of $O \in \mathbf{C}^{\mu-1}$ (Fig. 3). This hypersurface (more precisely, its germ at O) is called the *function bifurcation set* for f.

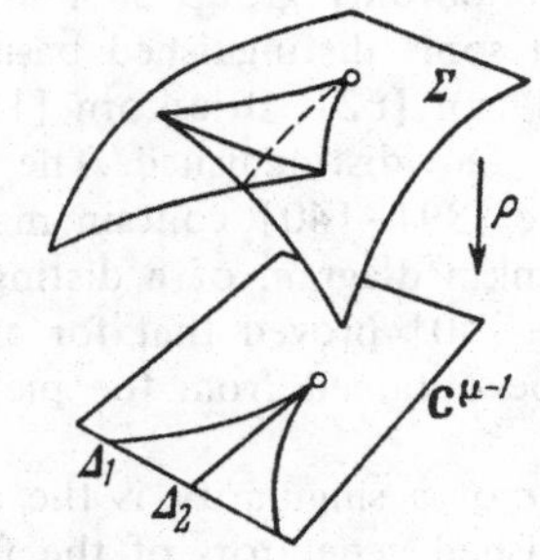

Fig. 3

THEOREM X (see [48], [36]). *The projection ρ of the base of an extended miniversal deformation onto the base of a restricted miniversal deformation, when restricted to the level bifurcation set Σ, gives a μ-sheeted covering of the complement to Δ. The group of this covering is the symmetric group S(μ).*

THEOREM XI (Lyashko [9], Looijenga [51]). *For simple functions the complement to the function bifurcation set,* $\mathbf{C}^{\mu-1} - \Delta$, *is a* $K(\pi, 1)$*-space, where π is a subgroup of finite index* $\nu = \mu! N^{\mu} W^{-1}$ *(where N is the Coxeter number, W the order of the Weyl group) in the Artin braid group on μ strings.*

The bifurcation set Δ splits in a natural way into two hypersurfaces Δ_1 and Δ_2: Δ_1 corresponds to functions with multiple critical points, Δ_2 to functions with coincident critical values.

The smooth mapping $\rho \circ \tau$: $T^{\mu-1} \to \mathbf{C}^{\mu-1}$ from the orbit transversal in $\mathfrak{m}^2$ to the base of a restricted deformation has Δ_1 as its set of critical values and gives a μ-sheeted cover of the complement to Δ_1 near O.

The hypersurface Δ_1 is called the *caustic*, and Δ_2 the *cut locus* or the *Maxwell stratum*.

§7. Lagrangian singularities and the classification of caustics

One can see caustics clearly on a wall, lit up by sun-rays reflected off a curved surface (for example, the inside of a cup). By moving the cup one can see that caustics in general position have only standard singularities, which do not disappear under a small perturbation (see, for example, [*144]).

The analysis of caustics is a part of the theory of *Lagrangian singularities* (see [6] and the articles of Guckenheimer [38] and Weinstein [85]), which is analogous to the usual theory of singularities of Whitney [86], Thom [72], and Mather [54].

We recall that a *symplectic manifold* is a smooth manifold M^{2n} with a closed non-degenerate 2-form ω. A *Lagrangian submanifold* of M^{2n} is one of dimension n on which $\omega = 0$. A fibration $p: M^{2n} \to B^n$ is said to be *Lagrangian* if its fibres are Lagrangian submanifolds. The standard example is the cotangent bundle $T^*B \to B$ (the "phase space" of classical mechanics).

Let $i: L \to M$ be the embedding of a Lagrangian submanifold into the total space of a Lagrange fibration $p: M \to B$. The projection $p \circ i: L \to B$ is called a *Lagrange mapping*, and the set of its critical values a *caustic*.

Two Lagrange mappings are *equivalent* if the corresponding Lagrangian manifolds are transformed into each other by a diffeomorphism of the fibration that preserves the 2-form; in this case the caustics are obviously diffeomorphic. A Lagrange mapping is *stable* at a point O if every nearby Lagrange mapping has, at some point near O, a germ equivalent to the germ of the given mapping at O.

The germ of a Lagrange mapping is *simple* if all nearby germs belong to finitely many equivalence classes. (A simple germ can be non-stable, and a stable germ need not be simple.)

THEOREM XII (see [6]). *Simple stable germs of Lagrange mappings are classified by the series A, D, and E. Every Lagrange mapping with $n < 6$ can be approximated by a mapping whose germ at each point is simple and stable.*

A description of the Lagrangian singularities A, D, and E is given below (§9) in coordinate form. From this description it follows, in particular, that the stable singularities of caustics in three-space are only the cuspidal edges (A_3), swallow tails (A_4), and the points of contact of three cuspidal edges, two of which may be imaginary ($D_4^{\pm}$) (Fig. 4). All other singularities split into these under a small perturbation.

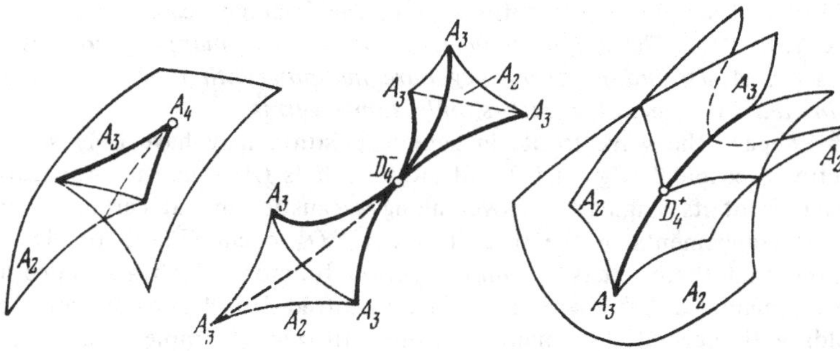

Fig. 4

§8. Legendre singularities and the classification of wave fronts

An example of a wave front is the curve that is obtained if we select a segment of length t on each interior normal to an ellipse. It is easy to check that a wave front may have singularities and that these singularities cannot be removed by a small deformation of the original ellipse.

The analysis of wave front singularities is a part of the theory of Legendre singularities (see [10], [11]).

The theory of Legendre singularities parallels that of Lagrangian singularities with the following differences: the role of the symplectic structure is taken by a contact structure, that of the affine structure by a projective structure, gradients by Legendre transformations, functions by hypersurfaces, etc.

The analogy between the two theories is none other than Hamilton's "optics-mechanics" analogy.

We recall that a *contact manifold* is an odd-dimensional manifold M^{2n+1} equipped with a maximally non-integrable "contact field" of hyperplanes (if the contact field is locally given by a form α, then the form $\alpha \wedge (d\alpha)^n$ is non-degenerate). The standard examples of contact manifolds are the projective cotangent bundle PT^*B^{n+1} and the manifold of 1-jets of functions $J^1(W^n, \mathbf{R})$ with their natural contact fields (given by the "integrability conditions").

The integral manifold of a contact field on M^{2n+1} is called a *Legendre submanifold* if it has the highest possible dimension (namely n). A fibration p: $M^{2n+1} \to B^{n+1}$ is called a *Legendre fibration* if its fibres are Legendre submanifolds (an example is the projective cotangent bundle $PT^*B \to B$). All the Legendre fibrations of the same dimension are locally Legendre equivalent.

If i: $L^n \to M^{2n+1}$ is an embedding of a Legendre submanifold in the total space of the Legendre fibration p: $M^{2n+1} \to B^{n+1}$, then $p \circ i$: $L^n \to B^{n+1}$ is called a *Legendre mapping* and the image is a *wave front.*

Stability and simplicity are defined as in the Lagrange case.

THEOREM XIII. *The simple stable germs of Legendre mappings are classified by the series A, D, and E. Every Legendre mapping with $n < 6$ can be approximated by a mapping with simple stable germs.*

In three-space the wave fronts in general position may have only self-intersections, cuspidal edges (A_2) and swallow tails (A_3); in the propagation of a wave front its singularity moves along a caustic, and at certain moments of time rearrangements of the three types A_4, $D_4^{\pm}$ occur (Fig. 5, p. 14).

We note that there exists a *symplectization* functor, which associates a symplectic manifold E^{2n+2} with a contact manifold M^{2n+1}, Lagrangian submanifolds with Legendre submanifolds, etc. However, symplectification takes Legendre singularities in general position to very special "conical" Lagrangian singularities. The correct way of obtaining Legendre singularities is by means of the *contactization* functor $M^{2n} \mapsto E^{2n+1}$ (the contactification functor is only defined for germs or for manifolds with integral class ω defining the symplectic structure).

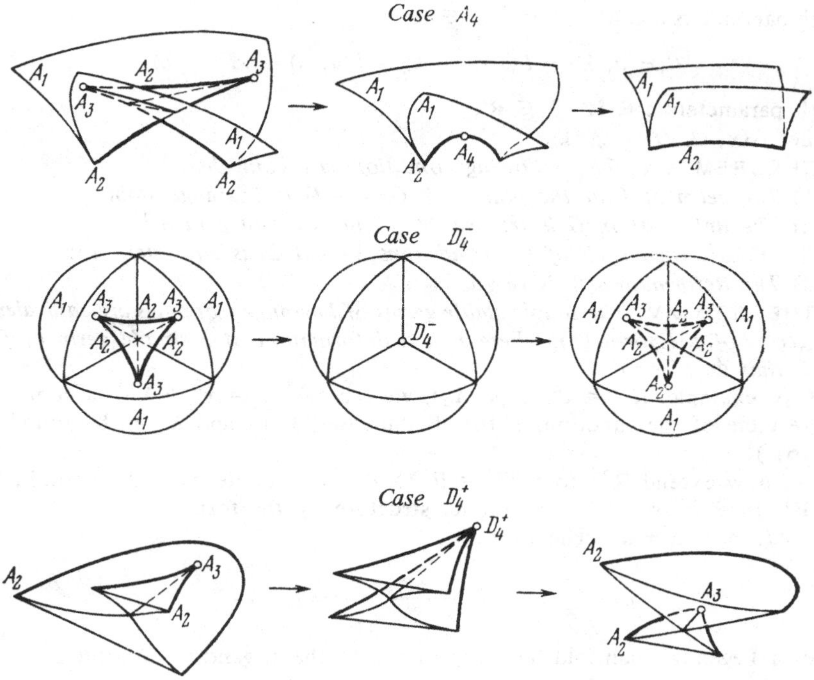

Fig. 5

§9. Normal forms for caustics and wave fronts

We use the old-fashioned coordinate notation. We consider a deformation $F(x, \lambda)$ of a function $f(x)$ of a variable $x \in \mathbf{R}^k$ with parameter $\lambda \in \mathbf{R}^l$. Suppose that $n = k + l$. We consider a symplectic space $\mathbf{R}^{2n}$ with coordinates $x \in \mathbf{R}^k$, $y \in \mathbf{R}^{k*}$, $\lambda \in \mathbf{R}^l$, $\varkappa \in \mathbf{R}^{l*}$ and with the 2-form

$$\omega = dx \wedge dy + d\varkappa \wedge d\lambda$$

(in intelligible shortened notation). The equations

$$(1) \qquad y = \frac{\partial F}{\partial x}, \quad \varkappa = -\frac{\partial F}{\partial \lambda}$$

define a Lagrangian submanifold, and together with the Lagrange fibration $(x, y, \lambda, \varkappa) \to (y, \lambda)$ they define a Lagrange mapping L.

With respect to the deformation F we construct two more families of functions of x:

$$\Phi(x;\ \lambda,\ y,\ z) = F(x,\ \lambda) - z - xy$$

with parameters $\lambda \in \mathbf{R}^l$, $y \in \mathbf{R}^{k*}$, $z \in \mathbf{R}$;

$$G(x;\ a,\ \lambda) = F(a + x,\ \lambda) - F(a,\ \lambda) - xF_a'(a,\ \lambda)$$

with parameters $a \in \mathbf{R}^k$, $\lambda \in \mathbf{R}^l$.

Let $G(x;\ O,\ O) = g(x)$.

THEOREM XIV. *The following conditions are equivalent*:

1) *The germ of L at the point $x = O$, $\lambda = O$ is Lagrange stable.*

2) *The deformation G is transversal to the orbit of g in m^2.*

If $f \in \mathfrak{m}^2$, then each of the conditions 1) *and* 2) *is equivalent to*:

3) *The deformation Φ is versal.*

THEOREM XV. *The simple stable germs of Lagrange mappings are equivalent to germs of the form* (1), *where F is a deformation of a simple germ of f such that Φ is versal.*

(For example, if $f = x^4$ (type A_3), we can take $F = x^4 + \lambda x^2$; a complete table of the functions F for all the cases A, D, and E can be found in [6].)

We now extend $\mathbf{R}^{2n}$ to $\mathbf{R}^{2n+1} = \mathbf{R}^{2n} \times \mathbf{R}^1$ and denote by z the coordinate in $\mathbf{R}^1$. In $\mathbf{R}^{2n+1}$ we define a contact structure by the form $\alpha = xdy + \varkappa\, d\lambda + dz$. The formula

$$(2) \qquad y = \frac{\partial F}{\partial x}, \qquad \varkappa = -\frac{\partial F}{\partial \lambda}, \qquad z = F - x\frac{\partial F}{\partial x}$$

gives a Legendre manifold, and together with the Legendre fibration $(x,\ y,\ \lambda,\ \varkappa\ ;\ z) - (y,\ \lambda;\ z)$ a Legendre mapping L'.

THEOREM XVI (see [11]). *The simple stable germs of Legendre mappings are equivalent to germs of the form* (2), *where F is the same as in the previous theorem.*

In addition to the group of diffeomorphisms of the argument space, the group of multiplication by non-vanishing functions acts on the space of functions of x and hence, the direct product of these two groups acts on that space. A deformation of the germ of a function is said to be *versal for levels* if it is transversal to the orbit of this (pseudo-) group.

As an example of such a deformation of a germ of f at O we can take

$$\lambda \mapsto f + \lambda_1 e_1 + \ldots + \lambda_r e_r,$$

where the e_s generate the quotient ring of the ring of power series at O modulo the ideal $(f,\ \partial f/\partial x)$ as an $\mathbf{R}$-module.

The direct product of the group of multiplications by non-vanishing germs at O and the group of diffeomorphisms leaving O fixed acts on $\mathfrak{m}^2$.

THEOREM XVII. *The following assertions are equivalent*:

1) *The germ of L' at the point $x = O$, $\lambda = O$ is Legendre stable.*

2) *The deformation G is transversal to the orbit of g under the action of the product group on $\mathfrak{m}^2$.*

If $f \in \mathfrak{m}^2$, *then each of the conditions* 1) *and* 2) *is equivalent to*:

3) *The deformation* Φ *is versal for levels.*

Comparing (1) and (2) with the results of §§5 and 6, we are led to the following conclusion:

THEOREM XVIII. *The mapping* τ *from the orbit transversal space to the base of the miniversal deformation gives a stable Legendre germ*; *the level bifurcation set is its wave front. The mapping* $\rho \circ \tau$ *from the orbit transversal space to the base of the restricted miniversal deformation defines a stable Lagrangian germ*; *the function bifurcation set* Δ_1 *is its caustic.*

Thus, the study of the topology of the bifurcation sets Σ and Δ_1 in the bases of versal deformations and that of the topology of caustics and wave fronts of stable Lagrangian and Legendre mappings are the same problem.

The above theorems may become clearer if we consider the germ of the *restricted critical set of the deformation* $F(x, \lambda)$, defined as

$$C = \{x, \lambda \colon \partial F/\partial x = 0, \quad F(x, \lambda) = 0\}.$$

If F is miniversal, then C is a germ of a smooth manifold of dimension $\mu - 1$. The natural projection $(x, \lambda) \mapsto \lambda$ defines a mapping $\pi \colon C \to \Sigma$. The coordinate system x defines a diffeomorphism $j \colon C \to T$ onto the orbit transversal (j is defined by the translation of the critical point to the origin). The diagram

$$\begin{array}{ccc} C & \xrightarrow{\;j\;} & T \\ & {}_{\tau}\searrow \quad \swarrow_{\pi} & \\ & \Sigma & \end{array}$$

is commutative. Therefore π, like τ, normalizes Σ, the mapping $\rho \circ \pi$ has the properties of $\rho \circ \tau$ and so on.

§10. Integrals of oscillatory functions

The determination of the intensity of light in a neighbourhood of a caustic leads to the problem of the asymptotic behaviour of integrals of rapidly oscillating functions of the type of Fresnel or Airy integrals

$$I(h, \lambda) = \int_{\mathbf{R}^k} e^{\frac{i}{h} F(x, \lambda)} \varphi \, dx, \qquad x \in \mathbf{R}^k, \qquad \lambda \in \mathbf{R}^l,$$

depending on a parameter λ, as the wave length tends to 0 (that is, as $h \to 0$). Here the parameter λ denotes the point of observation, φ a function with compact support, F a smooth real "phase function". Of course, similar integrals also turn up in other domains of physics and mathematics, for example, in number theory and the theory of partial differential equations (see [81], [55], [42]).

If the light is intense enough to destroy the medium, the destruction will begin at the points of caustics with the largest I. Thus, the problem arises of determining the asymptotic behaviour, as $h \to 0$, of the largest values of I in λ that occurs intrinsically for generic phase functions F. The classification of simple singularities was found as a by-product during the solution of this problem, which was communicated to the author by V. P. Maslov (see [56], [5]).

According to the stationary phase principle the fundamental contribution of the asymptotics is given by the critical points of F in x (for constant λ). In the absence of parameters ($l = 0$) for a function in general position the critical points are Morse (that is, non-degenerate), and I decreases like $h^{k/2}$ as $h \to 0$ (Fresnel [31]).

If there is a parameter, then degenerate critical points become irremovable, and for isolated "caustic" values of the parameter, the integral decreases more slowly than $h^{\frac{k}{2}-\beta}$. The so defined number β is called the *index of singularity* of the corresponding critical point.

Strictly speaking, for singularities with finite multiplicity the integral $I(h)$ has an asymptotic expansion

$$I \sim \sum_{\alpha,\varkappa} c_{\alpha,\varkappa} h^{\frac{k}{2}-\alpha} \ln^{\varkappa} h,$$

for fixed λ, where α runs through finitely many rational arithmetic progressions, $0 \leqslant \varkappa \leqslant k-1$ (see articles by I. N. Bernstein and Gel'fand [16], Atiyah [15], I. N. Bernstein [17], Malgrange [52], [53] and [*53a]).[1] The index of singularity β is the minimum of the α for which $c_{\alpha,\varkappa} \neq 0$ for some φ concentrated in an arbitrarily small neighbourhood of the critical point x in question.

THEOREM XIX (see [5], [7]). *For simple singularities* $\beta = \frac{1}{2} - \frac{1}{N}$, *where* N *is the Coxeter number*:

$N(A_{k+1}) = k+1$, $N(D_k) = 2k-2$, $N(E_6) = 12$, $N(E_7) = 18$, $N(E_8) = 30$. *For parabolic and hyperbolic singularities* $\beta = \frac{1}{2}$.

For all other singularities it appears that $\beta > \frac{1}{2}$.

We define the Coxeter number N of any singularity by the formula $\beta = \frac{1}{2} - \frac{1}{N}$, where β is the index of singularity.

THEOREM XX (see [7]). *For* $l \leqslant 10$ *and* $k \geqslant 3$, *the maximum index of singularity that inherently occurs in generic families has the form*

$$\beta_l = \frac{1}{2} - \frac{1}{N},$$

where the number N *is given by the table*

l	0	1	2	3	4	5	6	7	8	9	10, $k=3$	11, $k=3$	10, $k>3$
N	+2	+3	+4	+6	+8	+12	∞	∞	−24	−16	−12	−8	−6

[1] (Translator's note.) These functions have also been studied by M. Sato, M. Kashiwara and their co-workers under the name of *b*-functions. For a summary of their work, see M. Kashiwara, Sur la *b*-fonction, Séminaire Goulaouic-Lions-Schwartz 1974/75, Exp. 25. For more detailed expositions (in Japanese), see the collection "*b*-functions and singular points of hypersurfaces", Proc. Conf. Research Inst. Math. Sciences, Kyôto, Japan, 27 September–10 October, 1973, Sûrikaisekikenkyuo Kôkyûroku 225 (1975).

All the numbers $\beta_l = \beta_l(k)$ *are rational and for sufficiently large* k *do not depend on* k.

The calculation of the β_l seems to be very difficult; it appears that $\beta_l \sim \sqrt{(2l/6)}$. It seems probable that β is semicontinuous in λ and that for λ near λ_0 we have the uniform estimate

$$|I(h, \lambda)| \leqslant C(\varepsilon, \varphi)\, h^{\frac{k}{2} - \beta(\lambda_0) - \varepsilon}$$

for any $\varepsilon > 0$.

This estimate has been proved by I. M. Vinogradov [81] for the simple singularities of the series A, and by Duistermaat [30] for all simple singularities.

§11. Semiquasihomogeneous functions and the Newton diagram

The original proofs of all the classification theorems were based on long computations, which can be replaced by a distinctive geometric technique connected with the Newton diagram.

A function $f(x_1, \ldots, x_n)$ is said to be *quasihomogeneous* of degree d with weights $(\alpha_1, \ldots, \alpha_n)$ if $f(t^{\alpha_1} x_1, \ldots, t^{\alpha_n} x_n) = t^d f(x_1, \ldots, x_n)$ identically in $t \in \mathbf{C}^*$. Here $0 < \alpha_s \leqslant \frac{1}{2}$ are rational numbers. (For more about quasihomogeneous functions, see [8] and also papers by Saito [65], [66] Milnor and Orlik [57], Orlik and Wagreich [60].)

A function f is *semiquasihomogeneous* if $f = f_0 + f'$, where f_0 is quasihomogeneous of degree 1 and has a singularity of finite multiplicity at O, and all the monomials of f' are of degree greater than 1.

THEOREM XXI (see [8]). *A semiquasihomogeneous function can be reduced by a diffeomorphism to the normal form* $f_0 + c_1 e_1 + \ldots + c_r e_r$, *where the* c_s *are numbers and the* e_s *are basis monomials of the local ring of the function* f_0 *of degree greater than* 1.

The Newton diagram of an arbitrary germ f is a convex polyhedron in $\mathbf{R}^n$ constructed from the exponents of the monomials occurring in the Taylor expansion of f; it contains much useful information. Here I only state one result of Kushnirenko, which allows us to compute the multiplicity.

We assume that the Newton diagram of the function $f(x_1, \ldots, x_n)$ has points on each of the coordinate axes (this is not a restriction, since f has a sufficient jet).

THEOREM XXII (Kushnirenko). *Let* V *denote the volume of the subdomain of the positive orthant in* $\mathbf{R}^n$ *lying below the Newton diagram,* V_i *the* $(n-1)$*-dimensional volume under the Newton diagram on the i-th coordinate hyperplane,* V_{ij} *the* $(n-2)$*-dimensional volume on the coordinate plane orthogonal to the i-th and j-th basis vectors, etc. Then for all functions* f *with a given Newton diagram*

$$\mu(f) \geqslant n!V - (n-1)! \sum_i V_i + (n-2)! \sum_{i<j} V_{i,j} - \ldots \pm 1,$$

with equality for almost all f.

For example, for almost all functions of two variables with a given Newton diagram $\mu = 2S - a - b + 1$, where S is the area under the diagram, and a and b are the coordinates of the points of the diagram on the axes (Fig. 6).

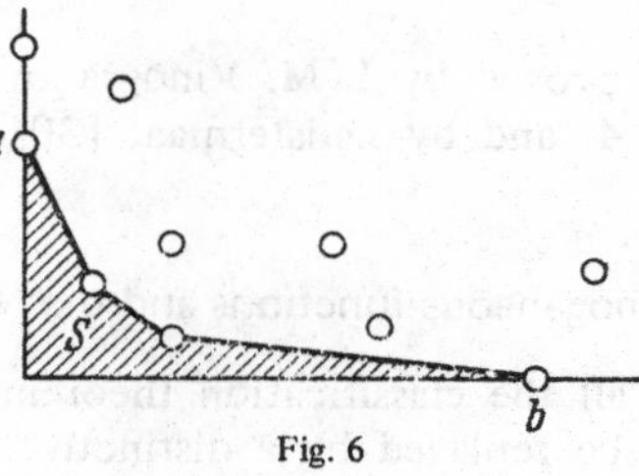

Fig. 6

Very recently D. N. Bernstein, Kushnirenko, and Khovanskii have obtained far-reaching generalizations of this theorem. In particular, the number of solutions of a system of polynomial equations $P_1(x_1, \ldots, x_n) = 0, \ldots, P_n(x_1, \ldots, x_n) = 0$, for which none of the coordinates x_s is zero, is equal to $n!$ times the mixed Minkowski volume of the convex hulls of the supports of the polynomials P_s (for almost all sets of polynomials P_s whose supports have given convex hulls). They have also obtained analogous formulae, expressing other numerical invariants of affine complete intersections in terms of the geometry of polyhedra. For example, the number of holomorphic forms on a hypersurface is equal to the number of lattice points inside the convex hull of the support (for almost all hypersurfaces whose supports have a given convex hull). (This formula seems to have been known to Hodge [*143], but according to Khovanskii, Hodge's arguments are not satisfactory.) D. N. Bernstein and Khovanskii have proved a formula for the Euler characteristics of complete intersections (spring 1975); another proof of this formula was given by Kushnirenko for the case of hypersurfaces. At the Bonner Arbeitstagung (June 1975) Varchenko discussed these problems with A'Campo and other Western mathematicians. In September 1975 Varchenko proved a formula for the ζ-function of the monodromy in terms of the Newton polyhedron. He uses the Bernstein-Khovanskii-Kushnirenko formula for Euler characteristics and A'Campo's results on the ζ-function of a monodromy [*145]. According to a letter from Brieskorn to Varchenko dated 30 December 1975, this formula was also found by A'Campo in October 1975 for functions of two variables, and by F. Ehlers in November 1975 for the general case. Varchenko has also found a formula for the index of singularity of oscillatory integrals in terms of Newton polyhedra. Among other things, he has proved that:

1) for functions of two variables with isolated singularities, the index of singularity can be calculated from the point of intersection of the Newton polygon with the diagonal in some system of coordinates, that is, for those systems of coordinates for which this point of intersection is as far from the origin as possible;

2) the same is true for functions of more variables if the coordinates of the point of intersection are greater than 1;

3) the index of singularity is not upper semicontinuous for some functions of three variables and for some Γ-non-degenerate functions of five variables for which the above formula fails (that is, the coordinates of the point of intersection are less than 1);

4) the index of singularity depends on the *real* form: it is different for some singularities of functions of three variables at isolated critical points even though the singularities are equivalent over **C** (this is the same example as in 3)).

Varchenko's paper on oscillatory integrals contains many other, more precise results, with fewer restrictions, but these are too long to be stated here [105].

Among the numerous unsolved problems in the theory of singularities there is the important problem of the smoothness of the stratum μ = const in the base of a miniversal deformation.

Lê and Ramanujam [50] have proved that for $n \neq 3$ neither the topological type of the singular level set nor the topological type of the "Milnor fibration" formed by the non-singular fibres varies. It also seems likely that neither the topological type of a function nor the index of singularity β varies. Pham [63] has noted the change of the topology of the bifurcation set along the stratum μ = const. Gabrielov [36] has proved that the dimension of the stratum μ = const is equal to the modality of the singularity. Using results of Teissier [71], Kushnirenko and Gabrielov have proved that for (semi)homogeneous singularities this dimension is equal to the number of generators of a monomial basis of the local ring on the Newton diagram and above it [37]. This result is probably true for (semi)quasihomogeneous functions.

In [8] I have expressed the conjecture that the modality of a function of two variables is equal to the number of (lattice) points between the Newton diagram and the coordinate rays passing through the point (2,2), counting the boundary points (for almost all functions with a given diagram); this was proved by Kushnirenko, using results of Varchenko and Lê Dũng Trang.

In this survey I have left to one side many important questions of the theory of critical points of functions, particularly, algebraic ones (see the ponderous tome "Singularités à Cargèse", [71]). From the results listed above we see the power of "transcendental" and topological methods, based on an analysis of the hierarchy of singularities, on a detailed study of the classes of low codimension, on considerations of semicontinuity and genericity, going back to Poincaré's theory of bifurcations [12], and formalized in Thom's transversality theorems. These ideas were first applied to the study of singular points of hypersurfaces by Galina N. Tyurina (see [76]–[78], [13]).

CHAPTER II

The hierarchy of singularities

§ 12. Normal forms

A large part of the results discussed in Chapter I is based on the classification theorems or is found by means of these theorems. However, to obtain these theorems there is no other approach so far than the experimental one. Below we state these experimental results (including a complete classification of the singularities with at most two moduli) and prove many of them (including the theorem on the classification of unimodal singularities).

Before proceeding to the technical statements of the results (see §§ 13 and 14), I indicate some conclusions of a general nature to which our list of singularities leads (§ 13). I do not state these conclusions in the form of a theorem, because I do not know the extent of the class of singularities to which they apply.

1. **Normal forms.** *For many classes of singularities there exist simple normal forms.*

The following definitions give a precise meaning to these words.

The group of germs (or jets) of diffeomorphisms of $\mathbf{C}^n$ leaving the critical point O fixed acts on the space of germs (or jets) of functions at O. A *class of singularities* is a subset of the space of germs (or jets) of functions that is invariant under this action. Orbits are examples of classes of singularities. Two germs (or jets) are said to be equivalent if they belong to the same orbit.

Another example of a class is the so-called μ = const *stratum.* The *multiplicity* (Milnor number) μ of a critical point $O \in \mathbf{C}^n$ of a function f is the index of the singular point O of the vector field grad f. The μ = const *stratum for* f is defined as the connected component containing f of the space of germs with fixed multiplicity μ at O.

To define a *normal form,* we regard the space of polynomials $M = \mathbf{C}\,[x_1, \ldots, x_n]$ as a subset of the space of germs of functions $f(x_1, \ldots, x_n)$ at O.

A *normal form for a class K of functions* is given by a smooth mapping $\Phi\colon B \to M$ of a finite-dimensional linear *space of parameters B* into the space of polynomials for which the following three conditions hold:

1) $\Phi(B)$ intersects all the orbits of K;

2) the inverse image in B of each orbit is finite;

3) the inverse image of the whole complement to K is contained in some proper hypersurface in B.

A normal form is said to be *polynomial* (respectively, *affine*) if the mapping Φ is polynomial (respectively, linear and inhomogeneous). An affine normal form is called *simple* if Φ has the form

$$\Phi(b_1, \ldots, b_r) = \varphi_0 + b_1\mathbf{x}^{\mathbf{m}_1} + \ldots + b_r\mathbf{x}^{\mathbf{m}_r},$$

where φ_0 is a fixed polynomial, the b_i are numbers and the $\mathbf{x}^{\mathbf{m}_i}$ are monomials. (In the applications the polynomial φ_0 is usually "simple" itself, that is, a sum of a few monomials.)

The existence of a unique normal form (at least a polynomial one) for the whole stratum μ = const is by no means obvious a priori. An astonishing conclusion from our calculations is the existence of such normal forms for all the singularities of the list in §13 (therefore, in particular, for all singularities with one and two moduli). The majority of our forms are simple; it is probable that all the singularities in §13 have simple normal forms. I do not know how extensive the class of functions is for which the stratum μ = const admits a simple (or at least polynomial) normal form (this question is naturally related to stable equivalence classes; see §12.5).

2. Series of singularities. In the list of §13 all the singularities are split into *series*, denoted by capital letters (we use light-face letters $A, \ldots, Z$ with various suffixes to denote the strata μ = const and bold-face letters $\boldsymbol{A}, \ldots, \boldsymbol{Z}$ with or without suffixes to denote classes of singularities that are unions of μ = const strata). Although the series undoubtedly exist, it is not at all clear what a series of singularities is.

For example, consider the series $\boldsymbol{A}$ and $\boldsymbol{D}$, which are formed by the orbits of the germs A_k: $f(x, y) = x^{k+1} + y^2$ and D_k: $f(x, y) = x^2y + y^{k-1}$. The classes A_k and D_k degenerate to each other as follows:

$$\begin{array}{ccccccccc} A_1 & \leftarrow & A_2 & \leftarrow & A_3 & \leftarrow & A_4 & \leftarrow & \ldots \\ & & & & \uparrow & & \uparrow & & \\ & & & & D_5 & \leftarrow & D_6 & \leftarrow & \ldots \end{array}$$

(A class of singularities L *degenerates* (or *is adjacent*) to a class K (notation: $K \leftarrow L$) if every function $f \in L$ can be deformed into a function of K by a suitably small perturbation.)

It is clear that there are two series in this example, $\boldsymbol{A}$ and $\boldsymbol{D}$. However, what is the formal meaning of this assertion, and does it exceed the limits of the arbitrary names?

Thus, to define the series $\boldsymbol{A}$ is to learn how to invert the arrows of the adherences, so that we go from A_k to A_{k+1}, without deviating through

D_{k+1}. In the given case this is not hard to do (the singularities of A have a second differential of corank $\leqslant 1$). In more complicated cases (see §§13, 14) we can also state a rule for inverting the arrows (in each case separately). As a result there arise series with one or several suffixes (for example, the three-suffix series $T_{k,l,m} = axyz + x^k + y^l + z^m$), and the functions of the series may depend on parameters.

As in the example of the series A and D above, so in all cases, after a series is found, we can define it. However, a general definition of a series of singularities is not known. It is only clear that the series are associated with singularities of infinite multiplicity (for example, $D \sim x^2y$, $T \sim xyz$), so that the hierarchy of series reflects the hierarchy of non-isolated singularities.

3. Periodicity. The partition of many classes of singularities into μ = const strata reveals a peculiar periodicity, which can be described as follows. The whole stratification (partition) is a chain of identical fragments (beasts). Each beast consists of points (strata), two of which (the head and the tail) are distinguished. Apart from the head and tail, the beast can contain the strata joining them by arrows of adjacency, and extremities (series of infinite length). The head of each beast adheres to the tail of the preceding beast.

For example, the stratification of singularities of corank 2 with 3-jet x^3 is given by a chain of beasts, each of which consists of 5 points and one infinite leg:

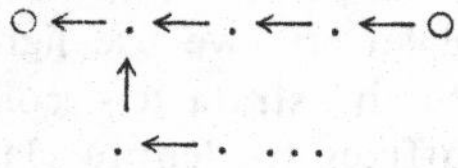

(J_k in the list of §13).

The form of the beasts corresponding to the other classes of singularities is also indicated in §13.

The reason for periodicity in the general case is not clear. A partial explanation has only been obtained for quasihomogeneous singularities by means of a technique resembling the fans of Enriques and Demazure [27] (see §16).

However, the periodicity becomes apparent not only under reduction of quasihomogeneous functions to normal forms, but also in all the calculations associated with the classifications (so that for all the calculations it suffices, in fact, to consider only one beast from a chain). This periodicity is an unexpected and, in the general case, unexplained experimental fact. Like the existence of series, periodicity suggests that there is some algebraic structure in the set of strata.

4. Classes of small modality. From the point of view of applications the most important characteristic of a class of singularities is its *codimension c* in the space of germs of functions with critical point O and critical value 0.

A generic function has only singularities of codimension $c = 0$ (non-degenerate). Degenerate singularities are irremovable only in the case of a *family* of functions depending on parameters. In this case the class of codimension c is irremovable by a small perturbation only if the number of parameters is $l \geqslant c$.

Thus, *in applications it is always necessary to study all classes up to codimension* l (that is, classes for which the complement to their union has codimension greater than l). This problem should not be confused with that of classifying singularities with orbital codimension $\leqslant l$ (that is, with $\mu \leqslant l + 1$). The latter problem occurs in applications only as a tool for solving the former.

From the topological point of view the most important characteristic of a singularity is the multiplicity μ of a critical point (equal to the number of simple critical points into which a composite point decomposes under a small perturbation).

The unexpected conclusion from these calculations is that *the algebraically most natural results are obtained not for the classification of classes of singularities up to a definite codimension* c *or multiplicity* μ, *but for the classification of classes of singularities of small modality* m.

I recall that the *modality* (or *number of moduli*) of an isolated critical point O of a function $f\colon (\mathbf{C}^n, O) \to (\mathbf{C}, 0)$ is the smallest number M such that a sufficiently small neighbourhood of any k-jet of f at O is covered by at most finitely many of at most M-parameter families of orbits of the action of the group of diffeomorphisms $(\mathbf{C}^n, O) \to (\mathbf{C}^n, O)$ on the space of jets of functions with critical point O and critical value 0 (see § 1).

The modality m is equal to the dimension of the stratum $\mu = \text{const}$ in the base of a versal deformation minus 1 (Gabrielov [36]). Therefore, the codimension c of the stratum $\mu = \text{const}$ in the space of germs of functions with critical point O and critical value 0, the multiplicity μ and the modality m are connected by the relation

$$\mu = c + m + 1.$$

At the present time the following are completely classified:

(1) all the singularities for which $c \leqslant 10$;

(2) all the singularities for which $\mu \leqslant 16$;

(3) all the singularities for which $m \leqslant 2$;

Singularities with $m = 0$, 1 and 2 are called, respectively, *simple, unimodal*, and *bimodal*. Lists of them are given in § 13. An analysis of these lists shows that

1) simple singularities are classified precisely by the Coxeter groups A_k, D_k. E_6, E_7, E_8 (that is, by the regular polyhedra in 3-space) [6];

2) unimodal singularities form a single infinite three-suffix series and 14 "exceptional" one-parameter families generated by quasihomogeneous singularities [7].

The quasihomogeneous unimodal singularities are obtained from automorphic functions connected with 14 distinguished triangles on the Lobachevskii plane and three distinguished triangles on the Euclidean plane in precisely the same way as simple singularities are connected with regular polyhedra (Dolgachev [28]).

3) bimodal singularities form 8 infinite series and 14 exceptional two-parameter families generated by quasihomogeneous singularities.

The quasihomogeneous bimodal singularities are associated with the 6 quadrilaterals and the 14 triangles on the Lobachevskii plane (in the latter case one must consider automorphic functions with automorphy factors corresponding to 2-, 3- or 5-sheeted coverings). (Dolgachev [29]).

All singularities with one and two moduli are classified precisely by the degenerations of elliptic curves, which were classified by Kodaira (Kulikov [89]). Kulikov's construction is that 1, 2 or 3 points are blown up in a minimal resolution of the degenerate fibre, after which the original fibre is blown down.

Unfortunately, all the results listed here are obtained by a comparison of independently proved classification theorems, none of which can (so far) be derived from another. Therefore all these results depend on our calculations.

5. Stable equivalence. I recall that two germs f: $(\mathbf{C}^n, O) \to (\mathbf{C}, 0)$ and g: $(\mathbf{C}^m, O) \to (\mathbf{C}, 0)$ are said to be *stably equivalent* if they become equivalent after adding non-degenerate quadratic forms (for example, $f(x) = x^3$ is stably equivalent to $g(x, y) = x^3 + y^2$). If two functions of the same number of variables are stably equivalent, then they are equivalent. Thus, passage to stable equivalence classes is not concerned with the number of variables of functions,[1] because nothing changes in the classification of functions of a fixed number of variables.

The number of stable μ-equivalence clasess for small μ that are powers of 2 is itself a power of 2:

μ	1	2	4	8	16
Number of classes	1	1	2	4	32

The *corank* of the germ of a function at a critical point is the dimension of the zero space of its second differential. Every function of corank r is stably equivalent to a function of r variables ("generalized Morse lemma"). In classification up to stable equivalence one usually takes normal forms with the number of variables equal to the corank.

[1] In § 13 we do not indicate the number of arguments of a function when all the arguments are explicitly included in the formula. For example, the stable equivalence class of the function x^3 includes the functions $f(x) = x^3$ and $g(x, y) = x^3 + y^2$, but not the function $f(x, y) = x^3$.

Calculations show, however, that in many cases more natural normal forms are obtained by passing to functions of a larger number of variables that are stably equivalent to the given function. The simplest example is the inclusion of the singularities $x^p + ax^2y^2 + y^q$ into the series T; a more interesting example is given by the series $W^{\#} \sim (x^2 + y^3)^2$.

I recall the definition of the *Newton polyhedron* of the series $f \in \mathbf{C}[[x_1, \ldots, x_n]]$. The *support* supp f of a series f is the set of all exponents $\boldsymbol{m} \in \mathbf{Z}^n$ of the monomials $\boldsymbol{x}^m$ that occur in f with non-zero coefficients. We assume the lattice $\mathbf{Z}^n$ to be embedded in $\mathbf{R}^n$. For each point $\boldsymbol{m} \in$ supp f we consider the positive orthant $\boldsymbol{m} + \mathbf{R}^n_+$. The Newton polyhedron Γ for f is the boundary of the convex hull of these orthants.

A function f is called Γ-non-degenerate if the multiplicity μ of its critical point O has the least possible value $\nu < \infty$ for functions with the Newton polyhedron Γ. An explicit formula for the Newton number $\nu(\Gamma)$ and a sufficient condition for Γ-non-degeneracy were found by Kushnirenko (Theorem XXII in § 11, and [90]).

A large part of our normal forms consists of Γ-non-degenerate functions. Is every function in a neighbourhood of a critical point of finite multiplicity stably equivalent to a Γ-non-degenerate function? If the answer to this question is positive, then the classification of singularities and the reduction to normal forms can be carried out in a rather explicit fashion: the study of Γ-non-degenerate functions leads to constructions basic to the stereometric constructions (see [8]).

The arrangement of material in this chapter is as follows. In § 13 we give lists of normal forms and some adherency diagrams. Here we list all the singularities with at most 2 moduli, all the singularities of corank 2 with non-zero 4-jet, all the singularities of corank 3 with 3-jet giving an irreducible cubic; we also indicate which singularities remain unclassified.

In § 14 we state 105 theorems, which form the main results of this chapter. These theorems, together with the list of singularities of § 13, constitute a determination of singularities, similar to botanical determination. Actually, in § 14 we describe an algorithm to find the place of an arbitrary singularity in the lists of § 13.

In § § 15-18 we prove the majority of the theorems of § 14. All these proofs are based on the geometric technique developed in [8]. The formulae of § § 13-14 can be interpreted as an abbreviated notation for the supports and the Newton polyhedra, and the proofs of the theorems consist not in calculations, but in the geometric manipulation of these supports and polyhedra.

The classification of quasihomogeneous singularities in § 16 is based on the general technique of the roots of a quasihomogeneous Lie algebra, developed here in greater detail than is necessary for proving the theorems of § 14.

The theorems in §§15–18 have as a consequence, in particular, the classification theorem for unimodal singularities (Theorem II of §1), which was announced in [7]. The basic part of the proof is contained in §18.

§13. Lists of singularities

(In this section the letters $A, \ldots, Z$ indicate *stable* equivalence classes of functions or families of functions.)

I. SINGULARITIES WITH $m = 0$, 1, and 2.

I.0. SIMPLE SINGULARITIES ($m = 0$). There are two infinite series A, D, and three exceptional singularities E_6, E_7, E_8:

$$A_k \ni x^{k+1} \quad (k \geqslant 1),$$
$$D_k \ni x^2 x + y^{k-1} \quad (k \geqslant 4);$$
$$E_6 \ni x^3 + y^4, \qquad E_7 \ni x^3 + xy^3, \qquad E_8 \ni x^3 + y^5.$$

The diagram of adjacencies:

$$\begin{array}{ccccccccc} A & \leftarrow & D & \leftarrow & E_6 & \leftarrow & E_7 & \leftarrow & E_8, \\ & & & & \uparrow & & \uparrow & & \uparrow \\ & & & & (P) & & (X) & & (J) \end{array} \qquad \begin{array}{ccccccccc} A = A_1 & \leftarrow & A_2 & \leftarrow & A_3 & \leftarrow & A_4 & \leftarrow & \ldots \\ & & & & \uparrow & & & & \\ & & & & (D) & & & & \\ D = D_4 & \leftarrow & D_5 & \leftarrow & D_6 & \leftarrow & \ldots & & \\ & & \uparrow & & & & & & \\ & & (E) & & & & & & \end{array}$$

The classes P, X and J are defined below: see II.

I.1. UNIMODAL SINGULARITIES ($m = 1$). There are 3 families of parabolic singularities, a three-suffix series of hyperbolic singularities and 14 families of exceptional singularities.

Parabolic:

P_8	$x^3 + y^3 + z^3 + axyz$	$a^3 + 27 \neq 0$
X_9	$x^4 + y^4 + ax^2y^2$	$a^2 \neq 4$
J_{10}	$x^3 + y^6 + ax^2y^2$	$4a^3 + 27 \neq 0$

Hyperbolic:

$$T_{p,q,r} \ni x^p + y^q + z^r + axyz, \quad a \neq 0, \quad \frac{1}{p} + \frac{1}{q} + \frac{1}{r} < 1.$$

14 *exceptional families*:

E_{12}	$x^3 + y^7 + axy^5$	E_{13}	$x^3 + xy^5 + ay^8$
E_{14}	$x^3 + y^8 + axy^6$	Z_{11}	$x^3y + y^5 + axy^4$
Z_{12}	$x^3y + xy^4 + ax^2y^3$	Z_{13}	$x^3y + y^6 + axy^5$
W_{12}	$x^4 + y^5 + ax^2y^3$	W_{13}	$x^4 + xy^4 + ay^6$
Q_{10}	$x^3 + y^4 + yz^2 + axy^3$	Q_{11}	$x^3 + y^2z + xz^3 + az^5$
Q_{12}	$x^3 + y^5 + yz^2 + axy^4$	S_{11}	$x^4 + y^2z + xz^2 + ax^3z$
S_{12}	$x^2y + y^2z + xz^3 + az^5$	U_{12}	$x^3 + y^3 + z^4 + axyz^2$

Some adjacencies of unimodal singularities:

$$\begin{array}{ccccccccccccc}
 & & & & J_{10}=T_{2,3,6} & \to & (E_8) & & & & & & \\
 & & & & \uparrow & & & & & & & & \\
\cdots & \to & T_{2,3,8} & \to & T_{2,3,7} & \leftarrow & E_{12} & \leftarrow & E_{13} & \leftarrow & E_{14} & \leftarrow & (\boldsymbol{J}_3)
\end{array}$$

The parentheses contain classes of singularities whose modality is not equal to 1.

$$\begin{array}{ccccccccccccc}
 & & & & X_9=T_{2,4,4} & \to & (E_7) & & & & & & \\
 & & & & \uparrow & & & & & & & & \\
\cdots & \to & T_{2,4,6} & \to & T_{2,4,5} & \leftarrow & Z_{11} & \leftarrow & Z_{12} & \leftarrow & Z_{13} & \leftarrow & (\boldsymbol{Z}_1^1) \\
 & & \uparrow & & \uparrow & & \uparrow & & \uparrow & & & & \\
\cdots & \rightrightarrows & T_{2,5,6} & \to & T_{2,5,5} & \leftarrow & W_{12} & \leftarrow & W_{13} & \leftarrow & (\boldsymbol{W}_{1,0};\ \boldsymbol{X}) & &
\end{array}$$

$$\begin{array}{ccccccccccccc}
 & & & & P_8=T_{3,3,3} & \to & (E_6) & & & & & & \\
 & & & & \uparrow & & & & & & & & \\
\cdots & \to & T_{3,3,5} & \to & T_{3,3,4} & \leftarrow & Q_{10} & \leftarrow & Q_{11} & \leftarrow & Q_{12} & \leftarrow & (\boldsymbol{Q}_2) \\
 & & \uparrow & & \uparrow & & \uparrow & & \uparrow & & & & \\
\cdots & \rightrightarrows & T_{3,4,5} & \to & T_{3,4,4} & \leftarrow & S_{11} & \leftarrow & S_{12} & \leftarrow & (\boldsymbol{S}_{1,0}) & & \\
 & & \uparrow & & \uparrow & & \uparrow & & & & & & \\
\cdots & \substack{\to\\ \to\\ \to} & T_{4,4,5} & \to & T_{4,4,4} & \leftarrow & U_{12} & \leftarrow & (\boldsymbol{U}_{1,0};\ \boldsymbol{V}) & & & & \\
 & & & & \uparrow & & & & & & & & \\
 & & & & (\boldsymbol{O}) & & & & & & & &
\end{array}$$

I.2. BIMODAL SINGULARITIES ($m = 2$). There are 8 infinite series and 14 exceptional families. Let $\boldsymbol{a} = a_0 + a_1 y$.

4 *infinite series of bimodal singularities of corank* 2:

Notation	Normal form	Restrictions	Multiplicity μ
$J_{3,0}$	$x^3+bx^2y^3+y^9+cxy^7$	$4b^3+27\neq 0$	16
$J_{3,p}$	$x^3+x^2y^3+\boldsymbol{a}y^{9+p}$	$p>0,\ a_0\neq 0$	$16+p$
$Z_{1,0}$	$x^3y+dx^2y^3+cxy^6+y^7$	$4d^3+27\neq 0$	15
$Z_{1,p}$	$x^3y+x^2y^3+\boldsymbol{a}y^{7+p}$	$p>0,\ a_0\neq 0$	$15+p$
$W_{1,0}$	$x^4+\boldsymbol{a}x^2y^3+y^6$	$a_0^2\neq 4$	15
$W_{1,p}$	$x^4+x^2y^3+\boldsymbol{a}y^{6+p}$	$p>0,\ a_0\neq 0$	$15+p$
$W^{\#}_{1,2q-1}$	$(x^2+y^3)^2+\boldsymbol{a}xy^{4+q}$	$q>0,\ a_0\neq 0$	$15+2q-1$
$W^{\#}_{1,2q}$	$(x^2+y^3)^2+\boldsymbol{a}x^2y^{3+q}$	$q>0,\ a_0\neq 0$	$15+2q$

4 infinite series of bimodal singularities of corank 3:

Notation	Normal form	Restrictions	Multiplicity μ
$Q_{2,0}$	$x^3+yz^2+ax^2y^2+xy^4$	$a_0^2 \neq 4$	14
$Q_{2,p}$	$x^3+yz^2+x^2y^2+ay^{6+p}$	$p>0,\ a_0 \neq 0$	$14+p$
$S_{1,0}$	$x^2z+yz^2+y^5+azy^3$	$a_0^2 \neq 4$	14
$S_{1,p}$	$x^2z+yz^2+x^2y^2+ay^{5+p}$	$p>0,\ a_0 \neq 0$	$14+p$
$S^{\#}_{1,2q-1}$	$x^2z+yz^2+zy^3+axy^{3+q}$	$q>0,\ a_0 \neq 0$	$14+2q-1$
$S^{\#}_{1,2q}$	$x^2z+yz^2+zy^3+ax^2y^{2+q}$	$q>0,\ a_0 \neq 0$	$14+2q$
$U_{1,0}$	$x^3+xz^2+xy^3+ay^3z$	$a_0(a_0^2+1) \neq 0$	14
$U_{1,2q-1}$	$x^3+xz^2+xy^3+ay^{1+q}z^2$	$q>0,\ a_0 \neq 0$	$14+2q-1$
$U_{1,2q}$	$x^3+xz^2+xy^3+ay^{3+q}z$	$q>0,\ a_0 \neq 0$	$14+2q$

14 *exceptional families*

E_{18}	$x^3+y^{10}+axy^7$	E_{19}	$x^3+xy^7+ay^{11}$
E_{20}	$x^3+y^8+axy^{11}$	Z_{17}	$x^3y+y^8+axy^6$
Z_{18}	$x^3y+xy^6+ay^9$	Z_{19}	$x^3y+y^9+axy^7$
W_{17}	$x^4+xy^5+ay^7$	W_{18}	$x^4+y^7+ax^2y^4$
Q_{16}	$x^3+yz^2+y^7+axy^5$	Q_{17}	$x^3+yz^2+xy^5+ay^8$
Q_{18}	$x^3+yz^2+y^8+axy^6$	S_{16}	$x^2z+yz^2+xy^4+ay^6$
S_{17}	$x^2z+yz^2+y^6+azy^4$	U_{16}	$x^3+xz^2+y^5+ax^2y^2$

All the functions of these families (under the restrictions indicated in the tables) are bimodal.

Some adjacencies of bimodal singularities:

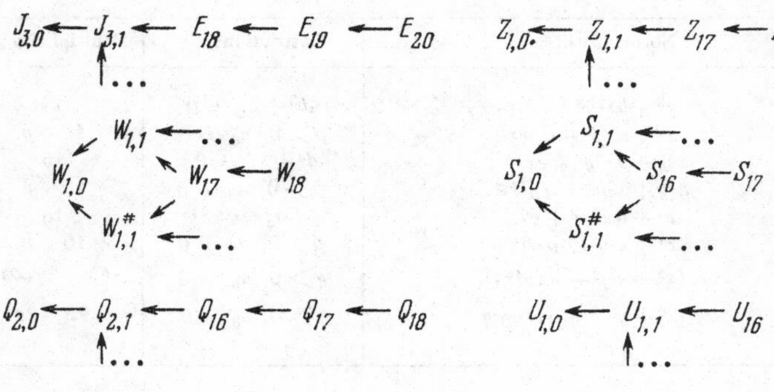

Pyramids of the exceptional singularities of modality 1 *and* 2:

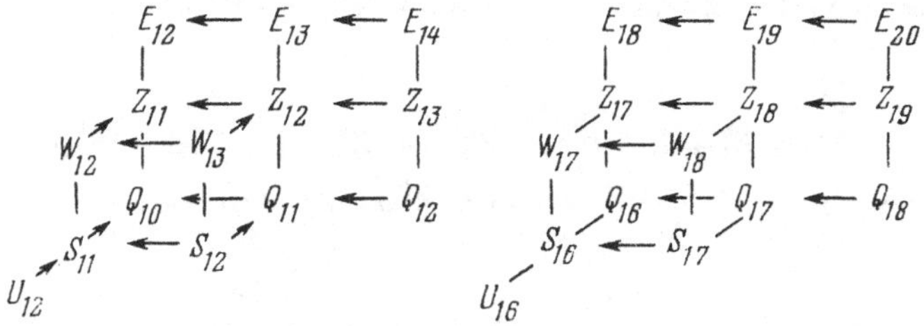

The vertical lines join singularities obtained from the same Kodaira class by means of the construction of Kulikov [89].

II. SINGULARITIES OF CORANK 2 WITH NON-ZERO 4-JET.

Throughout this section $a = a_0 + \ldots + a_{k-2}y^{k-2}$ ($a = 0$ for $k = 1$).

II.1. SINGULARITIES OF CORANK 2 WITH NON-ZERO 3-JET. Besides the simple singularities A, D, E_6, E_7, E_8 there is yet another infinite series of classes

$$\boldsymbol{J} = \leftarrow \boldsymbol{J}_2 \leftarrow \boldsymbol{J}_3 \leftarrow \ldots,$$

where

$$\begin{array}{l} \boldsymbol{J}_{\mathbf{k}} = J_{k,0} \leftarrow J_{k,1} \leftarrow E_{6k} \leftarrow E_{6k+1} \leftarrow E_{6k+2} \leftarrow (\boldsymbol{J}_{\mathbf{k}+1}) \\ \qquad\qquad\qquad\quad \uparrow \\ \qquad\qquad\qquad J_{k,2} \leftarrow J_{k,3} \leftarrow \ldots \end{array}$$

Notation	Normal form	Restrictions	Multiplicity μ	Modality m
$J_{k,0}$	$x^3+bx^2y^k+y^{3k}+cxy^{2k+1}$	$k>1, 4b^3+27\neq 0$	$6k-2$	$k-1$
$J_{k,i}$	$x^3+x^2y^k+ay^{3k+i}$	$k>1,\ i>0,\ a_0\neq 0$	$6k-2+i$	$k-1$
E_{6k}	$x^3+y^{3k+1}+axy^{2k+1}$	$k\geqslant 1$	$6k$	$k-1$
E_{6k+1}	$x^3+xy^{2k+1}+ay^{3k+2}$	$k\geqslant 1$	$6k+1$	$k-1$
E_{6k+2}	$x^3+y^{3k+2}+axy^{2k+2}$	$k\geqslant 1$	$6k+2$	$k-1$

Here $c = c_0 + \ldots + c_{k-3}y^{k-3}$ for $k > 2$; for $k = 2$ we set $c = 0$.

II.2. SINGULARITIES OF CORANK 2 WITH ZERO 3-JET AND NON-ZERO 4-JET. All such singularities form a single infinite series of classes

$$\boldsymbol{X} = \leftarrow \boldsymbol{X}_1 \leftarrow \boldsymbol{X}_2 \leftarrow \ldots,$$

where

$$\boldsymbol{X}_{\mathbf{k}} = X_{k,0} \leftarrow \boldsymbol{X}_{\mathbf{k}}^{*} \begin{array}{c} \swarrow\ \boldsymbol{Y}_{\mathbf{k}}\ \nwarrow \\ \nwarrow\ \boldsymbol{Z}_{\mathbf{k}}\ \swarrow \end{array} \boldsymbol{W}_{\mathbf{k}} \leftarrow (\boldsymbol{X}_{\mathbf{k}+1}),$$

and

$$X_k^* = \; \leftarrow X_{k,1} \leftarrow X_{k,2} \leftarrow \ldots, \quad Y_k = \; \leftarrow Y_{1,1}^k \leftarrow Y_{2,1}^k \Leftarrow \ldots,$$

$$Z_k = \; \leftarrow Z_0^k \leftarrow Z_1^k \leftarrow \ldots, \quad Z_0^k = \; \leftarrow Z_{12k-1}^k \leftarrow Z_{12k}^k \leftarrow Z_{12k+1}^k \leftarrow,$$

$$Z_i^k = \; \leftarrow Z_{i,0}^k \leftarrow Z_{i,1}^k \leftarrow Z_{12k+6i-1}^k \leftarrow Z_{12k+6i}^k \leftarrow Z_{12k+6i+1}^k \leftarrow (Z_{i+1}^k)$$

$$Z_{i,2}^k \leftarrow Z_{i,3}^k \leftarrow \ldots \qquad (i > 0),$$

$$W_{k,1} \leftarrow W_{k,2} \leftarrow \ldots$$

$$W_k = \; \leftarrow W_{12k} \leftarrow W_{12k+1} \leftarrow W_{k,0} \quad W_{12k+5} \leftarrow W_{12k+6} \leftarrow (X_{k+1}).$$

$$W_{k,1}^{\#} \leftarrow W_{k,2}^{\#} \leftarrow \ldots$$

Singularities of the classes X and Y:

Nota-tion	Normal form ($k>1$)	Restrictions	Multiplicity μ	Modality m
$X_{k,0}$	$x^4 + bx^3y^k + ax^2y^{2k} + xy^{3k}$	$\Delta \neq 0,\; a_0b_0 \neq 9$	$12k-3$	$3k-2$
$X_{k,p}$	$x^4 + ax^3y^k + x^2y^{2k} + by^{4k+p}$	$a_0^2 \neq 4,\; b_0 \neq 0,\; p>0$	$12k-3+p$	$3k-2$
$Y_{r,s}^k$	$[(x+ay^k)^2 + by^{2k+s}](x^2+y^{2k+r})$	$1 \leqslant s \leqslant r,\; a_0 \neq 0 \neq b_0$	$12k-3+r+s$	$3k-2$

In the case $k = 1$ these formulae need some modification:

$X_{1,0}$	$x^4 + a_0x^2y^2 + y^4$	$a_0^2 \neq 4$	9	1
$X_{1,p}$	$x^4 + x^2y^2 + a_0y^{4+p}$	$a_0 \neq 0$	$9+p$	1
$Y_{r,s}^1$	$x^{4+r} + a_0x^2y^2 + y^{4+s}$	$a_0 \neq 0$	$9+r+s$	1

Finally, $X_{1,0} = X_9$, $X_{1,p} = T_{2,4,4+p}$, $Y_{r,s}^1 = T_{2,4+r,4+s}$ (see I). Here

$$\Delta = 4(a_0^3 + b_0^3) - a_0^2b_0^2 - 18a_0b_0 + 27, \quad b = b_0 + \ldots + b_{2k-2}y^{2k-2}.$$

Singularities of the class Z.

The singularities $Z_{i,0}^k$ and Z_μ^k ($k > 1$) have a normal form of the kind $f = (x + ay^k)f_2$, where $a_0 \neq 0$ and f_2 is given by the following table:

Notation	f_2	Restrictions ($k>1$)	Multiplicity μ	Modality m
$Z_{i,0}^k$	$x^3 + dx^2y^{k+i} + cxy^{2k+2i+1} + y^{3k+3i}$	$4d^3 + 27 \neq 0,\; i>0$	$12k+6i-3$	$3k+i-2$
$Z_{12k+6i-1}^k$	$x^3 + bxy^{2k+2i+1} + y^{3k+3i+1}$	$i \geqslant 0$	$12k+6i-1$	$3k+i-2$
Z_{12k+6i}^k	$x^3 + xy^{2k+2i+1} + by^{3k+3i+2}$	$i \geqslant 0$	$12k+6i$	$3k+i-2$
$Z_{12k+6i+1}^k$	$x^3 + bxy^{2k+2i+2} + y^{3k+3i+2}$	$i \geqslant 0$	$12k+6i+1$	$3k+i-2$

The singularity $Z^k_{i,p}$ $(k > 1,\ i > 0,\ p > 0)$ has the normal form $Z^k_{i,p} \ni (x^2 + axy^k + by^{2k+i})(x^2 + y^{2k+2i+p})$, $a_0 \neq 0$, $b_0 \neq 0$; its multiplicity is $\mu = 12k + 6i + p - 3$, its modality is $m = 3k + i - 2$.

For $k = 1$ the preceding formulae are modified in the following way:

1) the upper index k is dropped;

2) the singularities $Z_{i,0}$, Z_{6i+11}, Z_{6i+12}, Z_{6i+13} $(i > 0)$ have normal forms of the kind $f = yf_2$, where f_2 is given by the preceding table;

3) $Z_{i,p} \ni y(x^3 + x^2y^{i+1} + by^{3i+p+3})$, $b_0 \neq 0, i > 0, p > 0; \mu = 9 + 6i + p$, $m = i + 1$.

Throughout the formulae above we have used

$$b = b_0 + \ldots + b_{2k+i-2}y^{2k+i-2}, \quad c = c_0 + \ldots + c_{2k+i-3}y^{2k+i-3}.$$

Singularities of the class W:

Notation	Normal form	Restrictions	Multiplicity μ	Modality m
W_{12k}	$x^4 + y^{4k+1} + axy^{3k+1} + cx^2y^{2k+1}$	$k \geqslant 1$	$12k$	$3k-2$
W_{12k+1}	$x^4 + xy^{3k+1} + ax^2y^{2k+1} + cy^{4k+2}$	$k \geqslant 1$	$12k+1$	$3k-2$
$W_{k,0}$	$x^4 + bx^2y^{2k+1} + axy^{3k+2} + y^{4k+2}$	$k \geqslant 1,\ b_0^2 \neq 4$	$12k+3$	$3k-1$
$W_{k,i}$	$x^4 + ax^3y^{k+1} + x^2y^{2k+1} + by^{4k+2+i}$	$i > 0,\ b_0 \neq 0$	$12k+3+i$	$3k-1$
$W^{\#}_{k,\ 2q-1}$	$(x^2 + y^{2k+1})^2 + bxy^{3k+1+q} + ay^{4k+2+q}$	$q > 0,\ b_0 \neq 0$	$12k+2+2q$	$3k-1$
$W^{\#}_{k,\ 2q}$	$(x^2 + y^{2k+1})^2 + bx^2y^{2k+1+q} + axy^{3k+2+q}$	$q > 0,\ b_0 \neq 0$	$12k+3+2q$	$3k-1$
W_{12k+5}	$x^4 + xy^{3k+2} + ax^2y^{2k+2} + by^{4k+3}$	$k \geqslant 1$	$12k+5$	$3k-1$
W_{12k+6}	$x^4 + y^{4k+3} + axy^{3k+3} + bx^2y^{2k+2}$	$k \geqslant 1$	$12k+6$	$3k-1$

In these formulae $b = b_0 + \ldots + b_{2k-1}y^{2k-1}$, $c = c_0 + \ldots + c_{2k-2}y^{2k-2}$; as elsewhere, $a = a_0 + \ldots + a_{k-2}y^{k-2}$ for $k > 1$ and $a = 0$ for $k = 1$.

III. SINGULARITIES OF CORANK 3 WITH A GIVEN 3-JET

Besides the unimodal singularities of the series T (see I.1) there are 3 infinite series of classes of such singularities, Q, S and U.

III.1. SERIES Q. Singularities with 3-jet $x^3 + yz^2$ form a single infinite series of classes

$$Q = \leftarrow Q_1 \leftarrow Q_2 \leftarrow \ldots, \text{ where } Q_1 = \leftarrow Q_{10} \leftarrow Q_{11} \leftarrow Q_{12} \leftarrow,$$

$$\underset{k>1}{Q_k} = \leftarrow Q_{k,0} \leftarrow \underset{\substack{\uparrow \\ Q_{k,2} \leftarrow Q_{k,3} \leftarrow \ \ldots}}{Q_{k,1}} \leftarrow Q_{6k+4} \leftarrow Q_{6k+5} \leftarrow Q_{6k+6} \leftarrow (Q_{k+1}).$$

Notation	Normal form	Restrictions	Multiplicity μ	Modality m
$Q_{k,0}$	$\varphi+bx^2y^k+xy^{2k}$	$k>1,\ b_0^2\neq 4$	$6k+2$	k
$Q_{k,i}$	$\varphi+x^2y^k+by^{3k+i}$	$k>1,\ b_0\neq 0$	$6k+2+i$	k
Q_{6k+4}	$\varphi+y^{3k+1}+bxy^{2k+1}$	$k\geqslant 1$	$6k+4$	k
Q_{6k+5}	$\varphi+xy^{2k+1}+by^{3k+2}$	$k\geqslant 1$	$6k+5$	k
Q_{6k+6}	$\varphi+y^{3k+2}+bxy^{2k+2}$	$k\geqslant 1$	$6k+6$	k

In these formulae $\varphi = x^3 + yz^2$, $b = b_0 + \ldots + b_{k-1}y^{k-1}$.

III.2. SERIES S. Singularities with 3-jet $x^2z + yz^2$ form a single infinite series of classes

$$S = \leftarrow S_1 \leftarrow S_2 \leftarrow \ldots,$$

where

$$\begin{array}{ccccccccc} & & & & S_{k,1}\leftarrow\ldots & & & & \\ & & & \swarrow & & \nwarrow & & & \\ S_k = \leftarrow S_{12k-1} & \leftarrow & S_{12k} \leftarrow & S_{k,0} & & S_{12k+4} & \leftarrow S_{12k+5} \leftarrow & S^*_{k+1} & \leftarrow (S_{k+1}). \\ & & & \nwarrow & & \nearrow & & & \\ & & & & S^{\#}_{k,1}\leftarrow\ldots & & & & \end{array}$$

Notation	Normal form	Restrictions	Multiplicity μ	Modality m
S_{12k-1}	$\varphi+y^{4k}+axy^{3k}+czy^{2k+1}$	—	$12k-1$	$3k-2$
S_{12k}	$\varphi+xy^{3k}+cy^{4k+1}+azy^{2k+1}$	—	$12k$	$3k-2$
$S_{k,0}$	$\varphi+y^{4k+1}+axy^{3k+1}+bzy^{2k+1}$	$b_0^2\neq 4$	$12k+2$	$3k-1$
$S_{k,i}$	$\varphi+x^2y^{2k}+ax^3y^k+by^{4k+1+i}$	$i>0,\ b_0\neq 0$	$12k+2+i$	$3k-1$
$S^{\#}_{k,2q-1}$	$\varphi+zy^{2k+1}+bxy^{3k+q}+ay^{4k+q+1}$	$q>0,\ b_0\neq 0$	$12k+2q+1$	$3k-1$
$S^{\#}_{k,2q}$	$\varphi+zy^{2k+1}+bx^2y^{2k+q}+axy^{3k+q+1}$	$q>0,\ b_0\neq 0$	$12k+2q+2$	$3k-1$
S_{12k+4}	$\varphi+xy^{3k+1}+azy^{2k+2}+by^{4k+2}$	—	$12k+4$	$3k-1$
S_{12k+5}	$\varphi+y^{4k+2}+axy^{3k+2}+bzy^{2k+2}$	—	$12k+5$	$3k-1$

In these formulae $\varphi = x^2z + yz^2$, $a = a_0 + \ldots + a_{k-2}y^{k-2}$ for $k > 1$, $a = 0$ for $k = 1$; $b = b_0 + \ldots + b_{2k-1}y^{2k-1}$, $c = c_0 + \ldots + c_{2k-2}y^{2k-2}$.

The singularities $S^*_k(k > 1)$ are divided into classes as follows:

$$\begin{array}{ccccc} & & SQ_k & & \\ & \swarrow & & \nwarrow & \\ S^*_k = \leftarrow S^*_{k,0} \leftarrow & SP_k & & & (S_k), \\ & \nwarrow & & \swarrow & \\ & & SR_k & & \end{array}$$

$$\mu(S^*_{k,0}) = 12k-4,\quad m(S^*_{k,0}, \ldots, SR_k) = 3k-2,\quad \operatorname{codim} S^*_k = 9k-3.$$

III. THE SERIES U. The singularities with 3-jet $x^3 + xz^2$ form a single infinite series of classes

$$U = \leftarrow U_1 \leftarrow U_2 \leftarrow \ldots,$$

where

$$\begin{array}{l} U_{\mathbf{k}} = \leftarrow U_{12k} \leftarrow U_{k,0} \leftarrow U_{k,1} \leftarrow U_{12k+4} \leftarrow U^*_{\mathbf{k}+1} \leftarrow (U_{\mathbf{k}+1}). \\ \qquad\qquad\qquad\qquad\quad \uparrow \\ \qquad\qquad\qquad\qquad\quad U_{k,2} \leftarrow \ldots \end{array}$$

Notation	Normal form	Restrictions	Multiplicity μ	Modality m
U_{12k}	$\varphi + y^{3k+1} + axy^{2k+1} + bzy^{2k+1} + dx^2y^{k+1}$	—	$12k$	$4k-3$
$U_{k,2q}$	$\varphi + xy^{2k+1} + ax^2y^{k+1} + by^{3k+2+q} + czy^{2k+1+q}$	$q \geqslant 0,\ c_0 \neq 0$	$12k+2+2q$	$4k-2$
$U_{k,2q-1}$	$\varphi + xy^{2k+1} + ax^2y^{k+1} + bzy^{2k+1+q} + cz^2y^{k+q}$	$q > 0,\ c_0 \neq 0$	$12k+1+2q$	$4k-2$
U_{12k+4}	$\varphi + y^{3k+2} + axy^{2k+2} + bzy^{2k+2} + cx^2y^{k+1}$	—	$12k+4$	$4k-2$

In these formulae $\varphi = x^3 + xz^2$; $c_0^2 + 1 \neq 0$ for $q = 0$; everywhere,

$$a = a_0 + \ldots + a_{k-2}y^{k-2} \quad \text{for} \quad k > 1, \quad a = 0 \quad \text{for} \quad k = 1;$$

$$b = b_0 + \ldots + b_{k-2}y^{k-2} \quad \text{for} \quad k > 1, \quad b = 0 \quad \text{for} \quad k = 1;$$

$$c = c_0 + \ldots + c_{2k-1}y^{2k-1}, \quad d = d_0 + \ldots + d_{2k-2}y^{2k-2}.$$

The singularities $U^*_{\mathbf{k}}$ $(k > 1)$ are divided into classes:

$$\begin{array}{ccccccc} U^*_{\mathbf{k}} = \leftarrow U^*_{k,0} \leftarrow & UP_{\mathbf{k}} & \leftarrow & UR_{\mathbf{k}} & \leftarrow & UT_{\mathbf{k}} \\ & \uparrow & & \uparrow & & \uparrow \\ & UQ_{\mathbf{k}} & \leftarrow & US_{\mathbf{k}} & \leftarrow & (U_{\mathbf{k}}), \end{array}$$

$$\mu(U^*_{k,0}) = 12k - 4, \quad m(U^*_{k,0}, \ldots, UT_{\mathbf{k}}) = 4k - 3,$$

$$\operatorname{codim}(U^*_{\mathbf{k}}) = 8k - 2.$$

IV. THE SERIES V.

The singularities of corank 3 with 3-jet x^2y are divided into classes

$$\begin{array}{ccccc} & & V_{1,1} \leftarrow V_{1,2} \leftarrow \ldots & & \\ & \swarrow & \nwarrow & & \\ V = V_{1,0} & & & V^*, & \\ & \nwarrow & \swarrow & & \\ & & V^\#_{1,1} \leftarrow V^\#_{1,2} \leftarrow \ldots & & \end{array}$$

,

where

Notation	Normal form	Restrictions	Multiplicity μ	Modality m
$V_{1,0}$	$x^2y+z^4+az^3y+bz^2y^2+zy^3$	$\Delta(a, b_0)\neq 0$	15	3
$V_{1,p}$	$x^2y+z^4+bz^3y+z^2y^2+ay^{4+p}$	$b^2\neq 4,\ a_0\neq 0$	$15+p$	3
$V^{\#}_{1,2q-1}$	$x^2y+z^3y+ay^2z^2+y^4+bxz^{2+q}$	$4a^3+27\neq 0,$ $b_0\neq 0$	$14+2q$	3
$V^{\#}_{1,2q}$	$x^2y+z^3y+ay^2z^2+y^4+bz^{4+q}$	$4a^3+27\neq 0,$ $b_0\neq 0$	$15+2q$	3

Here $p > 0$, $q > 0$, $a = a_0 + a_1y$, $b = b_0 + b_1z$. The singularities of the class V^* satisfy the conditions $\mu(V^*) \geqslant 17$, $m(V^*) \geqslant 3$, $c(V^*) = 13$.

V. OTHER SINGULARITIES

All the singularities whose normal forms are not given above belong to the following seven classes:

Notation	Corank	Adjacencies	Determination	$c \geqslant$	$\mu \geqslant$	$m \geqslant$	Theorem
N	2	$N \to W_{13}$	$j_4=0$	12	16	3	47—49
S^*	3	$S^*_k \to S_{12k-7}$	see Theorem 77	15	20	4	77—81
U^*	3	$U^*_k \to U_{12k-8}$	see Theorem 90	14	20	5	90—96
V^*	3	$V^* \to V_{1,1} \cup V^{\#}_{1,1}$	see Theorem 98	13	17	3	97—102
V'	3	$V' \to V$	$j_3=x^3$	13	18	4	103
V''	3	$V'' \to V'$	$j_3=0$	16	27	10	104
O	>3	$O \to T_{4,4,4}$	corank $\geqslant 4$	10	16	5	105

Here $k \geqslant 2$. The normal form for singularities of the class O, excepting a set of codimension $c = 11$, is indicated in [7].

§ 14. Determination of singularities

To shorten the statements of the theorems in § 14 we use the following *notation*:

f is the germ of a holomorphic function at an *isolated* critical point O of multiplicity μ, or its Taylor series at O, or a formal series in the variables x, y or x, y, z having finite μ.

$f \sim g$ means that the germs f and g at O are equivalent (that is, there exists a germ of a diffeomorphism or a formal series h such that $f = g \circ h$).

$\Rightarrow$ means "implies".

$\mapsto$ means "see" (references of the form $\mapsto i$ are not parts of the statements of the theorems; they indicate the number of the theorem where the singularities of that class are classified).

$j_k f$ means the k-jet of f at O (or the Taylor polynomial of order k at O).

A, . . . , **Z** are the stable equivalence classes of germs of functions that are defined in §13.

$m(f)$ is the modality of the germ f at O.

$c(f)$ is the codimension of the stratum μ = const of the germ of f in the space of germs of functions with critical point O and critical value 0.

$c(\mathbf{K})$ is the codimension of the class **K** in that space.

$j_{\{x^{m_i}\}}f$ is the quasijet of f at O, defined by the monomials x^{m_i} (or the corresponding Taylor polynomial).

$j_{\{x^{m_i}\}}f \approx g$ denotes quasihomogeneous equivalence of jets or Taylor polynomials.

j^*, φ – the meaning of these symbols for Theorems **58–65**, **66–81**, **82–89**, **98–102** is explained before the first theorem of each group.

Δ is the discriminant. In Theorems **36**, **37**, **47**, **48**, **98**, and **99**, $\Delta = 4(a^3 + b^3) + 27 - a^2b^2 - 18ab$.

Clarifications. A system of n monomials $\{x^{m_i}\}$ in $x_1, \ldots, x_n$ with independent exponents $m_i \in \mathbf{Z}^n \subset \mathbf{R}^n$ defines a hyperplane $\Gamma \subset \mathbf{R}^n$, $\Gamma = \{m\colon (\alpha, m) = 1\}$. If all the components α_i of the vector α are positive, then α is called the *quasihomogeneity type*, and the number (α, m) is the *order* of the monomial x^m. The polynomial $f = \Sigma f_m x^m$ is quasihomogeneous of degree d and type α if $(\alpha, m) = d \;\forall\, m{:}f_m \neq 0$.

The quasihomogeneity type defines a grading and a decreasing ring filtration $\mathscr{E}_0 \supset \ldots$, where

$$\mathscr{E}_d = \{f\colon (\alpha, m) \geqslant d \,\forall m\colon f_m \neq 0\}.$$

The factor-space $\mathscr{E}_0 / \bigcup \mathscr{E}_d$, $d > 1$, is called a *space of quasijets* defined by the monomials $\{x^{m_i}\}$ (or defined by the quasihomogeneity type α). For a fixed coordinate system quasijets can be identified with polynomials whose monomials are all of order at most 1 (that is, whose exponents lie on Γ or on the same side of Γ as O).

Quasihomogeneous diffeomorphisms are diffeomorphisms of $\mathbf{C}^n$ that preserve the grading of the ring $\mathbf{C}[[x_1, \ldots, x_n]]$. The Lie group of quasihomogeneous diffeomorphisms acts on the spaces of quasijets and on the spaces of quasihomogeneous polynomials. *Quasihomogeneous equivalence* refers to membership in a single orbit of this action.

DETERMINATION

1. $\mu(f) < \infty \Rightarrow$ one of four cases:

$$\begin{aligned} \text{corank } f \leqslant 1 &\models\!\Rightarrow \mathbf{2};\\ = 2 &\models\!\Rightarrow \mathbf{3};\\ = 3 &\models\!\Rightarrow \mathbf{50};\\ > 3 &\models\!\Rightarrow \mathbf{105}.\end{aligned}$$

2. corank $f \leqslant 1 \Rightarrow f \in A_k$ $(k \geqslant 1)$.

In Theorems 3–49, $f \in \mathbf{C}[[x, y]]$.

3. $j_2 f = 0 \Rightarrow$ one of four cases:

$$\begin{aligned} j_3 f &\approx x^2y + y^3 &&\models\!\Rightarrow \mathbf{4};\\ &\approx x^2y &&\models\!\Rightarrow \mathbf{5};\\ &\approx x^3 &&\models\!\Rightarrow \mathbf{6_1};\\ &= 0 &&\models\!\Rightarrow \mathbf{13}. \end{aligned}$$

4. $j_3 f = x^2y + y^3 \Rightarrow f \in D_4$.

5. $j_3 f = x^2 y \Rightarrow f \in D_k$ $(k > 4)$.

In Theorems 6–9 the number $k \geqslant 1$.

$6_{\mathbf{k}}$. $j_{x^3, y^{3k}} f(x, y) = x^3 \Rightarrow$ one of four cases:

$$\begin{aligned} j_{x^3,\, y^{3k+1}} f &\approx x^3 + y^{3k+1} &&\models\!\Rightarrow \mathbf{7_k};\\ j_{x^3,\, xy^{2k+1}} f &\approx x^3 + xy^{2k+1} &&\models\!\Rightarrow \mathbf{8_k};\\ j_{x^3,\, y^{3k+2}} f &\approx x^3 + y^{3k+2} &&\models\!\Rightarrow \mathbf{9_k};\\ j_{x^3,\, y^{3k+2}} f &= x^3 &&\models\!\Rightarrow \mathbf{10_{k+1}}. \end{aligned}$$

$7_{\mathbf{k}}$. $j_{x^3, y^{3k+1}} f = x^3 + y^{3k+1} \Rightarrow f \in E_{6k}$.

$8_{\mathbf{k}}$. $j_{x^3, y^{2k+1}} f = x^3 + xy^{2k+1} \Rightarrow f \in E_{6k+1}$.

$9_{\mathbf{k}}$. $j_{x^3, y^{3k+2}} f = x^3 + y^{3k+2} \Rightarrow f \in E_{6k+2}$.

In Theorems 10–12, $k > 1$.

$10_{\mathbf{k}}$. $j_{x^3, y^{3k-1}} f = x^3 \Rightarrow$ one of three cases:

$$\begin{aligned} j_{x^3,\, y^{3k}} f &\approx x^3 + ax^2y^k + y^{3k},\ 4a^3 + 27 \neq 0 &&\models\!\Rightarrow \mathbf{11_k};\\ &\approx x^3 + x^2y^k &&\models\!\Rightarrow \mathbf{12_k};\\ &\approx x^3 &&\models\!\Rightarrow \mathbf{6_k}. \end{aligned}$$

$11_{\mathbf{k}}$. $j_{x^3, y^{3k}} f = x^3 + ax^2y^k + y^{3k}$, $4a^3 + 27 \neq 0 \Rightarrow f \in J_{k,0}$.

$12_{\mathbf{k}}$. $j_{x^3, y^{3k}} f = x^3 + x^2y^k \Rightarrow f \in J_{k,p}$ $(p > 0)$.

The series X_1

13. $j_3 f(x, y) = 0 \Rightarrow$ one of six cases:

$$\begin{aligned} j_4 &\approx x^4 + ax^2y^2 + y^4,\ a^2 \neq 4 &&\models\!\Rightarrow \mathbf{14};\\ &\approx x^4 + x^2y^2 &&\models\!\Rightarrow \mathbf{15};\\ &\approx x^2y^2 &&\models\!\Rightarrow \mathbf{16};\\ &\approx x^3y &&\models\!\Rightarrow \mathbf{17};\\ &\approx x^4 &&\models\!\Rightarrow \mathbf{25};\\ &= 0 &&\models\!\Rightarrow \mathbf{47}. \end{aligned}$$

14. $j_4 f = x^4 + ax^2y^2 + y^4$, $a^2 \neq 4 \Rightarrow f \in X_9 = X_{1,0} = T_{2,4,4}$.

15. $j_4 f = x^4 + x^2y^2 \Rightarrow f \in X_{1,p} = T_{2,4,4+p}$ $(p > 0)$.

16. $j_4 f = x^2y^2 \Rightarrow f \in Y^1_{p,q} = T_{2,4+p,4+q}$ $(p \geqslant q > 0)$.

17. $j_4 f = x^3y \Rightarrow j_{x^3y, y^4} f = x^3y \Rightarrow \mathbf{18_1}$.

In Theorems 18–21, $p \geqslant 1$.

$18_{\mathbf{p}}$. $j_{x^3y, y^{3p+1}} f = x^3y \Rightarrow$ one of four cases:

$$j_{x^3y,\, y^{3p+2}} f \approx x^3y + y^{3p+2} \;|\!\Rightarrow \mathbf{19_p};$$
$$j_{x^3y,\, xy^{2p+2}} f \approx x^3y + xy^{2p+2} \;|\!\Rightarrow \mathbf{20_p};$$
$$j_{x^3y,\, y^{3p+3}} f \approx x^3y + y^{3p+3} \;|\!\Rightarrow \mathbf{21_p};$$
$$j_{x^3y,\, y^{3p+3}} f = x^3y \;|\!\Rightarrow \mathbf{22_{p+1}}.$$

19_p. $j_{x^3y, y^{3p+2}} f = x^3y + y^{3p+2} \Rightarrow f \in Z_{6p+5}$.

20_p. $j_{x^3y, xy^{2p+2}} f = x^3y + xy^{2p+2} \Rightarrow f \in Z_{6p+6}$.

21_p. $j_{x^3y, y^{3p+3}} f = x^3y + y^{3p+3} \Rightarrow f \in Z_{6p+7}$.

In Theorems 22–24, $p > 1$.

22_p. $j_{x^3y, y^{3p}} f = x^3y \Rightarrow$ one of three cases:

$$j_{x^3y,\, y^{3p+1}} f \approx y(x^3 + bx^2y^p + y^{3p}), \quad 4b^3 + 27 \neq 0 \;|\!\Rightarrow \mathbf{23_p};$$
$$\approx y(x^3 + x^2y^p) \;|\!\Rightarrow \mathbf{24_p};$$
$$\approx x^3y \;|\!\Rightarrow \mathbf{18_p}.$$

23_p. $j_{x^3y, y^{3p+1}} f = y(x^3 + bx^2y^p + y^{3p}), 4b^3 + 27 \neq 0 \Rightarrow f \in Z_{p-1,0}$.

24_p. $j_{x^3y, y^{3p+1}} f = y(x^3 + x^2y^p) \Rightarrow f \in Z_{p-1,r} (r > 0)$.

The series W.

25. $j_4 f(x, y) = x^4 \Rightarrow j_{x^4, y^4} f = x^4 \Rightarrow 26_1$.

In Theorems 26–35, $k \geqslant 1$.

26_k. $j_{x^4, y^{4k}} f = x^4 \Rightarrow$ one of three cases:

$$j_{x^4,\, y^{4k+1}} f \approx x^4 + y^{4k+1} \;|\!\Rightarrow \mathbf{27_k};$$
$$j_{x^4,\, xy^{3k+1}} f \approx x^4 + xy^{3k+1} \;|\!\Rightarrow \mathbf{28_k};$$
$$j_{x^4,\, xy^{3k+1}} f = x^4 \;|\!\Rightarrow \mathbf{29_k}.$$

27_k. $j_{x^4, y^{4k+1}} f = x^4 + y^{4k+1} \Rightarrow f \in W_{12k}$.

28_k. $j_{x^4, xy^{3k+1}} f = x^4 + xy^{3k+1} \Rightarrow f \in W_{12k+1}$.

29_k. $j_{x^4, xy^{3k+1}} f = x^4 \Rightarrow$ one of four cases:

$$j_{x^4,\, y^{4k+2}} f \approx x^4 + bx^2y^{2k+1} + y^{4k+2}, \; b^2 \neq 4 \;|\!\Rightarrow \mathbf{30_k};$$
$$\approx x^4 + x^2y^{2k+1} \;|\!\Rightarrow \mathbf{31_k};$$
$$\approx (x^2 + y^{2k+1})^2 \;|\!\Rightarrow \mathbf{32_k};$$
$$= x^4 \;|\!\Rightarrow \mathbf{33_k}.$$

30_k. $j_{x^4, y^{4k+2}} f = x^4 + bx^2y^{2k+1} + y^{4k+2}, b^2 \neq 4 \Rightarrow f \in W_{k,0}$.

31_k. $j_{x^4, y^{4k+2}} f = x^4 + x^2y^{2k+1} \Rightarrow f \in W_{k,i} \; (i > 0)$.

32_k. $j_{x^4, y^{4k+2}} f = (x^2 + y^{2k+1})^2 \Rightarrow f \in W_{k,i} \; (i > 0)$.

33_k. $j_{x^4, y^{4k+2}} f = x^4 \Rightarrow$ one of three cases:

$$j_{x^4,\, xy^{3k+2}} f \approx x^4 + xy^{3k+2} \;|\!\Rightarrow \mathbf{34_k};$$
$$j_{x^4,\, y^{4k+3}} f \approx x^4 + y^{4k+3} \;|\!\Rightarrow \mathbf{35_k};$$
$$j_{x^4,\, y^{4k+3}} f = x^4 \;|\!\Rightarrow \mathbf{36_{k+1}}.$$

34_k. $j_{x^4, xy^{3k+2}} f = x^4 + xy^{3k+2} \Rightarrow f \in W_{12k+5}$.

35_k. $j_{x^4, y^{4k+3}} f = x^4 + y^{4k+3} \Rightarrow f \in W_{12k+6}$.

In Theorems 36–46, $k > 1$.

The series X_k

36$_k$. $j_{x^4, y^{4k-1}} f = x^4 \Rightarrow$ one of five cases:

$$j_{x^4, y^{4k}} f \approx x^4 + bx^3y^k + ax^2y^{2k} + xy^{3k}, \ \Delta \neq 0, \ ab \neq 9 \models \mathbf{37_k};$$
$$\approx x^2(x^2 + axy^k + y^{2k}), \ a^2 \neq 4 \models \mathbf{38_k};$$
$$\approx x^2(x + y^k)^2 \models \mathbf{39_k};$$
$$\approx x^3(x + y^k) \models \mathbf{40_k};$$
$$\approx x^4 \models \mathbf{26_k}.$$

37$_k$. $j_{x^4, y^{4k}} f = x^4 + bx^3y^k + ax^2y^{2k} + xy^{3k}$, $\Delta \neq 0 \Rightarrow f \in X_{k,0}$.

38$_k$. $j_{x^4, y^{4k}} f = x^2(x^2 + axy^k + y^{2k})$, $a^2 \neq 4 \Rightarrow f \in X_{k,p}$ $(p > 0)$.

39$_k$. $j_{x^4, y^{4k}} f = x^2(x^2 + y^k)^2 \Rightarrow f \in Y^k_{r,s}$ $(1 \leqslant s \leqslant r)$.

40$_k$. $j_{x^4, y^{4k}} f = (x + y^k)x^3 \Rightarrow f \sim f_1 f_2$, where

$$j_{x, y^k} f_1 \approx x + y^k, \quad j_{x^3, y^{3k}} f_2 = x^3 \models \mathbf{41_k}.$$

In Theorems 41–44, $i \geqslant 0$, $p > 0$.

41$_k$. $j_{x^3, y^{3k}} f_2 = x^3 \Rightarrow$ one of five cases:

$$f_2 \in E_{6(k+i)} \models \mathbf{42_{k,\,i}};$$
$$f_2 \in E_{6(k+i)+1} \models \mathbf{43_{k,\,i}};$$
$$f_2 \in E_{6(k+i)+2} \models \mathbf{44_{k,\,i}};$$
$$f_2 \in J_{k+i+1,\,0} \models \mathbf{45_{k,\,i+1}};$$
$$f_2 \in J_{k+i+1,\,p} \models \mathbf{46_{k,\,i+1,\,p}}.$$

In Theorems 42–46, $f(x, y) = f_1 f_2$, where $j_{x,y^k} f_1 \approx x + y^k$ and $j_{x^3,y^{3k}} f_2 = x^3$.

42$_{k,i}$. $f_2 \in E_{6(k+i)} \Rightarrow f \in Z^k_{12k+6i-1}$.

43$_{k,i}$. $f_2 \in E_{6(k+i)+1} \Rightarrow f \in Z^k_{12k+6i}$.

44$_{k,i}$. $f_2 \in E_{6(k+i)+2} \Rightarrow f \in Z^k_{12k+6i+1}$.

In Theorems 45–46, $i \geqslant 1$, $p > 0$.

45$_{k,i}$. $f_2 \in J_{k+i,0} \Rightarrow f \in Z^k_{i,0}$.

46$_{k,i,p}$. $f_2 \in J_{k+i,p} \Rightarrow f \in Z^k_{i,p}$.

47. $j_4 f = 0 \Rightarrow$ one of two cases:

$$j_5 f \approx x^4y + ax^3y^2 + bx^2y^3 + xy^4, \ \Delta \neq 0, \ ab \neq 9 \models \mathbf{48};$$
$$j_5 f \text{ is degenerate } \models \mathbf{49}.$$

48. $j_5 f = x^4y + ax^3y^2 + bx^2y^3 + xy^4$, $\Delta \neq 0 \Rightarrow f \in N_{16}$, i.e.
$f \sim x^4y + ax^3y^2 + bx^2y^3 + xy^4 + cx^3y^3$, $\Delta \neq 0$, $ab \neq 9$; $\mu(f) = 16$, $m(f) = 3$, $c(f) = 12$.

49. $j_5 f$ is degenerate $\Rightarrow \mu(f) > 16$, $m(f) > 2$, $c(f) > 12$.

SINGULARITIES OF CORANK 3.

In Theorems 50–104, $f \in \mathbf{C}[[x, y, z]]$.

50. $j_2 f(x, y, z) = 0 \Rightarrow$ one of ten cases:

$$
\begin{aligned}
j_3 f &\approx x^3+y^3+z^3+axyz, \quad a^3+27\neq 0 \models\!\!\Rightarrow 51;\\
&\approx x^3+y^3+xyz \models\!\!\Rightarrow 52 \text{ (series } \mathbf{P});\\
&\approx x^3+xyz \quad \models\!\!\Rightarrow 54 \text{ (series } \mathbf{R});\\
&\approx xyz \quad \models\!\!\Rightarrow 56 \text{ (series } \mathbf{T});\\
&\approx x^3+yz^2 \quad \models\!\!\Rightarrow 58 \text{ (series } \mathbf{Q});\\
&\approx x^2z+yz^2 \quad \models\!\!\Rightarrow 66 \text{ (series } \mathbf{S});\\
&\approx x^3+xz^2 \quad \models\!\!\Rightarrow 82 \text{ (series } \mathbf{U});\\
&\approx x^2y \quad \models\!\!\Rightarrow 97 \text{ (class } \mathbf{V});\\
&\approx x^3 \quad \models\!\!\Rightarrow 103;\\
&= 0 \quad \models\!\!\Rightarrow 104.
\end{aligned}
$$

Series T

51. $j_3 f = x^3+y^3+z^3+axyz,\ a^3+27\neq 0 \Rightarrow f\in P_8 = T_{3,3,3}$.

52. $j_3 f = x^3+y^3+xyz \Rightarrow f\sim x^3+y^3+xyz+\alpha(z),\ j_3(\alpha)=0 \mapsto$ **53**.

53. $f = x^3+y^3+xyz+\alpha(z),\ j_3(\alpha)=0 \Rightarrow f\in P_{p+5} = T_{3,3,3+p}\ (p>3)$.

54. $j_3 f = x^3+xyz \Rightarrow f\sim x^3+xyz+\alpha(y)+\beta(z),\ j_3(\alpha,\beta)=0 \mapsto$ **55**.

55. $f = x^3+xyz+\alpha(y)+\beta(z),\ j_3(\alpha,\beta)=0 \Rightarrow f\in R_{p,q} = T_{3,p,q}\ (q\geqslant p>3)$.

56. $j_3 f = xyz \Rightarrow f\sim xyz+\alpha(x)+\beta(y)+\gamma(z),\ j_3(\alpha,\beta,\gamma)=0 \mapsto$ **57**.

57. $f = xyz+\alpha(x)+\beta(y)+\gamma(z),\ j_3(\alpha,\beta,\gamma)=0 \Rightarrow f\in T_{p,q,r}\ (r\geqslant q\geqslant p>3)$.

Series Q. In Theorems **58–65**, $\varphi = x^3+yz^2,\ j^*_\lambda = j_{yz^2,x^3,\lambda}$ (λ is a monomial).

58. $j_3 f = \varphi \Rightarrow f = \varphi+\alpha(y)+x\beta(y),\ j_3(\alpha, x\beta)=0 \mapsto$ **59_1**.

In Theorems **59–62**, $k\geqslant 1$.

59_k. $f = \varphi+\alpha(y)+x\beta(y),\ j^*_{y^{3k}} f = \varphi \Rightarrow$ one of four cases:

$$
\begin{aligned}
j^*_{y^{3k+1}} f &\approx \varphi+y^{3k+1} \models\!\!\Rightarrow 60_k;\\
j^*_{xy^{2k+1}} f &\approx \varphi+xy^{2k+1} \models\!\!\Rightarrow 61_k;\\
j^*_{y^{3k+2}} f &\approx \varphi+y^{3k+2} \models\!\!\Rightarrow 62_k;\\
j^*_{y^{3k+2}} f &= \varphi \models\!\!\Rightarrow 63_{k+1}.
\end{aligned}
$$

60_k. $j^*_{y^{3k+1}} f = \varphi + y^{3k+1} \Rightarrow f\in Q_{6k+4}$.

61_k. $j^*_{xy^{2k+1}} f = \varphi + xy^{2k+1} \Rightarrow f\in Q_{6k+5}$.

62_k. $j^*_{y^{3k+2}} f = \varphi + y^{3k+2} \Rightarrow f\in Q_{6k+6}$.

In Theorems **63–65**, $k>1$.

63_k. $f = \varphi+\alpha(y)+\beta(y),\ j^*_{y^{3k-1}} f = \varphi \Rightarrow$ one of three cases:

$$
\begin{aligned}
j^*_{y^{3k}} f &\approx \varphi+ax^2y^k+xy^{2k}, \quad a^2\neq 4 \models\!\!\Rightarrow 64_k;\\
&\approx \varphi+x^2y^k \models\!\!\Rightarrow 65_k;\\
&= \varphi \models\!\!\Rightarrow 59_k.
\end{aligned}
$$

64_k. $j^*_{y^{3k}} f = \varphi + ax^2y^k + xy^{2k},\ a^2\neq 4 \Rightarrow f\in Q_{k,0}$.

65_k. $j^*_{y^{3k}} f = \varphi + x^2y^k \Rightarrow f\in Q_{k,i}\ (i>0)$.

Series S. In Theorems **68–81**, $\varphi = x^2z+yz^2,\ j^*_\lambda = j_{x^2z,yz^2,\lambda}$ (λ is a monomial).

66. $j_3 f = \varphi \Rightarrow f \sim \varphi + \alpha(y) + x\beta(y) + z\gamma(y),\ j_3(\alpha, x\beta, z\gamma) = 0 \Rightarrow 67_1$.

In Theorems 67–76, $k \geqslant 1$.

67_k. $f = \varphi + \alpha(y) + x\beta(y) + z\gamma(y),\ j^*_{y^{4k-1}} f = \varphi \Rightarrow$ one of 3 cases:

$$j^*_{y^{4k}} f \approx \varphi + y^{4k} \quad |\Rightarrow 68_k;$$
$$j^*_{xy^{3k}} f \approx \varphi + xy^{3k} \quad |\Rightarrow 69_k;$$
$$j^*_{xy^{3k}} f = \varphi \quad |\Rightarrow 70_k.$$

68_k. $j^*_{y^{4k}} f = \varphi + y^{4k} \Rightarrow f \in S_{12k-1}$.

69_k. $j^*_{xy^{3k}} f = \varphi + xy^{3k} \Rightarrow f \in S_{12k}$.

70_k. $f = \varphi + \alpha(y) + x\beta(y) + z\gamma(y),\ j^*_{xy^{3k}} f = \varphi \Rightarrow$ one of four cases:

$$j^*_{y^{4k+1}} f \approx \varphi + y^{4k+1} + bzy^{2k+1}, \quad b^2 \neq 4 \ |\Rightarrow 71_k;$$
$$\approx \varphi + x^2 y^{2k} \quad |\Rightarrow 72_k;$$
$$\approx \varphi + zy^{2k+1} \ |\Rightarrow 73_k;$$
$$= \varphi \quad |\Rightarrow 74_k.$$

71_k. $j^*_{y^{4k+1}} f = \varphi + y^{4k+1} + bzy^{2k+1},\ b^2 \neq 4 \Rightarrow f \in S_{k,0}$.

72_k. $j^*_{y^{4k+1}} f = \varphi + x^2 y^{2k} \Rightarrow f \in S_{k,i}\ (i > 0)$.

73_k. $j^*_{y^{4k+1}} f = \varphi + zy^{2k+1} \Rightarrow f \in S^\#_{k,i}\ (i > 0)$.

74_k. $f = \varphi + \alpha(y) + x\beta(y) + z\gamma(y),\ j^*_{y^{4k+1}} f = \varphi \Rightarrow$ one of three cases:

$$j^*_{xy^{3k+1}} f = \varphi + xy^{3k+1} \ |\Rightarrow 75_k;$$
$$j^*_{y^{4k+2}} f = \varphi + y^{4k+2} \quad |\Rightarrow 76_k;$$
$$j^*_{y^{4k+2}} f = \varphi \quad |\Rightarrow 77_{k+1}.$$

75_k. $j^*_{xy^{3k+1}} f = \varphi + xy^{3k+1} \Rightarrow f \in S_{12k+4}$.

76_k. $j^*_{y^{4k+2}} f = \varphi + y^{4k+2} \Rightarrow f \in S_{12k+5}$.

In Theorems 77–81, $k > 1$.

77_k. $f = \varphi + \alpha(y) + x\beta(y) + z\gamma(y),\ j^*_{y^{4k+2}} f = \varphi \Rightarrow$ one of five cases:

$$j^*_{y^{4k-1}} f \approx \varphi + ax^2 y^{k-1} + bxy^k z + xy^{3k-1}, \quad \Delta \neq 0 \ |\Rightarrow 78_k;$$
$$\approx \varphi + xy^k z + ax^3 y^{k-1}, \quad a^2 \neq a \ |\Rightarrow 79_k;$$
$$\approx \varphi + x^3 y^{k-1} \ |\Rightarrow 80_k;$$
$$\approx \varphi + xy^k z \quad |\Rightarrow 81_k;$$
$$= \varphi \quad |\Rightarrow 67_k.$$

78_k. $j^*_{y^{4k-1}} f = \varphi + ax^2 y^{k-1} + bxy^k z + xy^{3k-1},\ \Delta \neq 0 \Rightarrow f \in S^*_{k,0}$; $\mu(f) = 12k-4$, $m(f) = 3k-2$, $c(S^*_{k,0}) = 9k-3$.

79_k. $j^*_{y^{4k-1}} f = \varphi + xy^k z + ax^3 y^{k-1},\ a^2 \neq a \Rightarrow f \in SP_k$; $\mu(f) \geqslant 12k-3$, $m(f) = 3k-2$, $c(SP_k) = 9k-2$.

80_k. $j^*_{y^{4k-1}} f = \varphi + x^3 y^{k-1} \Rightarrow f \in SQ_k$; $\mu(f) \geqslant 12k-2$, $m(f) = 3k-2$, $c(SQ_k) = 9k-1$.

81_k. $j^*_{y^{4k-1}} f = \varphi + xy^k z \Rightarrow f \in SR_k$; $\mu(f) \geqslant 12k-2$, $m(f) = 3k-2$, $c(SR_k) = 9k-1$.

The *series U*.

In Theorems **82–89**, $\varphi = x^3 + xz^2$, $j^*_\lambda = j_{x^3, z^3, \lambda}$ (λ is a monomial).

82. $j_3 f = \varphi \Rightarrow f \sim \varphi + \alpha(y) + x\beta(y) + z\gamma(y) + x^2\delta(y)$, $j_3(\alpha, x\beta, z\gamma, x^2\delta) = 0$ $\Rightarrow \mathbf{83_1}$.

In Theorems **83–89**, $k \geqslant 1$.

$\mathbf{83_k}$. $f = \varphi + \alpha(y) + x\beta(y) + z\gamma(y) + x^2\delta(y)$, $j^*_{y^{3k}} f = \varphi \Rightarrow$ one of 2 cases:

$$\begin{aligned} j^*_{y^{3k+1}} f &\approx \varphi + y^{3k+1} &&\models \mathbf{84_k}, \\ &= \varphi &&\models \mathbf{85_k}. \end{aligned}$$

$\mathbf{84_k}$. $j^*_{y^{3k+1}} f = \varphi + y^{3k+1} \Rightarrow f \in U_{12k}$.

$\mathbf{85_k}$. $f = \varphi + \alpha(y) + x\beta(y) + z\gamma(y) + x^2\delta(y)$, $j^*_{y^{3k+1}} f = \varphi \Rightarrow$ one of 3 cases:

$$\begin{aligned} j^*_{xy^{2k+1}} f &\approx \varphi + xy^{2k+1} + czy^{2k+1}, \quad c(c^2+1) \neq 0 &&\models \mathbf{86_k}; \\ &\approx \varphi + xy^{2k+1} &&\models \mathbf{87_k}; \\ &= \varphi &&\models \mathbf{88_k}. \end{aligned}$$

$\mathbf{86_k}$. $j^*_{xy^{2k+1}} f = \varphi + xy^{2k+1} + czy^{2k+1}$, $c(c^2+1) \neq 0 \Rightarrow f \in U_{k,0}$.

$\mathbf{87_k}$. $j^*_{xy^{2k+1}} f = \varphi + xy^{2k+1} \Rightarrow f \in U_{k,p}$ $(p > 0)$.

$\mathbf{88_k}$. $f = \varphi + \alpha(y) + x\beta(y) + z\gamma(y) + x^2\delta(y)$, $j^*_{xy^{2k+1}} f = \varphi \Rightarrow$ one of 2 cases:

$$\begin{aligned} j^*_{y^{3k+2}} f &\approx \varphi + y^{3k+2} &&\models \mathbf{89_k}; \\ &= \varphi &&\models \mathbf{90_{k+1}}. \end{aligned}$$

$\mathbf{89_k}$. $j^*_{y^{3k+2}} f = \varphi + y^{3k+2} \Rightarrow f \in U_{12k+4}$.

In Theorems **90–96**, $k \geqslant 2$; $\varphi = x^2 z + xz^2$, $j^*_\lambda = j_{x^3, z^3, \lambda}$ (λ is a monomial).

$\mathbf{90_k}$. $f = x^3 + xz^2 + \alpha(y) + x\beta(y) + z\gamma(y) + x^2\delta(y)$, $j^*_{y^{3k-1}} f = x^3 + xz^2 \Rightarrow$ one of seven cases:

$$\begin{aligned} j^*_{y^{3k}} f &\approx \varphi + ax^2y^k + bxy^kz + y^kz^2 + cxy^{2k}, \quad \Delta \neq 0 &&\models \mathbf{91_k}; \\ &\approx \varphi + ax^2y^k + bxy^kz + y^kz^2, 4a \neq b^2, \quad a(a+1-b) \neq 0 &&\models \mathbf{92_k}; \\ &\approx x^3 + ax^2z + xz^2 + y^kz^2, \quad a^2 \neq 4 &&\models \mathbf{93_k}; \\ &\approx \varphi + x^2y^k + axy^kz, \quad a^2 \neq a &&\models \mathbf{94_k}; \\ &\approx \varphi + x^2y^k &&\models \mathbf{95_k}; \\ &\approx \varphi + xy^kz &&\models \mathbf{96_k}; \\ &= \varphi &&\models \mathbf{83_k}. \end{aligned}$$

$\mathbf{91_k}$. $j^*_{y^{3k}} f = \varphi + ax^2y^k + bxy^kz + y^kz^2 + cxy^{2k}$, $\Delta \neq 0 \Rightarrow f \in U^*_{k,0}$; $\mu(f) = 12k - 4$, $m(f) = 4k - 3$, $c(U^*_{k,0}) = 8k - 2$.

$\mathbf{92_k}$. $j^*_{y^{3k}} f = \varphi + ax^2y^k + bxy^kz + y^kz^2$, $4a \neq b^2$, $a(a+1-b) \neq 0 \Rightarrow f \in \boldsymbol{UP_k}$; $\mu(f) \geqslant 12k - 3$, $m(f) = 4k - 3$, $c(\boldsymbol{UP_k}) = 8k - 1$.

$\mathbf{93_k}$. $j^*_{y^{3k}} f = x^3 + ax^2z + xz^2 + y^kz^2$, $a^2 \neq 4 \Rightarrow f \in \boldsymbol{UQ_k}$; $\mu(f) \geqslant 12k - 2$, $m(f) = 4k - 3$, $c(\boldsymbol{UQ_k}) = 8k$.

$\mathbf{94_k}$. $j^*_{y^{3k}} f = \varphi + x^2y^k + axy^kz$, $a^2 \neq a \Rightarrow f \in \boldsymbol{UR_k}$; $\mu(f) \geqslant 12k - 2$, $m(f) = 4k - 3$, $c(\boldsymbol{UR_k}) = 8k$.

$\mathbf{95_k}$. $j^*_{y^{3k}} f = \varphi + x^2y^k \Rightarrow f \in \boldsymbol{US_k}$, $\mu(f) \geqslant 12k - 2$, $m(f) = 4k - 3$, $c(\boldsymbol{US_k}) = 8k + 1$.

96_k. $j^*_{y^{3k}} f = \varphi + xy^k z \Rightarrow f \in UT_k$, $\mu(f) \geqslant 12k - 1$, $m(f) = 4k - 3$, $c(UT_k) = 8k + 1$.

The *Class V*.

97. $j_3 f(x, y, z) = x^2 y \Rightarrow f \sim x^2 y + \alpha(y, z) + x\beta(z) \Rightarrow$ 98.

In Theorems 98 and 102, φ is one of the following 10 polynomials

$$z^4 + z^3 y,\ z^3 y + z^2 y^2,\ z^2 y^2 + zy^3,\ z^4 + z^2 y^2,\ z^4,\ z^3 y,\ z^2 y^2,\ zy^3,\ y^4,\ 0.$$

98. $f = x^2 y + \alpha(y, x) + x\beta(z)$, $j_3 f = x^2 y \Rightarrow$ one of four cases:

$$\begin{aligned} j_{x^2y,\, y^4,\, z^4} f &\approx x^2y + z^4 + az^3y + bz^2y^2 + zy^3,\ \Delta \neq 0,\ ab \neq 9 &&\models\Rightarrow 99;\\ &\approx x^2y + z^4 + bz^3y + z^2y^2,\ b^2 \neq 4 &&\models\Rightarrow 100;\\ &\approx x^2y + z^3y + az^2y^2 + y^4,\ 4a^3 + 27 \neq 0 &&\models\Rightarrow 101;\\ &\approx x^2y + \varphi &&\models\Rightarrow 102. \end{aligned}$$

99. $j_{x^2y, y^4, z^4} f = x^2 y + z^4 + az^3 y + bz^2 y^2 + zy^3$, $\Delta \neq 0 \Rightarrow f \in V_{1,0}$.

100. $j_{x^2y, y^4, z^4} f = x^2 y + z^4 + bz^3 y + z^2 y^2$, $b^2 \neq 4 \Rightarrow f \in V_{1,p}$ $p > 0$.

101. $j_{x^2y, y^4, z^4} f = x^2 y + z^3 y + az^2 y^2 + y^4$, $4a^3 + 27 \neq 0 \Rightarrow f \in V^{\#}_{1,p}$ $p > 0$.

102. $j_{x^2y, y^4, z^4} f = x^2 y + \varphi \Rightarrow \mu(f) \geqslant 17$, $m(f) \geqslant 3$, $c(f) \geqslant 13$.

103. $j_3 f(x, y, z) = x^3 \Rightarrow \mu(f) \geqslant 18$, $m(f) \geqslant 4$, $c(f) \geqslant 13$.

104. $j_3 f(x, y, z) = 0 \Rightarrow \mu(f) \geqslant 27$, $m(f) \geqslant 10$, $c(f) \geqslant 16$.

105. corank $f > 3 \Rightarrow \mu(f) \geqslant 16$, $m(f) \geqslant 5$, $c(f) \geqslant 10$.

REMARKS. Proofs of Theorems 1–105 can be found in the following places:

§15–6, 18, 26, 33, 59, 67, 74, 83, 88.

§16–3, 10, 13, 22, 29, 36, 47, 50, 52, 54, 56, 58, 63, 66, 70, 77, 82, 85, 90, 97, 98.

§17–2, 4, 5, 7, 8, 9, 11, 14, 19, 20, 21, 23, 27, 28, 30, 34, 35, 37, 48, 51, 60, 61, 62, 64, 68, 69, 71, 75 ,76, 78, 84, 86, 89, 91, 99.

§18–53, 55, 57.

Theorems 1, 17 and 25 are obvious. Theorems 12_1, 15, 16 are proved in [8].

Theorems 49 and 102–105 are proved by the methods of [8], but we do not give the proofs.

The proofs of Theorems 12_k (k > 1), 24, 31, 32, 38–46, 65, 72, 73, 79, 80, 81, 87, 92–96, 100, 101 are based on new methods compared with those of [8] (using a spectral sequence) and are not carried out in the present article (see [91], [92]). In the classification of unimodal singularities these theorems are not used. A proof of the theorem on the classification of unimodal singularities is obtained by a juxtaposition of the following theorems:

1–5, $6_{1,2}$–$9_{1,2}$, 10_2, 11_2, 13–17, 18_1–21_1, 22_2, 23_2, 25, 26_1–30_1, 36_1, 37_1, 47, 48, 50–58, 59_1–62_1, 63_2, 64_2, 66, 67_1–71_1, 82, 83_1–86_1. 97, 98, 105.

Here Theorems 10, 22, 29, 36, 47, 63, 79, 85, 98 are not used in their entirety (only the first case of each of them is needed).

The classification of bimodal singularities uses, in addition, Theorems **6_3–11_3, 12_2, 18_2–21_2, 22_3, 23_3, 24_2, 31_1–35_1, 36_2, 37_2, 59_2–62_2, 63_3, 64_3, 65_2, 72_1–76_1, 77_2, 78_2, 87_1, 89_1, 90_2, 91_2, 99.**

Here Theorems **10_3, 22_3, 36_2, 63_3, 77_2, 90_2** are not used in their entirety (only the first case of each of them is needed).

§ 15. Proofs of the theorems on the rotation of a ruler (Theorems 6, 18, 26, 33, 59, 67, 74, 83, 88)

The method of the rotation of a ruler, by which these theorems are proved, is due to Newton [59]. Let us prove, for example, Theorem **6_k** According to the hypothesis of this theorem, the support of the function lies strictly above a line joining the points (3,0) and $(0,3k)$, excluding (3,0). We place a ruler on that line, and, holding one end at (3,0), we move the other end to the side, until we meet the exponent of a monomial occurring with a non-zero coefficient (Fig. 7).

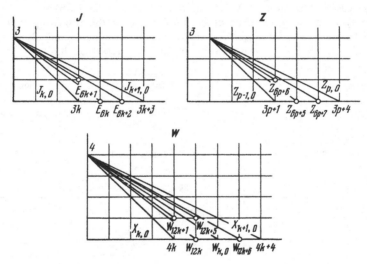

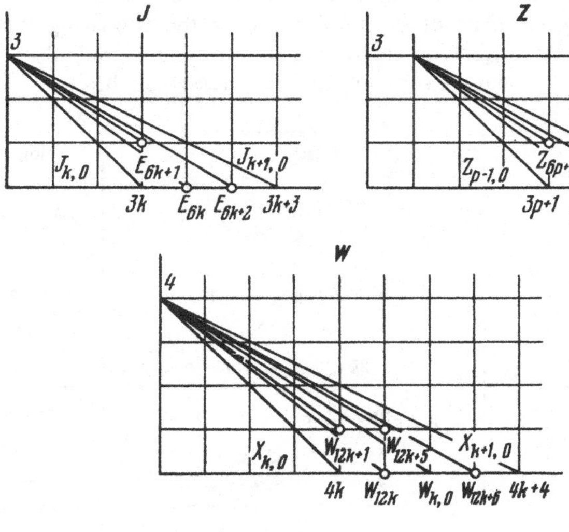

Fig. 7

Inside the triangle between the lines joining (3,0) to $(0,3k)$ and to $(0,3(k+1))$ there are in all three lattice points (in the positive quadrant), namely, $(0,3k+1)$, $(1, 2k+1)$ and $(0,3k+2)$.

If the coefficient of y^{3k+1} is different from 0, then we have the first case of Theorem **6_k**. If the coefficient of y^{3k+1} is 0, but that of xy^{2k+1} is different from zero, then we have the second case. If both monomials y^{3k+1} and xy^{2k+1} do not occur in f, but y^{3k+2} does, then we have the third case. If the coefficients of all three monomials y^{3k+1}, xy^{2k+1}, y^{3k+2}

are 0, then we have the fourth case.

In exactly the same manner the remaining theorems also reduce to the enumeration of the lattice points of one kind or another in the corresponding domains. The edges of the domains in the cases of corank 2 are as follows:

Theorem	Series	Near edge	Far edge	Monomials
6_k	***J***	(3, 0) (0, $3k$)	(3, 0) (0, $3k+3$)	y^{3k+1}, xy^{2k+1}, y^{3k+2}
18_p	***Z***	(3, 1) (0, $3p+1$)	(3, 1) (0, $3p+4$)	y^{3p+2}, xy^{2p+2}, y^{3p+3}
26_k	***W***	(4, 0) (0, $4k$)	(4, 0) (0, $4k+2$)	y^{4k+1}, xy^{3k+1}
33_k	***W***	(4, 0) (0, $4k+2$)	(4, 0) (0, $4k+4$)	xy^{3k+2}, y^{4k+3}

(see Fig. 7; following Newton, we put the exponent of x on the vertical, and that of y on the horizontal axis).

In the cases of corank 3 (Theorems **59**, **67**, **74**, **83** and **88**) we assume that f only contains monomials of definite types. In the following table we indicate the near and far edges for each type of monomial, and also monomials of those types with exponents in the prescribed limits.

Theorem	Series	Type	Edges	Monomials
59	***Q***	y^q xy^q	$3k < q < 3k+3$ $2k < q < 2k+2$	y^{3k+1}, y^{3k+2} xy^{2k+1}
67	***S***	y^q xy^q zy^q	$4k-1 < q < 4k+1$ $3k-1 < q < 3k+\frac{1}{2}$ $2k < q < 2k+1$	y^{4k} xy^{3k} —
74	***S***	xy^q y^q zy^q	$3k+\frac{1}{2} < q < 3k+2$ $4k+1 < q < 4k+3$ $2k+1 < q < 2k+2$	xy^{3k+1} y^{4k+2} —
83	***U***	y^q xy^q, zy^q x^2y^q	$3k < q < 3k+\frac{3}{2}$ $2k < q < 2k+1$ $k < q < k+\frac{1}{2}$	y^{3k+1} — —
88	***U***	y^q xy^q, zy^q x^2y^q	$3k+\frac{3}{2} < q < 3k+3$ $2k+1 < q < 2k+2$ $k+\frac{1}{2} < q < k+1$	y^{3k+2} — —

Let us prove, for example, Theorem **67**$_k$. According to the hypothesis of the theorem, on the plane passing through the exponents of the monomials x^2z, yz^2 and y^{4k-1} and closer to it than to the origin there are no points of the support of f, except two, namely the exponents of the monomials x^2z and yz^2. We join these two points by a line and we rotate the plane around this line, moving the point t of intersection with the axis of y-exponents to the side until we meet the exponent of the support of f. According to the hypothesis of the theorem, f contains only monomials of one of the three types y^q, xy^q and zy^q (excluding x^2z, yz^2). When t moves from $4k - 1$ to $4k + 1$, the plane (which we denote by Π_t) successively passes through the exponents of precisely two monomials of our types, y^{4k} and xy^{3k} (Fig. 8 shows the plane of x- and y-exponents; the line around which we rotate the plane Π_t, intersects it at $(4, -1)$).

If the coefficient of y^{4k} in f is not zero, then we have the first case of Theorem **67**; if it is zero, but the coefficient of xy^{3k} is not 0, then we have the second case; and if the coefficients are both zero, then we have the third case. This proves Theorem **67**; the other theorems (**59**, **74**, **83**, **88**) are proved in the same way.

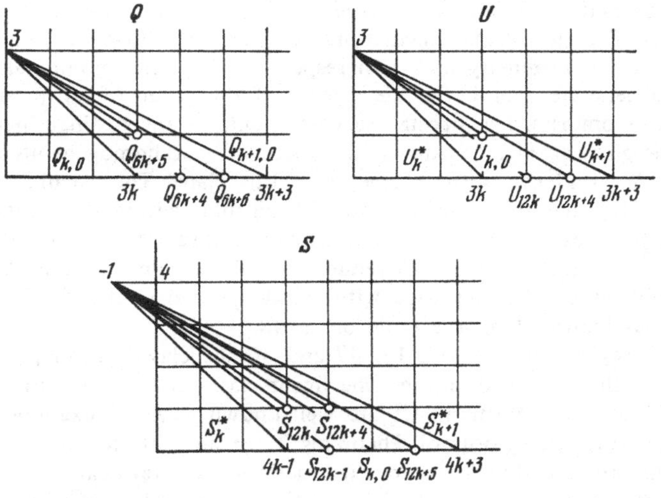

Fig. 8

REMARK. Actually, we have proved somewhat more than is stated in Theorems **6**, . . . , **88**. Namely, in each of these theorems there is a *last case*; in this case we prove that the *support of f is located no closer than the far edge* indicated in the table. (Explanation. In the cases of corank 2 in the table we have indicated two points on the far edge. In the cases of corank 3 the far edge is the plane Π_t, where $t = q_{\max}$ is the upper bound for the exponent q of the monomial y^q indicated in the table.)

EXAMPLE. In proving Theorem 67 we have proved that in the third case the support of f has no points lying on the same side as the origin of the plane passing through the exponents of the monomials x^2z, yz^2, y^{4k+1}.

In accordance with this remark, the proofs of Theorems 10, 22, 29, 36, 63, 70, 77, 85, 90 lead to the classification of quasihomogeneous singularities, for which all the exponents of the monomials lie on the corresponding far edge.

Analogous (but somewhat simpler) arguments with moving a ruler lead to the classification of quasihomogeneous singularities of the proof of Theorems 3, 49, 50, 98.

§16. Normal forms of quasihomogeneous singularities (proofs of Theorems 3, 10, 13, 22, 29, 36, 47, 50, 52, 54, 56, 58, 63, 66, 70, 77, 82, 85, 90, 97, 98)

The remark at the end of §15 reduces the proofs of these theorems to the classification of orbits of the action of the group of quasihomogeneous diffeomorphisms on the space of quasihomogeneous functions with fixed quasihomogeneity type and with fixed coefficients for certain *distinguished* monomials. The run of the calculations is as follows. First we compute the exponents of homogeneity and with respect to them the monomial vector fields that generate over **C** the Lie algebra of the group of all quasihomogeneous diffeomorphisms. Fixing the coefficients of the distinguished monomials gives an affine plane in the space of quasihomogeneous functions. We find the isotropy algebra of this plane. The calculations show (unexpectedly) that the actions of the Lie algebras within the limits of each of our series of singularities are affinely equivalent to each other. Thus, our classification theorems reduce to a finite number of problems. These problems can be identified with simple geometric classification problems, as indicated by the table on p. 48.

In the cases of Theorems 3, 13, 47 and 50 the classification problems coincide by their very definition. The linear classifications of 3-forms in $\mathbf{C}^2$ and $\mathbf{C}^3$ and of 4-forms in $\mathbf{C}^2$ are well known (see, for example, [45], [64]), therefore, the proofs of Theorems 3, 13 and 50 are not given here. I only remark that Theorem 50 describes the stratification by singularities shown in Fig.9 on p.48 of the 10-dimensional space of cubic forms.

The stratifications of the spaces of cubic curves in Theorems 63, 70, 77, 85 and 90 use different fragments of this diagram.

The identification of our classification problems with the geometric problems is based on two general theorems on quasihomogeneous functions (Theorems A and B below). To state these theorems we introduce some definitions and notation.

Theorem No.	Series	Geometric problem
3	D	Linear classification of 3-forms in $\mathbf{C}^2$
10, 22	J, Z	Affine classification of triples of points in $\mathbf{C}^1$
13	X	Linear classification of 4-forms in $\mathbf{C}^2$
29	W	Linear classification of pairs of points in $\mathbf{C}^1$
36	X	Affine classification of quadruples of points in $\mathbf{C}^1$
47	N	Linear classification of 5-forms in $\mathbf{C}^2$
50	P	Linear classification of 3-forms in $\mathbf{C}^3$
63	Q	Affine classification of cubic curves in $\mathbf{C}^2$ with a finite cusp
70	S	Affine classification of reducible cubic curves in $\mathbf{C}^2$ having a finite double point
77	S^*	Affine classification of cubic curves in $\mathbf{C}^2$ having a simple tangent at infinity
85	U	Affine classification of centrally-symmetric cubic curves in $\mathbf{C}^2$ with precisely three points at infinity
90	U^*	Affine classification of cubic curves in $\mathbf{C}^2$ with precisely three points at infinity
98	V	Affine classification of polynomials of degree $\leqslant 4$ in one variable in $\mathbf{C}^1$

Let $\alpha = (\alpha_1, \ldots, \alpha_n)$ be a set of positive rational numbers. We consider the arithmetic space $\mathbf{C}^n$ with fixed coordinates $x_1, \ldots, x_n$.

DEFINITION. A *quasihomogeneous function* of degree d and type $\boldsymbol{\alpha}$ is a polynomial $\Sigma a_m x^m \in \mathbf{C}[x_1, \ldots, x_n]$ for which $(m, \boldsymbol{\alpha}) = d$ for all m.

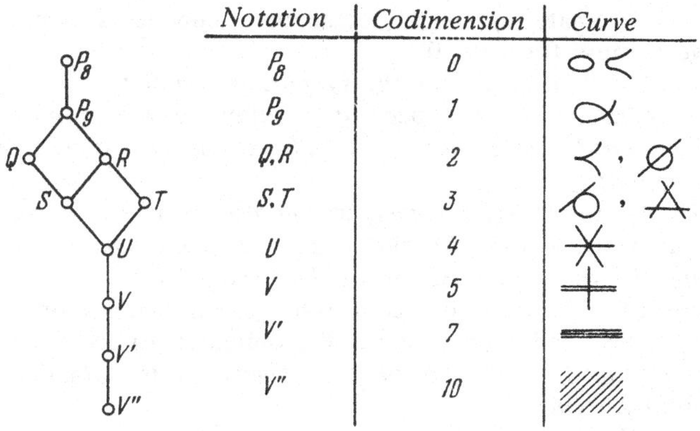

Fig. 9

DEFINITION. The *group of quasihomogeneous diffeomorphisms* of type $\boldsymbol{\alpha}$ is the group of germs of diffeomorphisms of $\mathbf{C}^n$ at O taking any quasihomogeneous function of type $\boldsymbol{\alpha}$ and degree d into a quasihomogeneous function of the same degree. The Lie algebra of this group is called the *quasihomogeneous algebra* and is denoted by $\mathfrak{a}(\alpha)$. For example, $\mathfrak{a}(1, 1) = \mathfrak{gl}(2, \mathbf{C})$.

DEFINITION. The *space of exponents* of functions of $x_1, \ldots, x_n$ is the arithmetic space $\mathbf{C}^n$ whose points m with non-negative integer coordinates are the exponents of the monomials x^m.

DEFINITION. The *support of quasihomogeneous functions of degree d and type* α is the set of all non-negative integer points m on the plane $(m, \alpha) = d$. The support is said to be *complete* if it is not contained in an affine subspace of $\mathbf{C}^n$ of dimension less than $n - 1$.

Quasihomogeneous functions can be regarded as functions given on the support ($\Sigma\, a_m x^m$ has the value a_m at m). All such functions form a linear space $\mathbf{C}^\nu$, where ν is the number of points of the support. The group of quasihomogeneous diffeomorphisms and the quasihomogeneous algebra $\mathfrak{a}(\alpha)$ act on this space $\mathbf{C}^\nu$. From the definitions it follows immediately that *the Lie algebra* $\mathfrak{a}(\alpha)$ *is generated as a* C*-linear space by all the monomial vector fields* $x^p\, \partial_i$ *for which* $(p, \alpha) = \alpha_i$ (here and later on, $\partial_i = \partial/\partial x_i$). For example, the n monomials $x_i \partial_i$ belong to $\mathfrak{a}(\alpha)$ for any α.

DEFINITION. The *roots* of the quasihomogeneous algebra $\mathfrak{a}(\alpha)$ are all the non-zero vectors m of the space of exponents that lie in the plane $(m, \alpha) = 0$ and have the form $m = p - \mathbf{1}_i$ (where $\mathbf{1}_i$ is the vector whose i-th component is equal to 1 and the remainder equal to 0, and the vector p has non-negative integer components).

In other words, m is a root if $x^p\, \partial_i$ is a monomial vector field from $\mathfrak{a}(\alpha)$ other than $x_i \partial_i$.

We observe that i can be regained from a root m, since m has precisely one negative coordinate $m_i = -1$ (not all the components of m can be non-negative because $(m, \alpha) = 0$).

THEOREM A. *We assume that the support is complete. Then the action of the Lie algebra* $\mathfrak{a}(\alpha)$ *on the space of functions on the support is uniquely determined with respect to the affine equivalence class of the pair (support, root system).*

THEOREM B. *The quasihomogeneous Lie algebra* $\mathfrak{a}(\alpha)$ *is determined, up to finitely many variants, by its root system (as a subspace of the linear space spanned by the roots) and by its dimension.*

In other words, if we do not distinguish algebras that are obtained from each other by direct addition of a trivial (commutative) algebra, then *there are only finitely many non-isomorphic Lie algebras* $\mathfrak{a}(\alpha)$ *with linearly equivalent root systems.*

In our examples this finite number is equal to 1, and I do not know whether it can be greater than 1.

REMARK. We must emphasize that the affine equivalences of supports and linear equivalences of root systems in Theorems A and B are not at all necessary to preserve either the coordinate simplex $m_i \geqslant 0$ on the plane $(m, \alpha) = d$ or the lattice of non-negative integral m in $\mathbf{C}^n$. We also observe that under the hypotheses of Theorems A and B the *groups* of quasihomogeneous diffeomorphisms and their orbits in spaces of quasihomogeneous functions do not necessarily coincide; however, the connected components of the orbits coincide.

PROOF OF THEOREM B. Let $M \subset \mathbf{C}^r \subset \mathbf{C}^n$ be a root system of $\mathfrak{a}(\alpha)$, generating the plane $\mathbf{C}^r$ in $\mathbf{C}^n$ $(0 \leqslant r \leqslant n-1)$. With each $m \in M$ we associate a basis vector e_m in $\mathbf{C}^\nu$ (where ν is the number of roots). We consider the r-dimensional linear space $H = \mathrm{Hom}(\mathbf{C}^r, \mathbf{C})$ and the direct sum $\mathfrak{b} = H \oplus \mathbf{C}^\nu$.

LEMMA. *The space* $\mathfrak{b} = H \oplus \mathbf{C}^\nu$ *can be given the following Lie algebra structure*:

(1) $[h_1, h_2] = 0$ $(\forall h_1, h_2 \in H)$;

(2) $[h, e_m] = (h, m)e_m$ $(\forall h \in H, m \in M)$;

(3) $[e_{m_1}, e_{m_2}] = N_{m_1, m_2} e_{m_1+m_2}$, *where* $m_1 + m_2 \neq 0$,

$$N_{m_1,m_2} \begin{cases} = 0 \textit{ if } m_1 + m_2 \textit{ is not a root;} \\ = -\max\{\lambda: m_1 + \lambda m_2 \text{ is a root}\} \textit{ if this maximum is } > 1; \\ = +\max\{\lambda: m_2 + \lambda m_1 \text{ is a root}\} \textit{ if this maximum is } > 1; \\ = \pm 1 \textit{ if both maxima} = 1 \textit{ (the case when both maxima are } > 1 \textit{ is} \end{cases}$$

impossible).

(4) $[e_m, e_{-m}] = h_m$, *where the function* $h_m \in H$ *changes sign under the reflection of* $\mathbf{C}^r$ *that preserves M and takes* m *into* $-m$, *normalized by the condition* $h_m(m) = 2$ *(such a reflection exists and is unique for any pair of opposite roots)*.

The quasihomogeneous Lie algebra $\mathfrak{a}(\alpha)$ *is isomorphic to the direct sum of the Lie algebra* $\mathfrak{b}$ *(for some choice of the sign* $\pm$ *in* (3)) *and a trivial (commutative) algebra*: $\mathfrak{a}(\alpha) \cong \mathfrak{b} \oplus \mathbf{C}^{n-r}$.

PROOF OF THE LEMMA. We consider a monomial basis for $\mathfrak{a}(\alpha)$ over $\mathbf{C}$ and write the basis monomials as follows:

$$h_i = x_i \partial_i, \quad e_m = x^m x_i \partial_i$$

(I recall that i is uniquely determined by the root m). We claim that these generators satisfy the commutation relations (1)–(4).

We regard the differentiations h_i and e_m as linear operators in the space of all functions on the lattice $\mathbf{Z}^n$ in the space of exponents $\mathbf{C}^n$. Then h_i is the operator of multiplication by the i-th coordinate. We denote the operator of multiplication by a function like that function. Thus,

$$(ha)(k) = h(k)\, a(k), \text{ where } k \in \mathbf{Z}^n, \ a: \mathbf{Z}^n \to \mathbf{C}.$$

The relation (1) is now proved, because multiplications by functions commute.

We denote by σ_m the action of translation by $m \in \mathbf{Z}^n$ on the function:

$$(\sigma_m a)(k) = a(k - m), \text{ where } k \in \mathbf{Z}^n, \ a: \mathbf{Z}^n \to \mathbf{C}.$$

Then $e_m = \sigma_m h_i$. Computing the commutator of multiplication by a linear function and a translation, we obtain

$$[h, \sigma_m] = h(m)\sigma_m.$$

From this we immediately get (2). Next, computing the commutator of e_{m_1} and e_{m_2} we obtain

$$e_{m_1}e_{m_2} = \sigma_{m_1}h_{i_1}\sigma_{m_2}h_{i_2} = \sigma_{m_1}\sigma_{m_2}(h_{i_1}h_{i_2} + h_{i_1}(m_2)h_{i_2}),$$
$$[e_{m_1}, e_{m_2}] = \sigma_{m_1+m_2}(h_{i_1}(m_2)h_{i_2} - h_{i_2}(m_1)h_{i_1}).$$

If $m_1 + m_2$ is not a root and not 0, then the operator on the right-hand side can belong to $\mathfrak{a}(\alpha)$ only if it is 0. In this case $[e_{m_1}, e_{m_2}] = 0$.

If $m_1 + m_2$ is a root, then this vector has precisely one negative component, which is equal to -1. The vectors m_1 and m_2 also have one negative component, equal to -1. Therefore $m_1 + m_2$ has either the same negative component as m_1 or as m_2. For example, let us assume that $m_1 + m_2$ has the same negative component as m_1.

Then the i_1-th and i_2-th components of the vectors m_1, m_2, and $m_1 + m_2$ have the form

	m_1	m_2	m_1+m_2	$m_1+\lambda m_2$
h_{i_1}	-1	0	-1	-1
h_{i_2}	$p>0$	-1	$p-1\geqslant 0$	$p-\lambda$

Consequently, $m_1 + \lambda m_2$ is a root for $\lambda \leqslant p = h_{i_2}(m_1)$, that is, $h_{i_2}(m_1) = \max\{\lambda: m_1 + \lambda m_2 \text{ is a root}\}$.

Thus, in case $h_{i_1}(m_1 + m_2) = -1$, we have

$$[e_{m_1}, e_{m_2}] = -h_{i_2}(m_1)\sigma_{m_1+m_2} = N_{m_1, m_2}e_{m_1+m_2},$$

where $N_{m_1,m_2} = h_{i_2}(m_1)$; and so we have proved (3).

Suppose that m and $-m$ are roots. Both these vectors have precisely one negative coordinate, equal to -1, so that $e_m = x_i\partial_j$, $e_{-m} = x_j\partial_i$. Consequently, the weights α_i and α_j are equal. The interchange of the coordinates i and j in $\mathbf{C}^n$ is a reflection taking the system of all roots into itself and changing the places of m and $-m$. Further, $[e_m, e_{-m}] = h_i - h_j$. If we regard $h_m = h_i - h_j$ as a function in $\mathbf{C}^n$, then it changes sign under the reflection switching i and j, and is equal to $+2$ on m. Thus, the operators h and e_m satisfy the relations (1)–(4).

Now we consider the subspace $\mathbf{C}^r$ spanned by the roots in the coordinate space $\mathbf{C}^n$ with the usual Hermitian metric, $\langle k, l\rangle = \Sigma\, k_i\bar{l}_i$. The space $\mathbf{C}^r$ and the metric are invariant relative to all permutations of the coordinates with equal weights ($\alpha_i = \alpha_j$). Therefore, the orthogonal complement $\mathbf{C}^{n-r}$ to $\mathbf{C}^r$ in $\mathbf{C}^n$ is also invariant. We represent the linear space of $\mathfrak{a}(\alpha)$ as $\mathfrak{a}(\alpha) = H^r \oplus K^\nu \oplus H^{n-r}$, where H^r consists of those linear functions h on $\mathbf{C}^n$ that vanish on $\mathbf{C}^{n-r}$; H^{n-r} of the linear functions h on $\mathbf{C}^n$ that vanish on $\mathbf{C}^r$; K^ν of linear combinations of the vectors e_m.

It follows from these commutation relations that H^{n-r} lies in the centre of $\mathfrak{a}(\alpha)$ and that $H^r \oplus K^\nu$ is an ideal isomorphic to $\mathfrak{b}$. Consequently,

$\mathfrak{a}(\alpha) \cong \mathfrak{b} \oplus H^{n-r}$, and the lemma is proved.

We can now complete the proof of Theorem B. The relations (1)–(4) express the commutators in $\mathfrak{b}$ and $\mathfrak{a}$ by the geometry of the roots, without reference to coordinates (excluding the choice of sign ± in one of the relations (3)). Thus, the set of roots, as vectors in the **C**-linear space $\mathbf{C}^r$, determines $\mathfrak{b}$ up to a finite number of possibilities, and this proves Theorem B.

REMARK. I do not know whether different choices of signs in (3) yield non-isomorphic algebras.

PROOF OF THEOREM A. We extend functions on the support onto the lattice $\mathbf{Z}^{n-1}$ of all integer points in the support plane, by setting them equal to 0 outside the support. The operators h_i, e_m, σ_m, h_m, defined in the proof of Theorem B on the functions on $\mathbf{Z}^n$, also act on the space of functions on the lattice $\mathbf{Z}^n$ in the support plane. These operators in the space of functions on $\mathbf{Z}^{n-1}$ will be denoted by the same letters. Thus, h_i and h_m are operators of multiplication by affine functions in the support plane, σ_m is the operator of translation by the root m and $e_m = \sigma_m h_i$.

Suppose that m is a root. We call a point k of the support a *base point* for m if $k + m$ does not belong to the support. The set of all base points of m is called the *base of the root* m (in the given support).

The base of the root $m = p - 1_i$ is formed by all the points of the support whose i-th coordinate is 0. Consequently, the entire base lies in an affine hyperplane $\mathbf{C}^{n-2}$ of the support plane.

The base of each root of the complete support belongs to precisely one affine hyperplane $\mathbf{C}^{n-2} \subset \mathbf{C}^{n-1}$. For, each point of the support is obtained from a point of the base by subtracting a non-negative integral multiple of m, so that if the base were contained in $\mathbf{C}^{n-3}$, the support would be contained in $\mathbf{C}^{n-2}$ and would not be complete.

Thus, there exists precisely one affine function in the support plane that is equal to 0 on the base of m and whose increment along m is -1. This function is h_i (restricted to the support plane). Consequently, the restriction of h_i to the support plane can be reconstructed uniquely from the support and the root m.

The action of e_m on the space of functions on the support can now be described in terms only of the geometry of the support and the roots: $e_m = \sigma_m h_i$, that is, $(e_m a)(k) = h_i(k - m)a(k - m)$ for any function a. We note that the operator e_m leaves invariant the space of functions that vanish outside the support, since h_i vanishes at base points.

The algebra $\mathfrak{a}(\alpha)$ acts on the space of functions on a lattice in the support plane, so that we have a representation φ: $\mathfrak{a}(\alpha) \to \mathscr{E}$, where $\mathscr{E}$ is the algebra of endomorphisms of this (infinite-dimensional) space of functions. We consider the image of the representation φ. *This image is determined by the geometry of the support and the roots.* More precisely, let $\mathfrak{a}_1 = \mathfrak{a}(\alpha_1)$ and $\mathfrak{a}_2 = \mathfrak{a}(\alpha_2)$ be two quasihomogeneous algebras, and

$S_1 \subset \mathbf{C}_1^{n-1}$, $S_2 \subset \mathbf{C}_2^{n-1}$ be complete supports. Let ψ: $\mathbf{C}_1^{n-1} \to \mathbf{C}_2^{n-1}$ be an affine mapping that takes S_1 into S_2 bijectively and the root system for $\mathfrak{a}_1$ into that for $\mathfrak{a}_2$. Then *the isomorphism* ψ^*, *induced by* ψ, *of the space of functions on* $\mathbf{C}_2^{n-1}$ *into the space of functions on* $\mathbf{C}_1^{n-1}$ *takes the Lie algebra* $\varphi(\mathfrak{a}_2)$ *isomorphically into the Lie algebra* $\varphi(\mathfrak{a}_1)$.

For $\mathfrak{a}(\alpha)$ is generated over $\mathbf{C}$ by the monomials $x_i\partial_i$ and $x^m x_i \partial_{i(m)}$. The images of the fields $\Sigma\, c_i x_i \partial_i$ in $\mathscr{E}$ are the operators of multiplication by all possible affine functions in the support plane. The images of the fields $x^m x_i \partial_{i(m)}$ are the operators e_m, defined by the geometry of the support and the roots. Consequently, $\psi^* e_{\psi(m)} = e_m \psi^*$ and, hence, ψ induces an isomorphism Ψ: $\varphi(\mathfrak{a}_2) \to \varphi(\mathfrak{a}_1)$.

The kernel of the Lie algebra homomorphism $\varphi: \mathfrak{a}(\alpha) \to \mathscr{E}$ *is* 0. For suppose that $(h + \Sigma\, c_m e_m)\, a = 0$ for all functions a. We choose a point k at which all the h_{i_m} are different from 0, and we apply $h + \Sigma\, c_m e_m$ to the function δ_k, which is equal to 1 only at k. We obtain $h(k)\delta_k + \Sigma c_m h_{i_m}(k)\, \delta_{k+m} = 0$, hence $c_m = 0$ and $h(k) = 0$. Consequently, $h \equiv 0$. Thus, the Lie algebra $\varphi(\mathfrak{a}(\alpha))$ is isomorphic to $\mathfrak{a}(\alpha)$.

The isomorphisms $\mathfrak{a}_2 \to \varphi(\mathfrak{a}_2) \overset{\Psi}{\to} \varphi(\mathfrak{a}_1) \to \mathfrak{a}_1$ show that both the Lie algebras $\mathfrak{a}_1$ and $\mathfrak{a}_2$, and their actions on the spaces of functions on S_1 and S_2 are isomorphic, and Theorem A is proved.

COROLLARY 1. *Suppose that the set of weights* α *and the degree* d *are such that there exists a quasihomogeneous function with an isolated critical point* O *with zero 2-jet. Then the root system and the support uniquely determine the Lie algebra* $\mathfrak{a}(\alpha)$ *and its action* φ.

PROOF. Under the assumptions the support is complete. For the isolation implies that for each i there is a monomial of the form $x_i^{a_i} x_{j(i)}$ $(j \in \{1, \ldots, n\}$, and by hypothesis $a_i > 1)$. We claim that the exponents of these n monomials that belong to the support are linearly independent. The system of equations $z_{j(i)} + a_i z_i = 0$ relative to z has only the zero solution for $a_i > 1$ (it is easy to see this by considering the cycles of an endomorphism of a finite set, $i \mapsto j$). Consequently, the determinant of its matrix is not 0, and hence, our support is complete. The assertion of the corollary now follows from Theorem A.

COROLLARY 2. *Let* γ: $\mathbf{C}_1^{n-1} \to \mathbf{C}_2^{n-1}$ *be an affine isomorphism of a complete support plane* S_1 *into a complete support plane* S_2, *taking* S_1 *into part of* S_2, *the roots of* $\mathfrak{a}_1$ *into part of the roots of* $\mathfrak{a}_2$ *and the bases of the roots of* $\mathfrak{a}_1$ *into part of the bases of the corresponding roots of* $\mathfrak{a}_2$. *Then* γ *induces an isomorphism of the action* φ *of* $\mathfrak{a}_1$ *onto functions on* S_1 *and an isomorphism of the action of a subalgebra of* $\mathfrak{a}_2$ *onto the space of functions on* S_2 *that vanish outside* $\gamma(S_1)$.

PROOF. The subalgebra indicated in the corollary is generated over $\mathbf{C}$ by the operators h of multiplication by all affine functions and by the operators

e_m, where m is the image of a root of the first algebra. The actions e_m are defined by the roots and the bases, and thus commute with the action of γ.

COROLLARY 3. *Suppose that under the hypotheses of Corollary 2 several points in the support S_1 are distinguished and the values of the functions are fixed at them. All the functions on S_1 with fixed values at these points form an affine plane P in the space of functions on S_1. Let $\mathfrak{a}_P$ be the isotropy algebra of this plane P. Then the isomorphism γ of Corollary 2 induces an isomorphism of the action of $\mathfrak{a}_P$ on P with the action of some subalgebra of the Lie algebra $\mathfrak{a}_2$, preserving the plane $\gamma^{-1}{}^*P$ in the space of functions on S_2.*

PROOF. The mapping of the first action into the second commutes with $\gamma^{-1}{}^*$, as required.

The *reduction of the classification Theorems* 10, . . . , 98 *to standard classification problems* is carried out by means of Corollary 3. In the table given below, for each theorem we have indicated: the weight α_i and the

Theorem	Type	Weights $\alpha; d$	Monomials of the support	e_m	Monomials of the homogeneous support	Images of e_m
10	J_k	$k,\ 1;\ 3k$	$x_*^3,\ x^2y^k,\ xy^{2k},\ y^{3k}$	$y^k\partial_x$	$X_*^3,\ X^2Y,\ XY^2,\ Y^3$	$Y\partial_X$
22	Z_{l-1}	$l,\ 1;\ 3l+1$	$x^3y_*,\ x^2y^{l+1},\ xy^{2l+1},\ y^{3l+1}$	$y^l\partial_x$	$X_3^*,\ X^2Y,\ XY^2,\ Y^3$	$Y\partial_X$
29	W_k	$2k+1, 2;\ 8k+4$	$x_*^4,\ x^2y^{2k+1},\ y^{4k+2}$	—	$X_*^2,\ XY,\ Y^2$	—
36	X_k	$k,\ 1;\ 4k$	$x_*^4,\ x^3y^k,\ x^2y^{2k},\ xy^{3k},\ y^{4k}$	$y^k\partial_x$	$X_*^4,\ X^3Y,\ X^2Y^2,\ XY^3,\ Y^4$	$Y\partial_X$
63	Q_k	$2k,\ 2,\ 3k-1;\ 6k$	$x_*^3,\ yz_*^2,\ x^2y^k,\ xy^{2k},\ y^{3k}$	$y^k\partial_x$	$X_*^3,\ Y^2Z_*,\ X^2Y,\ XY^2,\ Y^3$	$Y\partial_X$
70	S_k	$2k+1,\ 2,\ 4k;\ 8k+2$	$x^2z_*,\ yz_*^2,\ x^2y^{2k},\ y^{2k+1}z,\ y^{4k+1}$	$y^{2k}\partial_z$	$XYZ_*,\ YZ_*^2,\ XY^2,\ Y^2Z,\ Y^3$	$Y\partial_Z$
77	S_k^*	$k, 1,\ 2k-1;\ 4k-1$	$x^2z_*,\ yz_*^2,\ x^3y^{k-1},\ xy^kz,\ x^2y^{2k-1},\ zy^{2k},\ xy^{3k-1},\ y^{4k-1}$	$y^k\partial_x$ $y^{2k-1}\partial_z$ $xy^{k-1}\partial_z$	$X^2Z_*,\ YZ_*^2,\ X^3,\ XYZ,\ X^2Y,\ Y^2Z,\ XY^2\ Y^3$	$Y\partial_X$ $Y\partial_Z$ $X\partial_Z$
85	U_k	$2k+1, 2,\ 2k+1;\ 6k+3$	$x_*^3,\ xz_*^2,\ x^2z_\circ,\ z_\circ^3,\ xy^{2k+1},\ zy^{2k+1}$	$x\partial_z$ $z\partial_x$	$X_*^3,\ XZ_*^2,\ X^2Z_\circ,\ Z_\circ^3,\ XY^2,\ Y^2Z$	$X\partial_Z$ $X\partial_X$
90	U_k^*	$k,\ 1,\ k;\ 3k$	$x^2z_*,\ xz_*^2,\ x_\circ^3,\ z_\circ^3,\ x^2y^k,\ xy^kz,\ z^2y^k,\ xy^{2k},\ zy^{2k},\ y^{3k}$	$x\partial_z$ $z\partial_x$ $y^k\partial_x$ $y^k\partial_z$	$X^2Z_*,\ XZ_*^2,\ X_\circ^3,\ Z_\circ^3,\ X^2Y,\ XYZ,\ Z^2Y,\ XY^2,\ ZY^2,\ Y^3$	$X\partial_Z$ $Z\partial_X$ $Y\partial_X$ $Y\partial_Z$
98	V	$3,\ 2,\ 2;\ 8$	$x^2y_*,\ z^4,\ y^4,\ y^3z,\ y^2z^2,\ yz^3$	$y\partial_z$	$XY_*^3,\ Z^4,\ Y^4,\ Y^3Z,\ Y^2Z^2,\ YZ^3$	$Y\partial_Z$

degree of quasihomogeneity d, the monomials of the support (those distinguished by the sign $*$ have the coefficient 1 and those by the sign $\circ$ have the coefficient 0), the root vector fields e_m for each quasihomogeneity type, and then their images in the support of homogeneous forms (all of weight 1) and in the Lie algebra of the general linear group (Fig. 10).

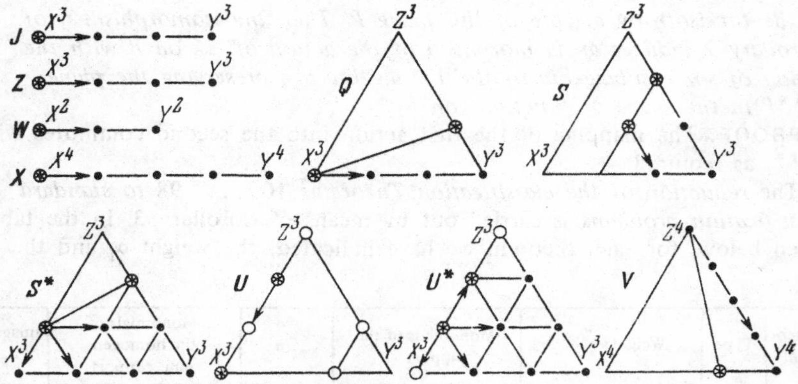

Fig. 10

According to Corollary 3, the orbits of the identity component of the group of quasihomogeneous diffeomorphisms that fix the plane P (consisting of functions on the support with the values 0 and 1 at the points distinguished by the signs $\circ$ and $*$) are transformed under a mapping onto the support of the homogeneous functions into orbits of the corresponding group of *linear* transformations. (This circumstance is a nice property of the roots in the relevant cases in Theorems **10–98**; in general, the images of the roots for a (α) in the homogeneous support need not also be roots of the linear group.)

The Lie algebra of this group of linear transformations is easy to describe: it is generated by a torus part, acting on a function on the support as multiplication by affine functions that are zero at the points of the support distinguished by the sign $*$ (at which a non-zero value of the function is fixed) and by the images of those root vectors e_m that do not take other points of the support into points distinguished by the sign $\circ$ (the last condition excludes only $X\partial_Z$ and $Z\partial_X$ in the cases U and U^*).

PROOF OF THEOREMS 10 AND 22. By Corollary 3 and the table, the problem reduces to the classification of the polynomials $p = X^3 + AX^2Y + BXY^2 + CY^3$ relative to the substitutions $X = X' + \lambda Y$, $Y = \gamma Y'$. By the substitution $X = X' + \lambda Y$ the polynomial p reduces to a form in which $B = 0$. If $C \neq 0$, the substitution $Y = \gamma Y'$ takes p to the form $X^3 + aX^2Y + Y^3$. The roots of the polynomial $X^3 + aX^2 + 1$ are distinct if and only if $4a^3 + 27 \neq 0$. If the roots have multiplicities 2 and 1, then the substitution $X = X' + \lambda Y$ takes p to the form $X^3 + AX^2Y$, $A \neq 0$. Using the substitution $Y = \gamma Y'$ we carry p to the form $X^3 + X^2Y$. If the multiplicity is 3, then $p = X'^3$. Theorems **10** and **22** are now proved.

PROOF OF THEOREM **29**. The problem reduces to the classification of the polynomials $X^2 + AXY + BY^2$ relative to the substitution $Y = \gamma Y'$. The four cases of Theorem **29** correspond to distinct non-zero roots, to one zero root, to a multiple non-zero root and to a multiple zero root of the polynomial $X^2 + AX + B$.

PROOF OF THEOREM **36**. The problem reduces to the classification of the polynomials $p = X^4 + AX^2Y + BX^2Y^2 + CXY^2 + DY^4$ with respect to the substitutions $X = X' + \lambda Y$ and $Y = \gamma Y'$. The first substitution changes the coefficients to

$$A' = 4\lambda + A, \quad B' = 6\lambda^2 + 3\lambda A + B, \quad C' = 4\lambda^3 + 3\lambda^2 A + 2\lambda B + C,$$
$$D' = \lambda^4 + A\lambda^3 + B\lambda^2 + C\lambda + D.$$

For λ we choose one of the roots of the polynomial D'. If these roots are simple, then $C' \neq 0$, so that the substitution $Y = \gamma Y'$ takes p to the form $X^4 + bX^3Y + aX^2Y^2 + XY^3$. The discriminant of $X^4 + bX^3 + aX^2 + X$ is $4(a^3 + b^3) - a^2b^2 - 18ab + 27$.

We observe that $ab - 9 = \dfrac{A'B' - 9C}{C'} = \dfrac{-12\lambda^3 + \ldots}{4\lambda^3 + \ldots}$ is a rational function with numerator of degree three in λ.

Consequently, if the roots of D' are simple, then at least one of the four roots λ of D' does not make the number $ab - 9$ vanish. Thus, in the case of simple roots, p reduces to the form $X^4 + bX^3Y + aX^2Y^2 + XY^3$, where $4(a^3 + b^3) + 27 \neq a^2b^2 + 18ab$, $ab \neq 9$.

If there is one root of multiplicity two, then the substitution $X = X' + \lambda Y$ allows us to reach $D' = C' = 0$, $B' \neq 0$, after which $Y = \gamma Y'$ gives $X^2(X^2 + aXY + Y^2)$, $a^2 \neq 4$.

If there are two double roots, one triple or one quadruple root we obtain $X^2(X + Y)^2$, $X^3(X + Y)$ and X^4, respectively. Theorem **36** is now proved.

REMARK. The restriction $ab \neq 9$ is necessary for the reduction to normal form of non-quasihomogeneous functions with principal quasi-homogeneous part of the form indicated here (that is, for the proof of Theorem 37). The possibility of reduction to a normal form with $ab \neq 9$ in Theorems **47** and **98** is proved in exactly the same way (and follows from the calculations given here and from Corollary 3).

PROOF OF THEOREM **63**. The problem reduces to the classification of the polynomials $p = X^3 + Y^2Z + AX^2Y + BXY^2 + CY^3$ relative to the substitution $X = X' + \lambda Y$, $Y = \gamma Y'$, $Z = \gamma^2 Z'$. We consider the roots λ of the polynomial $\lambda^3 + A\lambda^2 + B\lambda + C$. If they are simple, the substitution $X = X' + \lambda Y$ yields $C = 0$, $B \neq 0$, after which a second substitution transforms p to $X^3 + Y^2Z + aX^2Y + XY^2$, $a^2 \neq 4$. Then we have the first case of Theorem 63. If λ is a double root, we obtain $X^3 + Y^2Z + X^2Y$. Under the conditions of the theorem ($A = 0$) a triple root can only occur if $B = C = 0$, and Theorem 63 is proved.

REMARK. It is not hard to check that we have classified plane cubic curves with a finite cusp ($X = 0$, $Y = 0$, $Z = 1$) relative to the affine group. The three cases we have met are distinguished by the nature of the points of intersection with the line at infinity ($Z = 0$), which can be 3, 2, or 1 in number.

PROOF OF THEOREM 70. The problem reduces to the classification of the polynomials $p = XYZ + YZ^2 + AXY^2 + BY^2Z + CY^3$ relative to the substitution $Z = Z' + \lambda Y$; $X = \gamma X'$, $Y = \gamma^{-2}Y'$, $Z = \gamma Z'$. Instead of (A, B, C) it is convenient to consider in the space of polynomials the coordinates $(A,\ E = B - 2A,\ D = 4C - B^2)$. The first substitution takes the point (A, E, D) to $(A + \lambda, E, D)$. Consequently, it suffices to consider the action of the second substitution on the plane $A = 0$, which carries the point (E, D) to $(\gamma^{-3}E, \gamma^{-6}D)$. Consequently, the orbits are 0, half the parabola $D = \varkappa E^2$ without 0, and the line $E = 0$ without 0. Also let $4\Delta = D + E^2$ (that is, $\Delta = A^2 + C - AB$). Four cases are possible:

	Condition	The orbit passes through the point with the coordinates	Consequently, the polynomial reduces to the form
1°	$D \neq 0,\ \Delta \neq 0$	$A = 0,\ B = b,\ C = 1$	$XYZ + YZ^2 + bY^2Z + Y^3,\ b^2 \neq 4$
2°	$D = 0,\ \Delta \neq 0$	$A = 1,\ B = C = 0$	$XYZ + YZ^2 + XY^2$
3°	$D \neq 0,\ \Delta = 0$	$A = C = 0,\ B = 1.$	$XYZ + YZ^2 + Y^2Z$
4°	$D = 0,\ \Delta = 0$	$A = B = C = 0$	$XYZ + YZ^2$

In the last case from the very start the polynomial has the form $XYZ + YZ^2$, since under the hypotheses of Theorem 70 the original polynomial has no monomial x^2y^{2k}, and hence $A = 0$. Theorem 70 is now proved.

REMARK. It is not hard to check that we have classified reducible plane cubic curves (with Y splitting off as factor), having at least two finite (with $X \neq 0$) double points ($Y = 0$, $XZ + Z^2 = 0$) relative to the affine group. The condition $D = 0$ corresponds to tangency with infinity, and $\Delta = 0$ is the complete degeneration of the cubic into 3 lines (Fig.11 on p.58).

PROOF OF THEOREM 77. The problem reduces to the classification of cubic forms $X^2Z + YZ^2 + 0XZ^2 + 0Z^3 + \ldots$ relative to the substitutions $X = X' + \alpha Y$, $Z = Z' + \beta Y + \gamma X$; $X = \lambda X'$, $Y = \lambda^4 Y'$, $Z = \lambda^{-2}Z'$.

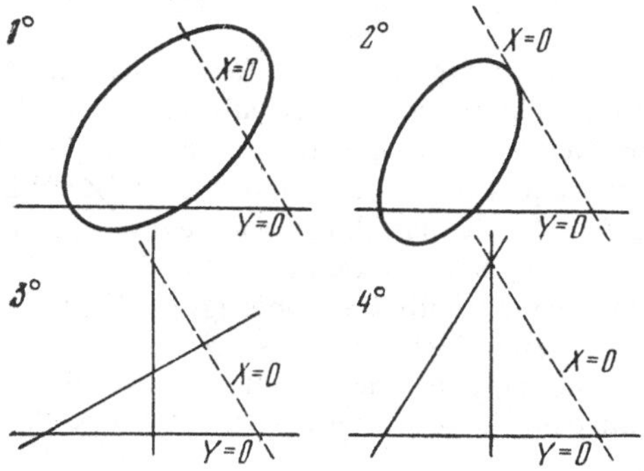

Fig. 11

A form of this kind gives a cubic curve with a simple tangent $Y = 0$ at the non-singular point $X = Y = 0$. Depending on the kind of singularities of this curve five cases are possible.

1°. If the curve is non-singular, then we pass a tangent to the curve different from $Y = 0$ through the point $X = Y = 0$, and set $X = X' + \alpha Y$, where $X' = 0$ is the equation of the new tangent. Then, joining the second point of intersection of the line $Y = 0$ with the curve to the point of contact of the line $Y = 0$ with the line $Z' = 0$ (where $Z = Z' + \beta Y + \gamma X$). Now the form is $X^2Z + YZ^2 + aX^2Y + bXYZ + cXY^2$. The point $X = Z = 0$ is non-singular, so that $c \neq 0$. By dilations we can achieve that $c = 1$. This is the first case of Theorem 77.

2°. Suppose that the curve has a singular point. This cannot lie on the line "at infinity" $Y = 0$. Moving $X' = Z' = 0$ to the singular point by the translations $X = X' + \alpha Y$, $Z = Z' + \beta Y$, we take the form into $X^2Z + YZ^2 + aX^2Y + bXYZ$.

Now by the substitution $Z = Z' + \gamma X$ we annihilate the term with X^2Y, and then X^3 arises and the form becomes $X^2Z + YZ^2 + bXYZ + aX^3$.

If the singular point is non-degenerate, then $b \neq 0$ and by dilations we may take $b = 1$. This is the second case of Theorem 77 (the case *SB*). The condition $a^2 \neq a$ guarantees the uniqueness of the singular point.

3°. If the singular point is degenerate, then $b = 0$. If it is a cusp, then $a \neq 0$ and by dilations we may take $a = 1$. This is the third case of Theorem 77 (case *SQ*).

4°. If there are two singular points, then both are non-degenerate and the curve splits. By the substitution $Z = Z' + \beta Y + \gamma X$ we reduce the equation of the linear factor to the form Z. By the substitution $X = X' + \alpha Y$ we make the point $X' = Z' = 0$ singular. The form is taken to $X^2Z + YZ^2 + aXYZ$. Since there are two singular points, we have

$a \neq 0$ and we may make $a = 1$. This is the fourth case of Theorem 77 (the case SR).

5°. The curve cannot completely split (since the axis $Y = 0$ is a *simple* tangent). Therefore, what remains to be analyzed is the case when the curve consists of an irreducible conic and a tangent to it. But by the hypothesis of Theorem 77, the equation of the curve was $X^2Z + YZ^2 + AY^3 + BXY^2 + CZY^2 = 0$. The line $Z = 0$ has third order contact at $Y = Z = 0$ with such a curve, and hence, belongs to the curve. Consequently, $A = B = 0$. The condition of tangency with the conic gives $C = 0$. Thus, in the last case $A = B = C = 0$, and Theorem 77 is proved.

REMARK. It is not hard to check that we have given an affine classification of all plane cubic curves having the line "at infinity" $Y = 0$ as a simple tangent.

PROOF OF THEOREM 85. The problem reduces to the classification of the forms $\varphi + Y^2\psi$ (where φ is a non-degenerate cubic form in X and Z, and ψ is an arbitrary linear form in X and Z) relative to linear substitutions of X and Z and dilations of Y.

Let $\psi \neq 0$. Then, depending on whether φ is divisible by ψ or not, by a linear transformation taking φ to the form $X^3 + XZ^2$, we can transform ψ either to the form $AX + CZ$, where $C^3 + A^2C \neq 0$, $A \neq 0$ or to the form AX. The substitution $Y = \lambda Y'$ makes $A = 1$ and we obtain either the first ($C^3 + C \neq 0$) or the second ($C = 0$) case of Theorem 85. Finally, if $\psi = 0$, then we have the third case.

REMARK. It is not hard to check that we have given an affine classification of all centrally-symmetric plane cubic curves having precisely three points at infinity ($Y = 0$). Here the three cases of Theorem 85 correspond to a non-degenerate curve, a curve split into a line and a conic, and a curve completely split (into three lines).

PROOF OF THEOREM 90. The problem reduces to the classification of the forms $\varphi_3 + Y\varphi_2 + Y^2\varphi_1 + Y^3\varphi_0$, where φ_i is a form of degree i with respect to (X, Z), and φ_3 is not degenerate. We admit linear substitutions of (X, Z), dilations of Y and the substitutions $X = X' + \alpha Y$, $Z = Z' + \beta Y$. In other words, the question is one of the classification of the equations of all cubic curves having exactly three points on the line at infinity $Y = 0$. Depending on the singularities of the curve, seven cases are possible.

1°. We assume that the curve is non-singular. We choose one of the points of intersection of the curve with the line at infinity and we pass a tangent to the curve through this point, so that the point of contact is finite (which, as is easily seen, is always possible). Then the contact is of second order (simple). We choose an origin of the affine coordinate system $(X/Y, Z/Y)$ at the point of contact. We direct the axis $X = 0$ along the tangent and the axis $Z = 0$ into one of the two remaining points at infinity.

In writing our form in this coordinate system, the terms

Y^3, Y^2Z, Z^3, X^3 are missing and YZ^2, X^2Z, XZ^2 are present with non-zero coefficients. By dilations we make the non-zero coefficients equal to 1, after which the form becomes $X^2Z + XZ^2 + aX^2Y + bXYZ + YZ^2 + cXY^2$. This form is non-degenerate if its discriminant, the polynomial $\Delta(a, b, c)$, is different from 0. This is the first case of Theorem **90**.

2°. We assume that the curve is the folium of Descartes. We choose an origin of the affine coordinate system $(X/Y, Z/Y)$ at the singular point and direct the axes $(X = 0, Z = 0)$ to the two points of the curve at infinity. Then in writing down the form, all the terms divisible by Y^2 are missing, as are the terms X^3 and Z^3, and X^2Z, XZ^2, X^2Y, Z^2Y are present with non-zero coefficients.

By dilatations we make the first three non-zero coefficients equal to 1, after which the form becomes $X^2Z + XZ^2 + aX^2Y + bXYZ + YZ^2$. This form yields the folium of Descartes if $b^2 \neq 4a$, $a(a + 1 - b) \neq 0$. This is the second case of Theorem **90** (case *UP*).

3°. If the curve has a cusp, then we choose the origin of the system $(X/Y, Z/Y)$ there and direct the axis $Z = 0$ along the tangent to the cusp, and the axis $X = 0$ to one of the points at infinity of the curve. After dilatations we reduce the form to $X^3 + aX^2Z + XZ^2 + Z^2Y$. This form yields a curve with a cusp $X = Z = 0$. It has three distinct points at infinity $(Y = 0)$ for $a^2 \neq 4$. This is the third case of Theorem **90** (case *UQ*).

4°. If the curve splits into a line and a non-singular conic, then we choose the origin of the coordinate system $(X/Y, Z/Y)$ at one of their points of intersection, and we direct the axis $X = 0$ along a line and the axis $Z = 0$ to one of the points of the conic at infinity. After dilatations the form becomes $X^2Z + XZ^2 + X^2Y + aXYZ$.

The conic does not split if $a \neq 1$. The conic is not tangent to the line $X = 0$ for $a \neq 0$. This is the fourth case of Theorem **90** (case *UR*).

5°. If the conic is tangent to the line under the conditions of 4°, then $a = 0$. This is the fifth case of Theorem **90** (case *US*).

6°. If the curve splits into three lines, then we make two of them the coordinate axes $X = 0$ and $Z = 0$. Then the form becomes $XZ(aX + bY + cZ)$, $ac \neq 0$. If the lines do not intersect at one point, then $b \neq 0$ and we reduce the form to $X^2Z + XZ^2 + XYZ$ by dilatations. This is the sixth case of Theorem **90** (case *UT*).

7°. If the curve given by the form $X^3 + XZ^2 + \alpha Y^3 + \beta XY^2 + \gamma ZY^2 + \delta YX^2$ splits into three lines intersecting at the same point, then necessarily $\alpha = \beta = \gamma = \delta = 0$. For let us consider the point $X = Y = 0$. The tangent to the curve at this point $(X = 0)$ must be contained in the curve. Consequently, $\gamma = \alpha = 0$. The curve $X^2 + Z^2 + \beta Y^2 + \delta XZ$ must have a singular point on the axis $X = 0$. Consequently, $\beta = \delta = 0$. This is the last case of Theorem **90**, which is therefore proved, since we have exhausted all the possible cases (by Theorem **50**).

PROOF OF THEOREM **98**. The problem reduces to the classification of the polynomials $aZ^4 + bZ^3 Y + cZ^2 Y^2 + dZY^3 + eY^4$ relative to the transformations $Z = \alpha Z' + \beta Y', Y = \gamma Y'$. This classification in the case $a \neq 0$ is carried out in the proof of Theorem 36, and in the case $a = 0$, $b \neq 0$ in the proof of Theorem 10. The case $a = b = 0$ is trivial, and Theorem 98 is proved.

PROOFS OF THEOREMS 52, 54, 56, 58, 66, 82, 97. In [8], 7.3 we proved the following lemma.

LEMMA 1. *Let f_0 be a quasihomogeneous function of degree d (not necessarily non-degenerate) and suppose that the monomials e_k form a basis of the local ring of the gradient mapping f_0. Then every series $f = f_0 + f_1$, where the series f_1 has a higher order of quasihomogeneity than d, reduces by a formal change of variables to the form $f_0 + \Sigma c_k e_k$, where the c_k are constants, and the e_k are monomials of the basis whose degree is greater than d.*

We verify the next result directly.

LEMMA 2. *The ideals generated by the partial derivatives of the cubic forms of Theorems 52–97 contain the following elements*:

Theorem No.	f_0	Elements of the ideal
52	x^3+y^3+xyz	$x^3,\ y^3,\ x^2z,\ y^2z,\ xz^2,\ yz^2$
54	x^3+xyz	$xy,\ xz,\ x^3,\ y^2z,\ yz^2$
56	xyz	$xy,\ xz,\ yz$
58	x^3+yz^2	$x^2,\ z^2,\ yz$
66	x^2z+yz^2	$xz,\ z^2,\ x^2+2yz$
82	x^3+xz^2	$xz,\ 3x^2+z^2,\ x^3,\ z^3$
97	x^2y	$xy,\ x^2$

The proofs of all these theorems follow from Lemmas 1 and 2.

§17. Proofs of theorems on the normal form of semiquasihomogeneous functions (Theorems 2, 4, 5, 7, 8, 9, 11, 14, 19, 20, 21, 23, 27, 28, 30, 34, 35, 37, 48, 51, 60, 61, 62, 64, 68, 69, 71, 75, 76, 78, 84, 86, 89, 91, 99)

In these theorems we are given a non-degenerate quasihomogeneous function f_0 of degree N and we find a normal form for the function $f_0 + f_1$, where f_1 has order greater than N. The normal form is like $f_0 + \Sigma c_s e_s$, where the e_s are monomials that generate the space of all functions of order greater than N over $\mathbf{C}$ together with the gradient ideal $I = \{(\partial f_0/\partial x_i)\}$ (see Lemma 1 in §16 and Theorem 7.2 in [8]).

Therefore, for the proof it suffices to verify that for each (weighted) degree $d > N$ the space of quasihomogeneous polynomials of degree d is generated over $\mathbf{C}$ by the polynomials $x^m \partial f_0/\partial x_i$ (where $x^m \partial_i$ are the "leading monomials" of degree $d - N$) and monomials e_s of degree d included in the normal form of §13. To verify the independence of the monomials e_s mod I it suffices to compare the number of these with the number p_d of basis monomials of degree d, which is given by the Poincaré polynomial [8]

$$p(t) = \Sigma p_d t^d = \Pi (t^{N-A_i} - 1)/\Pi (t^{A_i} - 1),$$

where the A_i are weights (so that the degree of x^m is equal to $\Sigma\, m_i A_i$, that of f_0 is N).

As an example we consider the proof of Theorem 37. In this case $f_0 = x^4 + bx^3y^k + ax^2y^{2k} + xy^{3k}$, so that $A_1 = k$, $A_2 = 1$; $N = 4k$. The degree of $x^p y^q$ is $d = kp + q$. The degree of the Poincaré polynomial is $6k - 2$. Consequently, the condition for f_0 to be non-degenerate is the same as that for all the monomials of degree $d = 6k - 1$ to belong to I.

In the following table, for the values of d in which we are interested (that is, for $4k < d \leqslant 6k - 1$) we have indicated the number $\nu(d)$ of monomials of degree d, the leading monomials $x^m \partial_i$, and the basis monomials e_s.

Here in the notation of the basis and leading monomials we do not indicate the degree of y, for example, in place of $xy^u \partial_y$ we write $x\partial_y$ (the quantity u can be recovered from d).

d	$4k \leqslant d \leqslant 5k-2$	$5k-1$	$5k \leqslant d \leqslant 6k-2$	$6k-1$
$\nu(d)$	5	5	6	6
$\{x^m \partial_i\}$	$x\partial_x,\ \partial_x,\ \partial_y$	$x\partial_x,\ \partial_x,\ x\partial_y,\ \partial_y$	$x^2\partial_x,\ x\partial_x,\ \partial_x,\ x\partial_y,\ \partial_y$	$x^2\partial_x,\ x\partial_x,\ \partial_x,\ x^2\partial_y,\ x\partial_y,\ \partial_y$
$\{e_s\}$	$x^2,\ x^3$	x^3	x^3	—

To check that $\{e_s\}$ and $\{x^m \partial_i f_0\}$ generate $\mathbf{C}^{\nu(d)}$, it suffices to verify that the "quasidiscriminant" is different from 0 (this is a determinant formed from the components of the expansions of these $\nu(d)$ functions in the basis of monomials of degree d). The four cases of the preceding table correspond to 4 determinants $\Delta_1 - \Delta_4$, which are, respectively,

$$\begin{vmatrix} 4 & & & & \\ 3b & 4 & kb & & 1 \\ 2a & 3b & 2ka & 1 & \\ 1 & 2a & 3k & & \\ & 1 & & & \end{vmatrix}, \quad \begin{vmatrix} 4 & & kb & & \\ 3b & 4 & 2ka & kb & 1 \\ 2a & 3b & 3k & 2ka & \\ 1 & 2a & & 3k & \\ & 1 & & & \end{vmatrix}, \quad \begin{vmatrix} 4 & & & & & \\ 3b & 4 & & kb & & \\ 2a & 3b & 4 & 2ka & kb & 1 \\ 1 & 2a & 3b & 3k & 2ka & \\ & 1 & 2a & & 3k & \\ & & 1 & & & \end{vmatrix}, \quad \begin{vmatrix} 4 & & & & & \\ 3b & 4 & & kb & & \\ 2a & 3b & 4 & 2ka & kb & \\ 1 & 2a & 3b & 3k & 2ka & kb \\ & 1 & 2a & & 3k & 2ka \\ & & 1 & & & 3k \end{vmatrix}.$$

Thus, $\Delta_1 = 12k$, $\Delta_2 = 4k^2(ab - 9)$, $\Delta_3 = 4\Delta_2$, $\Delta_4 = -64k^3\Delta_5$, where Δ_5 is the discriminant of the polynomial $x^3 + bx^2 + ax + 1$. The non-degeneracy condition $\Delta_5 \neq 0$ has the form $4(a^3 + b^3) + 27 \neq a^2b^2 + 18ab$.

If $\Delta_5 \neq 0$, $ab \neq 9$, then all four determinants are different from 0, and hence, f reduces to the normal form

$$f_0+\sum c_s e_s = x^4 + bx^3y^k + ax^2y^{2k} + xy^{3k},$$

where $a = a_0 + \ldots + a_{k-2}y^{k-2}$, $b = b_0 + \ldots + b_{2k-2}y^{2k-2}$.

But by Theorem 36 every non-degenerate quasihomogeneous function $x^4 + bx^3y^k + ax^2y^{2k} + xy^{3k}$ is equivalent to one for which $ab \neq 9$. Theorem 37 is now proved.

The difficulty in the proof of this and the remaining theorems of the present section is that we have to guess the monomials e_s that are suitable simultaneously for all the values of the parameters of the quasihomogeneous principal part. The existence of such monomials is not a priori obvious.

It is also appropriate to remark that the classification of semiquasihomogeneous functions, like that of quasihomogeneous polynomials, reduce to calculations that actually do not depend on k. However, in contrast to the quasihomogeneous case, in the semiquasihomogeneous case (and a fortiori in the general case of an arbitrary Newton diagram), this independence from k has not yet been explained satisfactorily.

In many cases we may avoid the determinants and guess the basis monomials with the help of the technique of "crosswords" from [8]. For example, let us prove Theorem 71.

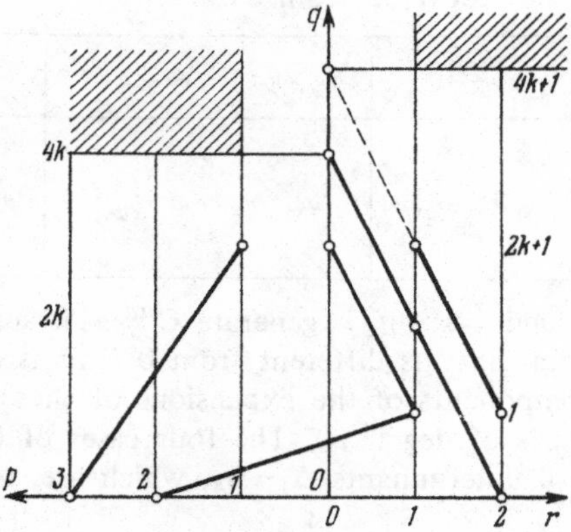

Fig. 12

In this theorem, $f_0 = x^2z + yz^2 + y^{4k+1} + bzy^{2k+1}$, $b^2 \neq 4$.

The gradient ideal I contains xz, so that the exponents of all the basis monomials $e_s = x^p y^q z^r$ lie on the planes (p, q) and (q, r), where it is also convenient to draw the crossword. The derivatives

$$\partial f_0/\partial y = z^2 + \alpha z y^{2k} + \beta y^{4k}, \quad \partial f_0/\partial z = x^2 + 2yz + by^{2k+1}$$

(where $\alpha = (2k+1)b$, $\beta = 4k+1$) correspond to chains of three points. Translating the chains, we can see that all the monomials on the planes

(p, q) and (q, r) are congruent mod I to combinations of monomials of the form xy^q, y^q, zy^q. Looking at the crossword we successively find (Fig. 12) on p. 63 $xy^4 \in I$, $x^3 + bxy^{2k+1} \in I$, $2yz^2 + bzy^{2k+1} \in I$, $yz^2 + \alpha zy^{2k+1} + \beta y^{4k+1} \in I$, $bzy^{2k+1} + 2y^{4k+1} \in I$, $2zy^{4k+1} + by^{6k+1} \in I$. For $b^2 \neq 4$, from the last two relations we conclude $y^{6k+1} \in I$, $zy^{4k+1} \in I$. Consequently, the monomials $xy^q (3k + 1 \leqslant q \leqslant 4k - 1)$ and $zy^q (2k + 1 < q \leqslant 4k)$ together with I generate the whole space of functions whose order is greater than the degree of f_0. Hence, for $b^2 \neq 4$ the quasihomogeneous function f_0 is non-degenerate.

To prove the independence of these new monomials mod I, we observe that $12k + 2$ of the monomials $xy^q (0 \leqslant q \leqslant 4k - 1)$, $y^q (0 \leqslant q \leqslant 4k)$, $zy^q (0 \leqslant q \leqslant 4k)$ together with I generate the whole ring $\mathbf{C}[[x, y, z]]$ over $\mathbf{C}$. On the other hand, the weights A_i and the degree N are equal to $(2k + 1, 2, 4k)$ and $8k + 2$. Hence, the dimension of the local ring $\mathbf{C}[[x, y, z]]/I$ over $\mathbf{C}$ is equal to

$$\mu = \Pi(N - A_i)/\Pi A_i = 12k + 2.$$

Consequently, the listed $12k + 2$ monomials are independent, as required.

The remaining theorems on semiquasihomogeneous singularities are verified similarly; the determinants only must be considered for the proof of Theorems **37**, **48**, **51**, **64**, **71**, **78**, **86**, **91**, **99**. Here the crossword gives not only the normal forms, but also all the generators of the local ring (and, consequently, a formula for the versal deformation).

§ 18. Proofs of theorems on the normal forms of singularities of the series T (Theorems 53, 55, 57)

We use the terminology of [8].

PROPOSITION. *Let* $f_0 = ax_1x_2x_3 + x_1^{p_1} + x_2^{p_2} + x_3^{p_3}$, *where* $3 \leqslant p_1 \leqslant p_2 \leqslant p_3 > 3, a \neq 0$. *Then*:

1) *the function* f_0 *has a critical point* O *of finite multiplicity* $\mu = p_1 + p_2 + p_3 - 1$;

2) *the system of monomials* $1, x_1x_2x_3, x_i^{s_i}$ $(0 < s_i < p_i,\ i = 1, 2, 3)$ *defines a regular basis of the local ring*;

3) *condition* **A** *holds for the Newton filtration defined by the four monomials of* f_0.

PROOF. We apply the technique of crosswords described in [8], §9.7.

LEMMA 1. *All the cycles of the admissible segments for* f_0 *are trivial.*

PROOF. This follows from the linear independence of the three basic segments, which is guaranteed by the conditions on p_i.

LEMMA 2. *The maximal admissible chains are the following*:

1) *each of the points* $x_i^{s_i}$ $(0 \leqslant s_i \leqslant p_i - 2)$ *has a trivial admissible chain from one of these points*;

2) *there are four finite maximal chains*

$$x_1x_2 - x_3^{p_3-1}, \quad x_2x_3 - x_1^{p_1-1}, \quad x_3x_1 - x_2^{p_2-1}, \quad x_1x_2x_3 - \{x_i^{p_i}\};$$

3) *infinite admissible chains issue from all the remaining points.*

Lemma 2 is geometrically obvious, but we give a formal proof.

PROOF OF LEMMA 2. We consider an arbitrary monomial $\Pi x_i^{q_i}$. It can be written in the form $x^\xi y^\eta z^\zeta$, where $\xi \leqslant \eta \leqslant \zeta$; in this notation (depending on q_i), $f_0 = axyz + x^p + y^q + z^r$. Then there are eight mutually exclusive possibilities:

0. $0 = \xi < \eta = \zeta = 1$. IV. $0 < \xi = \eta = \zeta = 1$.
I. $\xi < \eta \leqslant \zeta \neq 1$. V. $0 = \xi = \eta, \quad \zeta > r$.
II. $0 < \xi = \eta < \zeta$. VI. $0 = \xi = \eta, \quad \zeta = r$.
III. $0 < \xi = \eta = \zeta > 1$. VII. $0 = \xi = \eta, \quad \zeta < r$.

Corresponding to this we distinguish eight types of monomials. If for a monomial $x^\xi y^\eta z^\zeta$ the condition $\zeta \leqslant \eta \leqslant \xi$ does not hold, then we define its type as that of a monomial with regular order of the exponents obtained by a permutation of the names of the variables.

The vector-exponent of a monomial of type i is called a point of type i. From each point of the types I, II, III, IV there issues an admissible x-, y- or z-segment, taking into one point the same of a higher Newton filtration. In the following table we indicate the ends of these segments and their type:

Type of start	Start	Type of segment	End	Type of end	Filtration
I	ξ, η, ζ	x	$\xi+p-1, \eta-1, \zeta-1$	I, II or III	$>$
II	ξ, ξ, ζ	y	$\xi-1, \xi+q-1, \zeta-1$	I	$=$
III	ξ, ξ, ξ	z	$\xi-1, \xi-1, \xi+r-1$	II	$=$
V	$0, 0, \zeta$	z	$1, 1, \zeta-r+1$	II	$=$

Consequently, from each point of each of the types I, II, III, V there issues an admissible chain of at most three segments, ending at a point of a strictly larger filtration of one of the same types. Thus, the maximal chains of points of these types are infinite.

The admissible chains of points $x_i^{s_i}$ $(0 \leqslant s_i \leqslant p_i - 2)$ of type VII are obviously trivial. There remain ten points, corresponding to monomials of the types 0, IV, VI, VII:

$$x_i^{p_i}, \quad x_i^{p_i-1}, \quad x_1x_2, \quad x_2x_3, \quad x_3x_1, \quad x_1x_2x_3.$$

The six admissible segments passing through them have no other vertices. Therefore the maximal chains of them are finite. Lemma 2 is now proved.

The fact that of the critical point O of f_0 is isolated can be verified directly. In each finite maximal chain of monomials we choose a higher filtration. We obtain a regular basis of $\mu = p_1 + p_2 + p_3 - 1$ monomials. Assertions 1 and 2 of our proposition are proved (see [8], §9.7).

All strictly upper monomials for f_0 have types I, II, III, V. Consequently, all of them belong to the ideal $\{\partial f_0/\partial x_i\}$. We check that condition *A* holds (see [8], §9.2).

We denote by φ the Newton filtration (φ assigns to each function the cardinal number d of the smallest term $\mathscr{E}_d$, of the filtration to which the function belongs; $\varphi(f_0) = \varphi(x_i^{p_i}) = \varphi(x_1x_2x_3) = 1$). We denote by ψ the filtration induced by in the space of vector fields:

$$\psi(\gamma) = \max\{\delta : \gamma\mathscr{E}_d \subset \mathscr{E}_{d+\delta} \forall d\}.$$

LEMMA 3. *Each upper monomial α admits the decomposition*

$$\alpha = \gamma f_0 + \alpha', \quad \varphi(\alpha') > \varphi(\alpha), \quad \psi(\gamma) = \varphi(\alpha) - 1.$$

PROOF OF LEMMA 3. In the proof of Lemma 2 we constructed an admissible chain (of at most three segments), joining α to a monomial of the higher filtration α' ($\alpha = \alpha_1; \alpha_2; \ldots; \alpha'$). By the definition of an admissible chain

$$\alpha_i = a_i\alpha_{i+1} + b_iv_i\partial f_0/\partial w_i, \tag{1}$$

where a_i and b_i are constants, v_i is a monomial, and w_i is one of the coordinates x_1, x_2, x_3. By construction, $\varphi(\alpha_{i+1}) \geqslant \varphi(\alpha_i)$, and at least one of the inequalities is strict.

We claim that $\psi(v_i\partial/\partial w_i) = \varphi(\alpha_i) - 1$. For we can extend a piecewise-linear function φ of the exponents of the monomials into a domain where one of the exponents is equal to -1, linearly through a plane on which this exponent is 0. Then

$$\psi(v_i\partial/\partial w_i) = \varphi(v_i/w_i).$$

The vector-exponents of the monomials $x_1x_2x_3$ and v_i/w_i lie together in one of three convex cones, in which φ depends linearly on the exponents. Therefore,

$$\varphi(v_ix_1x_2x_3/w_i) = \varphi(v_i/w_i) + \varphi(x_1x_2x_3) = \psi(v_i\partial/\partial w_i) + 1.$$

On the other hand, we compute v_i and w_i for each of the groups constructed in the proof of Lemma 2. If α_i is of type I, II or III, we obtain $\varphi(v_ix_1x_2x_3/w_i) = \varphi(\alpha_i)$, and if of type V, $\varphi(v_ix_1x_2x_3/w_i) = \varphi(\alpha_{i+1}) = \varphi(\alpha_i)$. Thus, in all cases $\psi(v_i\partial/\partial w_i) = \varphi(\alpha_i) - 1$.

Since $\varphi(\alpha_i) \geqslant \varphi(\alpha_1)$, we obtain $\psi(v_i\partial/\partial w_i) \geqslant \varphi(\alpha_1) - 1$. Now we express according to (1) α_1 by α_2, α_2 by α_3 etc. to α', with $\varphi(\alpha') > \varphi(\alpha)$. This gives us the required decomposition and proves the proposition.

Now we can deduce Theorems 53, 55, 57. We claim that if *f has a critical point O of finite multiplicity and*

$$f = xyz + \alpha(x) + \beta(y) + \gamma(z), \quad j_2(\alpha, \beta) = 0, \quad j_3(\gamma) = 0,$$

then in a neighbourhood of O,

$$f \sim f_0 = axyz + x^p + y^q + z^r, \quad a \neq 0, \quad 3 \leqslant p \leqslant q \leqslant r > 3.$$

For suppose that the first non-zero terms of the Taylor series for α, β, γ have the exponents p, q, r. Changing the notation, if necessary, we obtain $3 \leqslant p \leqslant q \leqslant r > 3$. By a dilatation of the axes we reduce f to the form $f_0 + f_1$, where $\varphi(f_1) > 1$. By the proposition proved above, all the monomials of f_1 lie in the ideal $\{\partial f_0/\partial x_i\}$ and f_0 satisfies condition A. By Theorem 9.5 of [8], $f \sim f_0$, as required.

References[1]

[1] N. A'Campo, Sur la monodromie des singularités isolées d'hypersurfaces complexes, Invent. Math. **20** (1973), 147–169. MR **49** # 3201.

[2] N. A'Campo, Le nombre de Lefschetz d'une monodromie, Indag. Math. **35** (1973), 113–118. MR **47** # 8903.

[3] V. I. Arnol'd, Singularities of smooth mappings, Uspekhi Mat. Nauk **23**:1 (1968), 3–44. MR **37** # 2243.
= Russian Math. Surveys **23**:1 (1968), 1–43.

[4] V. I. Arnol'd, On some topological invariants of algebraic functions, Trudy Moskov. Mat. Obshch. **21** (1970), 27–46. MR **43** # 225.
= Trans. Moscow Math. Soc. **21** (1970), 30–52 (1972).

[5] V. I. Arnol'd, Integrals of rapidly oscillating functions and singularities of the projections of Lagrangian manifolds, Funktsional. Anal. i Prilozhen. **6**:3 (1972), 61–62. MR **50** # 8594.
= Functional Anal. Appl. **6** (1973), 222–224.

[6] V. I. Arnol'd, Normal forms of functions close to degenerate critical points, the Weyl groups A_k, D_k, E_k, and Lagrangian singularities, Funktsional. Anal. i Prilozhen. **6**:4 (1972), 3–25. MR **50** # 8595.
= Functional Anal. Appl. **6** (1973), 254–272.

[7] V. I. Arnol'd, Remarks on the stationary phase method and Coxeter numbers, Uspekhi Mat. Nauk 28:5 (1973), 17–44. MR **53** # 1635.
= Russian Math. Surveys **28**:5 (1973), 19–48.

[8] V. I. Arnol'd, Normal forms of functions in a neighbourhood of a degenerate critical point, Uspekhi Mat. Nauk **29**:2 (1974), 11–49.
= Russian Math. Surveys **29**:2 (1974), 10–50.

[9] V. I. Arnol'd, Critical points of functions and the classification of caustics, Uspekhi Mat. Nauk **29**:3 (1974), 243–244 (*summary of a lecture, with errata to [6] and [8]).

[10] V. I. Arnol'd, *Matematicheskoi metody klassicheskoi mekhaniki,* Izdat. Nauka, Moscow 1974.
*Translation: *Mathematical methods of classical mechanics,* Springer-Verlag, Berlin-Heidelberg-New York and Izdat. Mir, Moscow, to appear.

[1] The references prefixed by an asterisk were added by the translator, while references [91] through [132] were added by the author for this translation.

[11] V. I. Arnol'd, Contact manifolds, Legendrian mappings, and singularities of wave fronts, Uspekhi Mat. Nauk **29**:4 (1974), 153–154 (*summary of a lecture)

[12] V. I. Arnol'd, Lectures on bifurcations in versal families, Uspekhi Mat. Nauk **27**:5 (1972), 119–184.
= Russian Math. Surveys, **27**:5 (1972), 54–123.

[13] V. I. Arnol'd, I. M. Gel'fand, Yu. I. Manin, B. G. Moishezon, S. P. Novikov and I. R. Shafarevich, Galina Nikolaevna Tyurina (obituary), Uspekhi Mat. Nauk **26**:1 (1971), 207–211. MR **44** # 6430.
= Russian Math. Surveys **26**:1 (1971), 193–197.

[14] M. Artin, On the solutions of analytic equations, Invent. Math. **5** (1968), 277–291. MR **38** # 344.

[15] M. F. Atiyah, Resolution of singularities and division of distributions, Comm. Pure Appl. Math. **23** (1970), 145–150. MR **41** # 815.

[16] I. N. Bernstein and S. I. Gel'fand, Meromorphy of the functions P^{λ}, Funktsional. Anal. i Prilozhen. **3**:1 (1969), 84–85. MR **40** # 723.
= Functional Anal. Appl. **3** (1969), 68–69.

[17] I. N. Bernstein, Analytic continuation of generalized functions with respect to a parameter, Funktsional. Anal. i Prilozhen. **6**:4 (1972), 26–40. MR **47** # 9269.
= Functional Anal. Appl. **6** (1973), 273–285.

[18] E. Brieskorn, Rationale Singularitäten komplexer Flächen, Invent. Math. **4** (1968), 336–358. MR **36** # 5136.

[19] E. Brieskorn, Beispiele zur Differentialtopologie von Singularitäten, Invent. Math. **2** (1966), 1–14. MR **34** # 6788.

[20] E. Brieskorn, Singular elements of semisimple algebraic groups, Actes Congrès Internat. Mathématiciens, Nice 1970, Tome 2, 279–284, Gauthier-Villars, Paris 197

[21] E. Brieskorn, Die Fundamentalgruppe des Raumes der regulären Orbite einer endlichen komplexen Spiegelungsgruppe, Invent. Math. **12** (1971), 57–61. MR **45** # 2692.
= Uspekhi Mat. Nauk **30**:6 (1975), 147–151.

[22] E. Brieskorn, Sur les groupes des tresses (d'après V. I. Arnol'd), Séminaire Bourbaki 1971–72, exp. 401, Lecture Notes in Math. **317**, Springer-Verlag, Berlin-Heidelberg New York 1973.

[23] E. Brieskorn and K. Saito, Artin-Gruppen unde Coxeter-Gruppen, Invent. Math. **17** (1972), 245–271. MR **48** # 2263.
= Matematika **18**:6 (1974), 56–79.

[24] E. Brieskorn, Die Monodromie der isolierten Singularitäten von Hyperflächen, Manuscripta Math. **2** (1970), 103–106. MR **42** # 2509.
= Matematika **15**:4 (1971), 130–160.

[25] J. Cerf, La stratification naturelle des espaces de fonctions différentiables réelles et le théorème de la pseudo-isotopie, Inst. Hautes Études Sci. Publ. Math. **39** (1970), 5–173. MR **45** # 1176.

[26] P. Deligne, Les immeubles des groupes de tresses généralisés, Invent. Math. **17** (1972) 273–302.

[27] M. Demazure, Sousgroupes algébriques de rang maximum du groupe de Cremona, Ann. Sci. Ecole Normale Sup. (4) **3** (1970), 507–588. MR **44** # 1672.

[28] I. V. Dolgachev, Factor-conical singularities of complex hypersurfaces, Funktsional. Anal. i Prilozhen. **8**:2 (1974), 75–76. MR **49** # 10700.
= Functional Anal. Appl. **8** (1974), 160–161.

[29] I. V. Dolgachev, Automorphic forms and quasihomogeneous singularities, Funktsional. Anal. i Prilozhen. **9**:2 (1975), 67–68.
= Functional Anal. Appl. **9** (1975), 149–151.

[30] J. J. Duistermaat, Oscillatory integrals, Lagrange immersions, and unfolding of singularities, Comm. Pure Appl. Math. **27** (1974), 207–281.

[31] A. Fresnel, Mémoire sur la diffraction de la lumière, Mém. Acad. Roy. Sci. Paris **5** (1821–22), 339–475 (*published in 1826, submitted in 1818 – the usual citation is incorrect).

[32] D. B. Fuks, Quillenization and bordism, Funktsional. Anal. i Prilozhen. **8**:1 (1974), 36–42. MR **49** # 8043.
= Functional Anal. Appl. **8** (1974), 31–36.

[33] D. B. Fuks, Cohomology of braid groups mod 2, Funktsional. Anal. i Prilozhen. **4**:2 (1970), 62–73; MR **43** # 226.
= Functional Anal. Appl. **4** (1970), 143–151.

[34] A. M; Gabrielov, Intersection matrices for certain singularities, Funktsional. Anal. i Prilozhen. **7**:3 (1973), 18–32. MR **48** # 2418.
= Functional Anal. Appl. **7** (1974), 182–193.

[35] A. M. Gabrielov, Dynkin diagrams of unimodal singularities, Funktsional. Anal. i Prilozhen. **8**:3 (1974), 1–6. MR **51** # 3516.
= Functional Anal. Appl. **8** (1975), 192–196.

[36] A. M. Gabrielov, Bifurcations, Dynkin diagrams, and the modality of isolated singularities, Funktsional. Anal. i Prilozhen. **8**:2 (1974), 7–12.
= Functional Anal. Appl. **8** (1974), 94–98.

[37] A. M. Gabrielov and A. G. Kushnirenko, Description of the deformations with constant Milnor number for homogeneous functions, Funktsional. Anal. i Prilozhen. **9**:4 (1975), 67–68.

[38] J. Guckenheimer, Catastrophes and partial differential equations, Ann. Inst. Fourier (Grenoble) **23**:2 (1973), 31–59. MR **51** # 1879.

[39] S. M. Gusein-Zade, Intersection matrices for some singularities of functions of two variables, Funktsional. Anal. i Prilozhen. **8**:1 (1974), 11–15. MR **49** # 3202.
= Functional Anal. Appl. **8** (1974), 10–13.

[40] S. M. Gusein-Zade, Dynkin diagrams for singularities of functions of two variables, Funktsional. Anal. i Prilozhen. **8**:4 (1974), 23–30.
= Functional Anal. Appl. **8** (1975), 295–300.

[41] A. E. Hatcher, Parametrized *h*-cobordism theory, Ann. Inst. Fourier (Grenoble) **23**:2 (1973), 61–74. MR **50** # 1267.

[*41a] A. E. Hatcher and J. Wagoner, *Pseudo-isotopies of compact manifolds,* Astérisque No. 6, Offilib, Paris 1973. MR **50** # 5821.

[42] L. Hörmander, Fourier integral operators. I, Acta Math. **127** (1971), 79–183. MR **52** # 9399.
= Matematika **16**:1 (1972), 17–61; **16**:2 (1972), 67–136.

[43] F. Hirzebruch, The topology of normal singularities of an algebraic surface (d'après Mumford), Séminaire Bourbaki, 15ème année, Exp. No. 250, 1962/1963, Benjamin, New York.

[44] F. Hirzebruch and K. H. Mayer, *O(n)-Mannigfaltigkeiten, exotische Sphären und Singularitäten,* Lecture Notes in Math. **57**, Springer-Verlag, Berlin-Heidelberg-New York, 1968. MR **37** # 4825.

[45] F. Klein, *Vorlesungen über die Entwicklung der Mathematik im 19. Jahrhundert*, Springer-Verlag, Berlin 1926; reprinted, Chelsea, New York 1967.

[46] N. Kuiper, Algebraic equations for non-smoothable 8-manifolds, Inst. Hautes Etudes Scient. Publ. Math. **33** (1967), 139–155. MR **39** # 3494.

[47] F. Latour, Stabilité des champs d'applications differentiables, généralisation d'un théorème de J. Mather, C. R. Acad. Sci. Paris **268** (1969), A1331–1334. MR **39** # 7617.

[48] F. Lazzeri, A theorem on the monodromy of isolated singularities (Singularités à Cargèse, 1972), Astérisque **7–8** (1973), 269–275, Soc. Math. France, Paris 1973.

[49] Lê Dũng Tráng, Les théorèmes de Zariski de type Lefschetz, preprint, Centre Maths. École Polytechnique, Paris 1971.

[*49a] (published version of [49]) Lê Dũng Tráng and Helmut A. Hamm, Un théorème de Zariski du type de Lefschetz, Ann. Sci. Ecole Norm. Sup. (4) **6** (1973), 317–355.

[50] Lê Dũng Tráng and C. P. Ramanujam, The invariance of the Milnor number implies the invariance of the topological type, preprint, Centre Maths. École Polytechnique, Paris 1973, Amer. J. Math. **98** (1976), 67–68.

[51] E. Locijenga, The complement of the bifurcation variety of a simple singularity, Invent. Math. **23** (1974), 105–116.

[52] B. Malgrange, Intégrales asymptotiques et monodromie, Sém. Leray au Collège de France, 1972/1973; *published version: Ann. Sci. Ecole Norm. Sup. (4) **7** (1974), 405–430. MR **51** # 8459.

[53] B. Malgrange, Sur les polynômes de I. N. Bernstein, Sém. Goulaouic-Schwartz 1973/74, École Polytechnique, Paris, Exp. No. 20.
= Uspekhi Mat. Nauk **29**:4 (1974), 81–88.
= Russian Math. Surveys **29**:4 (1974), 81–88.

[*53a] B. Malgrange, Le polynôme de Bernstein d'une singularité isolée, in: Fourier integral operators and partial differential equations, Lecture Notes in Math. **459**, 98–119, Springer-Verlag, Berlin-Heidelberg-New York 1975.

[54] J. N. Mather, Stability of C^∞-mappings I–VI: Ann. of Math.(2) **87** (1968), 89–104. MR **38** # 726; ibid. **89** (1969), 254–291. MR **41** # 4582; Inst. Hautes Etudes Sci. Publ. Math. **35** (1969), 127–156 (279–308). MR **43** # 1215a; ibid. **37** (1970), 223–248. MR **43** # 1215b; Advances in Math. **4** (1970), 301–335. MR **43** # 1215c; Proc. Liverpool Conf. on Singularities, Vol. I, Lecture Notes in Math. **192**, 207–253, Springer-Verlag, Berlin-Heidelberg-New York 1971. MR **45** # 2747.
= (I) (translation of a preliminary version). In the collection: *Osobennosti differentsiruemykh otobrazhenii* (Singularities of differentiable mappings), 198–215, Izdat. Mir, Moscow 1968. MR **39** # 3511.
= (II) (translation of a preliminary version) In the collection *Osobennosti differentsiruemykh otobrazhenii* (Singularities of differentiable mappings), 217–267, Izdat. Mir, Moscow 1968. MR **40** # 4966.
= (III) Matematika **14**:1 (1970), 146–175.
= (IV) Uspekhi Mat. Nauk **28**:6 (1973), 165–190.
= (V) Uspekhi Mat. Nauk **29**:1 (1974), 99–128.
= (VI) Uspekhi Mat. Nauk **29**:1 (1974), 129–158.

[55] V. P. Maslov, *Teoriya vozmushchenii i asimptoticheskie metody* (Perturbation theory and asymptotic methods), Izdat. Moskov. Gos. Univ., Moscow 1965. Translation: *Théorie des perturbations et méthodes asymptotiques*, Dunod, Paris, 1972.

[56] V. P. Maslov, On the focusing of energy in a crystal lattice, Uspekhi Mat. Nauk **27**:6 (1972), 224 (*summary of a lecture).

[57] J. Milnor, *Singular points of complex hypersurfaces,* Annals of Math. Studies **61**, Princeton Univ. Press, Princeton, N.J., 1968. MR **39** # 969.
Translation: *Osobye tochki kompleksnykh giperpoverkhnostei*, Izdat. Mir, Moscow 1971. MR **50** # 8560.

[58] J. Milnor and P. Orlik, Isolated singularities defined by weighted homogeneous polynomials, Topology **9** (1970), 385–393. MR **45** # 2757.

[59] I. Newton, The method of fluxions (1671), Mathematical papers, Vol. 3, Cambridge University Press, Cambridge 1969, 49 ff.

[60] P. Orlik and P. Wagreich, Isolated singularities of algebraic surfaces with C^*-action, Ann. of Math. (2) **93** (1971), 205–228. MR **44** # 1662.

[61] V. P. Palamodov, The multiplicity of a holomorphic mapping, Funktsional. Anal. i Prilozhen. **1**:3 (1967), 54–65. MR **38** # 4720.
= Functional Anal. Appl. **1** (1967), 218–226.

[62] F. Pham, Formules de Picard-Lefschetz généralisées et ramifications des intégrales, Bull. Soc. Math. France **93** (1968), 333–367. MR **33** # 4064.

[63] F. Pham, Remarque sur l'equisingularité universelle, Preprint, Faculté des Sciences, Nice (1970), 1–24.

[64] H. Poincaré, Sur les formes cubiques ternaires et quaternaires. I, J. École Polytechnique **50** (1881), 190–253 = Oeuvres V, 28–72.
= *Izbrannye trudy*, Tom II, Izdat. Nauka, Moscow 1973, 829–836 (*These pages correspond to §4 of the original paper.)

[65] K. Saito, Einfach-elliptische Singularitäten, Invent. Math., **23** (1974), 289–325. MR **50** # 7147.

[66] K. Saito, Quasihomogene isolierte Singularitäten von Hyperflächen, Invent. Math. **14** (1971), 123–142. MR **45** # 3767.

[67] A. M. Samoilenko, On the equivalence of a smooth function with a Taylor polynomial in a neighbourhood of a critical point of finite type. Funktsional. Anal. i Prilozhen. **2**:4 (1968), 63–69. MR **42** # 5132.
= Functional Anal. Appl. **2** (1969), 318–323.

[68] G. Segal, Configuration-spaces and iterated loop-spaces, Invent. Math. **21** (1973), 213–222. MR **48** # 9710.

[69] D. Siersma, The singularities of C^∞-functions of right codimension smaller or equal than eight, Indag. Math. **35**:1 (1973), 31–37. MR **51** # 11574.

[*69a] D. Siersma, *Classification and deformation of singularities,* Academisch Proefschrift, Univ. of Amsterdam, 1974. MR **50** # 3267.

[70] *Singularités à Cargèse,* Astérisque 7/8, Soc. Math. France, Paris 1973. MR **48** # 11548 (table of contents review).

[71] B. Teissier, Cycles évanescents, sections planes et conditions de Whitney, Astérisque 7/8, 285–362, Soc. Math. France, Paris 1973. MR **51** # 10682.

[72] R. Thom and H. Levine, Stability of differentiable mappings. I, Bonner Math. Schriften **6**, Univ. Bonn 1959.
= Proc. Liverpool Singularities Symposium. I, 1–89, Lecture Notes in Math. **192**, Springer-Verlag, Berlin-Heidelberg-New York 1971.
= in the collection: *Osobennosti differentsiruemykh otobrazhenii* (Singularities of differentiable mappings), 9–101, Izdat. Mir, Moscow 1968. MR **40** # 896.

[73] R. Thom and M. Sebastiani, Un résultat sur la monodromie, Invent. Math. **13** (1971), 90–96. MR **45** # 2201.

[74] R. Thom, *Stabilité structurelle et morphogénèse,* Benjamin, Reading, Mass., 1972. Translation: *Structural stability and morphogenesis,* Benjamin, Reading, Mass., 1974.

[75] R. Thom, The bifurcation subset of a space of maps, in: *Manifolds*, Amsterdam, 1970, Lecture Notes in Math. **197**, 202–208, Springer-Verlag, Berlin-Heidelberg-New York 1971. MR **43** # 6941.
= Uspekhi Mat. Nauk **27**:5 (1972), 51–58.

[76] G. N. Tyurina, The topological properties of isolated singularities of complex spaces of codimension one, Izv. Akad. Nauk SSSR Ser. Mat. **32** (1968), 605–620. MR **37** # 3053.
= Math. USSR-Izv. **2** (1968), 557–571.

[77] G. N. Tyurina, Locally semi-universal flat deformations of isolated singularities of complex spaces, Izv. Akad. Nauk SSSR Ser. Mat. **33** (1969), 1026–1058. MR **40** # 5903.
= Math. USSR-Izv. **3** (1969), 967–999.

[78] G. N. Tyurina, Resolution of singularities of flat deformations of rational double points, Funktsional. Anal. i Prilozhen. **4**:1 (1970), 77–83. MR **42** # 2031.
= Functional Anal. Appl. **4** (1970), 68–73.

[79] J. C. Tougeron, Idéaux de fonctions différentiables, Ann. Inst. Fourier (Grenoble) **18**:1 (1968), 177–240. MR **39** # 2171.

[80] A. N. Varchenko, Theorems of topological equisingularity of families of algebraic varieties and a family of polynomial mappings, Izv. Akad. Nauk SSSR Ser. Mat. **36** (1972), 957–1019. MR **49** # 2725.
= Math. USSR-Izv. **6** (1972), 949–1008.

[81] I. M. Vinogradov, *Metod trigonometricheskikh summ v teorii chisel* (The method of trigonometric sums in number theory), Izdat. "Nauka", Moscow 1971.

[82] I. A. Volodin, Generalized Whitehead groups and pseudo-isotopies, Uspekhi Mat. Nauk **27**:5 (1972), 229–230. MR **52** # 15501.

[83] P. Wagreich, Singularities of complex surfaces with solvable fundamental groups, Topology **11** (1972), 51–72. MR **44** # 2754.

[84] J. B. Wagoner, Algebraic invariants for pseudoisotopies, Proc. Liverpool Singularities Symposium. II, 164–190. Lecture Notes in Math. **209**, Springer-Verlag, Berlin-Heidelberg–New York 1971. MR **50** # 1266.

[85] A. Weinstein, Singularities of families of functions, in: *Differentialgeometrie im Grossen, Oberwolfach* 1971, Ber. Math. Forschungsinst. Oberwolfach, Heft 4, BI, Mannheim 1971.

[86] H. Whitney, On singularities of mappings of Euclidean spaces. I, Ann. of Math. (2) **62** (1955), 374–410. MR **17**–518.

[87] V. M. Zakalyukin, A theorem of versality, Funktsional. Anal. i. Prilozhen. **7**:2 (1973), 28–32. MR **47** # 9670.
= Functional Anal. Appl. **7** (1973), 110–112.

[88] O. Zariski, On the Poincaré group of a projective hypersurface, Ann. of Math. (2) **38** (1937), 131–141.

[89] V. S. Kulikov, Degenerations of elliptic curves and resolutions of unimodal and bimodal singularities, Funktsional. Anal. i Prilozhen. **9**:1 (1975), 72–73.
= Functional Anal. Appl. **9** (1975), 69–70.

[90] A. G. Kushnirenko, Newton polygon and Milnor numbers, Funktsional. Anal. i Prilozhen. **9**:1 (1975), 74–75.
= Functional Anal. Appl. **9** (1975), 71–72.

[91] V. I. Arnol'd, A spectral sequence to reduce functions to a normal form, Funktsional. Anal. i Prilozhen. **9**:3 (1975), 81–82. MR **52** #4332.

[92] V. I. Arnol'd, A spectral sequence to reduce functions to a normal form, I. G. Petrovskii memorial volume, Acad. Sci. USSR, 1976, to appear.

[93] V. I. Arnol'd, On the theory of envelopes, Uspekhi Mat. Nauk **31**:5 (1976), to appear.

[94] V. I. Arnol'd, Evolution of wave fronts and equivariant Morse lemma, Comm. Pure Appl. Math. 1976, to appear in a volume dedicated to Siegel.

[95] V. I. Arnol'd, Some unsolved problems in the theory of singularities, Trudy Sobolev Seminar **1**, to appear (1976).

[96] A. G. Kushnirenko, The Newton polyhedron and the number of solutions of a system of *k* equations in *k* unknowns, Uspekhi Mat. Nauk **30**:2 (1975), 266–267 (*summary of a lecture)

[97] A. G. Kushnirenko, The Newton polyhedron and Bezout's theorem, Funktsional. Anal. i Prilozhen. **10**:3 (1976), to appear

[98] A. G. Kushnirenko (Kouchnirenko), Polyhèdres de Newton et nombres de Milnor, Invent. Math., **32** (1976), 1–31.

[99] A. G. Kushnirenko (Kouchnirenko), Recepte pour diviner le nombre de solutions d'un système d'équations polynomiales à polyhèdre de Newton donnée, C. R. Acad. Sci., to appear (1976).

[100] D. N. Bernstein, The number of solutions of a system of equations, Funktsional. Anal. i Prilozhen. **9**:3 (1975), 1–4.

[101] D. N. Bernstein, The number of lattice points in polyhedra with integer vertices, Funktsional. Anal. i Prilozhen. **10**:3 (1976), to appear.

[102] V. M. Zakalyukin, On Lagrangian and Legendre singularities, Funktsional. Anal. i Prilozhen. **10**:1 (1976), 37–45.

[103] V. M. Zakalyukin, Rearrangements in wave fronts depending on a parameter, Funktsional. Anal. i Prilozhen. **10**:2 (1976), 69–70.

[104] A. N. Varchenko, ζ-functions of monodromy and the Newton diagram, Invent. Math., to appear (1976).

[105] A. N. Varchenko, Oscillatory integrals and the Newton polyhedron, Funktsional. Anal. i Prilozhen. **10**:3 (1976), to appear.

[106] A. N. Varchenko, Oscillatory integrals and the Newton polyhedron, Notes prepared for Bonner Arbeitstagung, June 1975.

[107] S. M. Gusein-Zade, On the characteristic polynomial of the monodromy for series of singularities, Funktsional. Anal. i Prilozhen. **10** (1976), to appear.

[107a] S. M. Gusein-Zade, Monodromy of isolated singularities, Uspekhi Mat. Nauk **31** (1976), to appear.

[108] O. V. Lyashko, Decompositions of simple singularities of functions, Funktsional. Anal. i Prilozhen. **10**:2 (1976), 49–56.

[109] S. I. Kaliman, The holomorphic universal covering of the space of polynomials without multiple roots, Funktsional. Anal. i Prilozhen. **9**:1 (1975), 71.

[110] V. S. Kulikov, Degenerations of elliptic curves and pseudotaut singularities, Uspekhi Mat. Nauk **30**:2 (1975), 215–216.

[111] V. S. Kulikov, Degenerations of elliptic curves and surface singularities, Ph.D. thesis, Moscow State University, Moscow 1975.

[112] J. Milnor, On the 3-dimensional Brieskorn manifolds $M(p, q, r)$, In: *Knots, groups, and 3-manifolds* (Papers dedicated to the memory of R.H. Fox), 175–225, Annals of Math. Study **84**, Princeton University Press, Princeton, N.J., 1975.

[113] J. Steenbrink, Limits of Hodge structures, Notes prepared for the Bonner Arbeitstagung, June 1975.

[*113a] J. H. M. Steenbrink, Limits of Hodge structures, Invent. Math. **31** (1976), 229–257.

[114] J. H. M. Steenbrink, Intersection form for quasi-homogeneous singularities, Univ. of Amsterdam, Mathematisch Instituut, Report No. 75–09.

[115] M. Golubitsky, Contact equivalence for Lagrangian submanifolds, in *Dynamical Systems – Warwick* 1974, Lecture Notes in Math. **468**, 71–72, Springer-Verlag, Berlin-Heidelberg-New York 1975.

[*115a] M. Golubitsky and V. W. Guillemin, Contact equivalence for Lagrangian manifolds, Advances in Math. **15** (1975), 375–387. MR **51** # 1868 (detailed version of above).

[116] K. Jänich, Caustics and catastrophes, in: *Dynamical Systems – Warwick* 1974, Lecture Notes in Math. **468**, 100–101, Springer-Verlag, Berlin-Heidelberg-New York 1975.

[117] E. C. Zeeman, Catastrophe theory in biology, in: *Dynamical Systems – Warwick* 1974, Lecture Notes in Math. **468**, 101–104, Springer-Verlag, Berlin-Heidelberg-New York 1975.

[118] John Guckenheimer, Solving a single conservation law, in: *Dynamical Systems – Warwick* 1974, Lecture Notes in Math. **468**, 108–134, Springer-Verlag, Berlin-Heidelberg-New York 1975.

[119] John Guckenheimer, Bifurcation and catastrophe, in: *Dynamical Systems* – Salvador 1971, 95–109, Academic Press, New York-London 1973. MR **49** # 9878.

[120] Gordon Wassermann, *Stability of unfoldings*, Lecture Notes in Math. **393**, Springer-Verlag, Berlin-Heidelberg-New York 1974.

[121] Gordon Wassermann, Stability of unfoldings in space and time, Acta Math. **135** (1975), 57–128.

[122] Gordon Wassermann, Stability of caustics, Math. Ann. **216** (1975), 43–50. MR **51** # 9116.

[123] John Guckenheimer, Caustics, in: Global analysis and its applications. II, 281–289, Internat. Atomic Energy Agency, Vienna 1974.

[124] John Guckenheimer, Caustics and non-degenerate Hamiltonians, Topology **13** (1974), 127–133. MR **50** # 11316.

[125] John Guckenheimer, Isochrons and phaseless sets, J. Math. Biol. **1** (1974/75), 259–273.

[126] John Guckenheimer, Shocks and rarefactions in two space dimensions, Arch. Rational Mech. Anal. **59** (1975), 281–291. MR **52** # 8668.

[127] E. C. Zeeman, Levels of structure in catastrophe theory illustrated by applications in the social and biological sciences, Proc. Internat. Congress Mathematicians, Vancouver 1974. II, 533–546, Canadian Math. Congress 1975.

[128] K. Jänich, Caustics and catastrophes, Math. Ann. **209** (1974), 161–180. MR **50** # 3273.

[129] K. Lamotke, Die Homologie isolierter Singularitäten, Math. Z. **143** (1975), 27–44. MR **52** # 11104.

[130] N. A'Campo, Le groupe de monodromie du déploiement des singularités isolées des courbes planes. I, Math. Ann. **213** (1975), 1–32. MR **51** # 13282.

[131] N. A'Campo, Le groups de monodromie du déploiement des singularités isolées des courbes planes. II, Proc. Internat. Congress Mathematicians, Vancouver 1974, Vol. I, 395–404, Canadian Math. Congress 1975.

[132] A. Durfee, Fibered knots and algebraic singularities, Topology **13** (1974), 47–59. MR **49** # 1523.

[*133] P. Gabriel, Unzerlegbare Darstellungen. I, Manuscripta Math. **6** (1972), 71–103. MR **48** # 11212.

[*134] P. Gabriel, Problèmes actuels de théorie de représentations, Enseigement Math. **20** (1974), 323–332. MR **51** # 3229.

[*135] P. Gabriel, Représentations indécomposables, Sém. Bourbaki, 1973/74, Exp. 444, Lecture Notes in Math. **431**, Springer-Verlag, Berlin-Heidelberg-New York 1975.

[*136] L. A. Nazarova and A. V. Roiter, Representations of partially ordered sets, Zap. Nauchn. Sem. Leningrad. Otdel. Mat. Inst. Steklov (LOMI) **28** (1972), 5–31. MR **49** # 4877.
= J. Soviet Math. 3 (1975), 585–606.

[*137] L. A. Nazarova, Representations of quivers of infinite type, Izv. Akad. Nauk SSSR Ser. Mat. **37** (1973), 752–791. MR **49** # 2785.
= Math. USSR-Izv. 7 (1973), 749–792.

[*138] I. M. Gel'fand and V. A. Ponomarev, Problems of linear algebra and classification of quadruples of subspaces in a finite-dimensional vector space. In: Hilbert space operators and operator algebras, Colloq. Math. Soc. János Bolyai **5**, 163–237, North-Holland, Amsterdam 1972. MR **50** # 9896.

[*139] I. N. Bernstein, I. M. Gel'fand and V. A. Ponomarev, Coxeter functors and a theorem of Gabriel, Uspekhi Mat. Nauk **28**:2 (1973), 19–33.
= Russian Math. Surveys 28:2 (1973), 17–32.

[*140] I. M. Gel'fand and V. A. Ponomarev, Free modular lattices and their representations, Uspekhi Mat. Nauk **29**:6 (1974), 3–58.
= Russian Math. Surveys **29**:6 (1975), 1–56.

[*141] V. Dlab and C. M. Ringel, On algebras of finite representation type, J. Algebra **33** (1975), 306–394. MR **50** # 9974.

[*142] V. Dlab and C. M. Ringel, Representations of graphs and algebras, Mem. Amer. Math. Soc. No. **173** (1976).

[*143] W. V. D. Hodge, The isolated singularities of an algebraic surface, Proc. London Math. Soc. (2) **36** (1929), 133–143.

[*144] M. V. Berry, Waves and Thom's theorem, Advances in Physics **25**:1 (1976), 1–26.

[*145] N. A'Campo, La fonction zêta d'une monodromie, Comment. Math. Helvet. **50** (1975), 233–248. MR **51** # 8106.

[*146] D. Siersma, Periodicities in Arnold's lists of singularities, Univ. of Amsterdam, Mathematisch Instituut, Report No. 76–01.

[*147] A. H. Durfee, Fourteen characterizations of rational double points and simple critical points, Inst. for Adv. Study, Princeton, N.J., 1976, preprint.

Received by the Editors, 26 December 1974

Translated by J. S. Joel

CRITICAL POINTS OF FUNCTIONS ON A MANIFOLD WITH BOUNDARY, THE SIMPLE LIE GROUPS B_k, C_k, AND F_4 AND SINGULARITIES OF EVOLUTES

V. I. Arnol'd

Contents

Introduction
§1. Classification of simple critical points of functions on the boundary of a manifold 208
§2. Versal deformations and bifurcation diagrams 209
§3. Vanishing cycles and half-cycles 211
§4. The form of intersections 212
§5. The monodromy group 213
§6. Elliptic singularities 214
§7. Classification of singularities 215
§8. The determinant of singularities 217
§9. Some corollaries 220
§10. Singularities of a family of evolutes 221
References 223

Introduction

In the investigation of degenerate critical points of functions, a useful instrument are the simple Lie groups whose Dynkin diagrams contain only ordinary edges, that is, the groups A_k, D_k, and E_k (see [1] and [9]). It turns out that the groups with double edges, that is, B_k, C_k, and F_4, are connected in a similar way with the theory of critical points of functions on a manifold with boundary.

The bifurcation diagrams of B_2 and B_3 appeared unexpectedly in the investigation of the geometry of families of evolutes of plane curves in general position. The classification given below of simple critical points of functions on a manifold with boundary is the result of an attempt to understand this observation.

Going over to Dynkin diagrams with edges of different lengths leads to a distinctive variant of the Picard–Lefschetz theory in which, along with vanishing cycles, "vanishing half-cycles" are involved. The example of simple singularities suggests in the general situation a new, rather unusual, variant of relative monodromy on the space of vanishing cycles and half-cycles. In particular, for the relative case we have succeeded in defining a form of intersections, and consequently, elliptic, parabolic, and hyperbolic singularities. Below we show that elliptic singularities on a manifold with boundary are precisely the simple singularities. We have found three one-parameter families of quasihomogeneous parabolic singularities (which apparently exhaust all the parabolic singularities). We also classify all unimodel singularities of functions of two variables.

The author wishes to thank V. I. Borzenko, V. V. Goryunov, V. I. Guts, V. I. Matov, D. B. Fuks, S. M. Husein-Zade, and V. P. Palamodov for useful discussions.

§1. Classification of simple critical points of functions on the boundary of a manifold

In what follows, a manifold with boundary is understood to be a smooth (real or complex) manifold with a fixed smooth hypersurface. Two functions on a manifold with boundary are called *equivalent* if one goes over into the other under a diffeomorphism of the manifold that takes the boundary into itself.

On the boundary we consider a distinguished point O. The group of germs of diffeomorphisms of a manifold with boundary at a distinguished point that keep the boundary fixed acts on the spaces of germs and jets of functions at the distinguished point for which this is a critical point with critical value zero.

DEFINITION. The *modality* of a critical point of a function with critical value zero is defined as the smallest number m such that a sufficiently small neighbourhood of the jet of this function in the space of jets of arbitrarily high order intersects at most finitely many of at most m-parameter families of orbits of the action indicated above.

A critical point is called *simple* if its modality is zero.

THEOREM 1. *The simple critical points of functions on the boundary of a manifold with boundary are exhausted, to within equivalence, by the following list of germs of functions $f(x, y)$ at the point $x = y = 0$ of the boundary $x = 0$:*

$$B_k = \pm x^k \pm y^2, \qquad C_k = xy \pm y^k, \qquad F_4 = \pm x^2 + y^3; \qquad k \geqslant 2.$$

Here equivalence of functions of different numbers of variables is understood to be stable equivalence, that is, equivalence of a function of a large number of variables to the sum of a function of a smaller number of variables and a non-degenerate quadratic form in the missing variables.

REMARK 1. The set of non-simple critical points has codimension 3 in the space of functions with critical point O and critical value 0.

REMARK 2. The codimensions of the sets of simple critical points of the types B_k, C_k, and F_k are equal to $k-2$.

REMARK 3. The singularities B_2 and C_2 are equivalent.

The proof of Theorem 1 is carried out in §8.

§2. Versal deformations and bifurcation diagrams

We carry over to the case of a manifold with boundary the basic ideas of the theory of critical points of functions (see [2] and [5]).

A versal deformation of the germ of a smooth function at a distinguished point of the boundary is understood henceforth to be a versal deformation for the group of diffeomorphisms that take the boundary into itself. For the germ at zero of a function $f(x_1, \ldots, x_n)$ at the point $x = 0$ of the boundary $x_1 = 0$ we can take as such a deformation one of the form

$$F(x, \lambda) = f(x) + \lambda_1 e_1 + \ldots + \lambda_\mu e_\mu,$$

where the numbers λ_k are the parameters of the deformation and the e_k are monomials in x that give a basis of the local ring

$$Q_{f\,|\,x_1} = \mathbf{C}\,[[x_1, \ldots, x_n]]/(x_1 f_1, f_2, \ldots, f_n), \qquad f_k = \partial f/\partial x_k$$

(for definiteness we consider the complex case). The space $\mathbf{C}^\mu$ with coordinates $\lambda_1, \ldots, \lambda_\mu$ is called the *base* of the deformation F.

We define the *multiplicity* of a critical point by the formula

$$\mu\,(f|x_1) = \dim_{\mathbf{C}} Q_{f\,|\,x_1}.$$

Critical points of infinite multiplicity form a set of infinite codimension. In what follows we assume that the multiplicities of the critical points under consideration are finite.

The non-singular local level manifolds and the bifurcation diagram of zeros for a critical point of finite multiplicity of a holomorphic function on a manifold with boundary $x_1 = 0$ are defined in the following way.

We fix a versal deformation F and choose a sufficiently small ball $B_\rho = \{x \in \mathbf{C}^n : |x| \leqslant \rho\}$. Depending on ρ we choose a sufficiently small δ and consider for a λ in the ball $|\lambda| \leqslant \delta$ the local level set $V_\lambda = \{x \in B_\rho : F(x, \lambda) = 0\}$.

DEFINITION. The local level set V_λ is called *non-singular* if 1) 0 is not a critical value for $F(\cdot, \lambda)$, 2) the manifold V_λ is transversal to the boundary (that is, the hyperplane $x_1 = 0$).

The set of singular values λ is denoted by

$$\Sigma_{f\,|\,x_1} = \{\lambda \in \mathbf{C}^\mu : |\lambda| \leqslant \delta, \quad V_\lambda \text{ singular}\}.$$

The germ of the hypersurface Σ at zero is called the *bifurcation diagram of zeros* for f. It is determined uniquely to within local biholomorphic diffeomorphisms of the base of the deformation at zero (and does not depend on the choice of F nor on ρ and δ).

REMARK. The bifurcation diagram of zeros of the germ of a function on a manifold with boundary consists of two hypersurfaces, corresponding to the two conditions of non-singularity 1) and 2): the level manifold may be non-smooth or non-transversal to the boundary.

The roots for the groups B_k, C_k, and F_4 (and the mirrors corresponding to them) are also of two unmixed types. This suggests the results presented below.

THEOREM 2. *The bifurcation diagram of zeros for a simple critical point* B_μ, C_μ, *or* F_μ, *embedded in* $\mathbf{C}^\mu$, *is biholomorphically equivalent to the manifold of irregular orbits of the corresponding group generated by reflections that acts on the complexification of Euclidean space.*

The complement of the bifurcation diagram is a space $K(\pi, 1)$, where π is Brieskorn's braid group [12], constructed with respect to the corresponding group generated by reflections.

PROOF. The versal deformations for B_μ (and $n = 1$) and for C_μ (and $n = 2$) can be taken in the form

$$B_\mu : F = x^\mu + \lambda_1 x^{\mu-1} + \ldots + \lambda_\mu,$$
$$C_\mu : F = xy + y^\mu + \lambda_1 y^{\mu-1} + \ldots + \lambda_\mu.$$

A local level set is smooth if the polynomial with the coefficients $\lambda_1, \ldots, \lambda_\mu$ has no multiple roots. It is transversal to the hyperplane $x = 0$ if zero is not a root of the polynomial.

The action of the group generated by reflections of B_μ (or C_μ) on the complexification $\mathbf{C}^\mu$ of Euclidean space reduces to arbitrary changes of sign of the coordinates z_k and permutations of them. The space of orbits is identified with the manifold $\mathbf{C}^\mu$ of polynomials with roots z_k^2. The space of irregular orbits (that is, the image of the mirrors $z_k = 0$, $z_i = \pm z_j$) under the factorization mapping $\mathbf{C}^\mu \to \mathbf{C}^\mu$, $z \mapsto \lambda$, is a space of polynomials with multiple ($z_i^2 = z_j^2$) or zero ($z_k = 0$) roots.

Information on the homotopy groups of the complement is now obtained in the usual way of [8] and [12]. In the cases B_μ and C_μ the theorem is proved.

In the case F_4 the proof proceeds by means of the theorems of Brieskorn [11] and Deligne [15] (see also a paper of Slodowy [19]).

REMARK. Versal deformations of quasihomogeneous singularities (and all the normal forms of simple singularities mentioned above are quasihomogeneous) can be taken to be quasihomogeneous. In the case of simple singularities the weights of the variables λ_k turn out to be positive; they are connected with the indices p_k of the corresponding Weyl groups and with their Coxeter numbers N just as in the cases A, D, and E ([2]).

More precisely, let $f(x_1, x_2)$ be a quasihomogeneous simple function on a manifold with boundary $x_1 = 0$ of degree 1 for weights $\deg x_k = \alpha_k$.

Then the Coxeter number N is defined by the formula

$$2/N = \alpha_1 + 2\alpha_2 - 1,$$

and the weights d_k of the basis monomials e_k are given by

$$Nd_k = p_k - 1 \qquad (k = 1, \ldots, \mu).$$

The numerical values are:

Type	α_1	α_2	N	e_k	$N \deg e_k$
B_μ	$1/\mu$	$1/2$	2μ	$1, x, \ldots, x^{\mu-1}$	$0, 2, \ldots, 2\mu-2$
C_μ	$(\mu-1)/\mu$	$1/\mu$	2μ	$1, y, \ldots, y^{\mu-1}$	$0, 2, \ldots, 2\mu-2$
F_4	$1/2$	$1/3$	12	$1, x, y, xy$	$0, 4, 6, 10$

It is easy to verify that the quasihomogeneous functions with critical point on a manifold with boundary $x_1 = 0$ for which the weights of all the λ_k are positive (that is, for which all the $d_k < 1$), are exhausted by the simple singularities B_μ, C_μ and F_4 (see [3]).

§3. Vanishing cycles and half-cycles

The bifurcation diagrams of the singularities B_μ and C_μ are the same, and the preceding arguments do not explain which of the series of singularities is connected with the root system of B_μ and which with that of C_μ. The distinction between short and long roots becomes apparent only in an attempt to associate (co)homology classes with the roots.

The fundamental group of the complement of the bifurcation diagram acts on the various homology groups that can be defined by means of a local non-singular level manifold V_λ and boundary $x_1 = 0$. We denote by V'_λ the intersection of V_λ with the hyperplane $x_1 = 0$ of $\mathbf{C}^n$.

THEOREM 3. *The factor-space V_λ/V'_λ has the homotopy type of a bouquet of μ spheres of dimension $n-1$.*

We use the following notation:

$$\mu = \mu(f|x_1) = \dim_{\mathbf{C}} \mathbf{C}[x_1, \ldots, x_n]]/(x_1f_1, f_2, \ldots, f_n);$$

$$\mu_1 = \mu(f) = \dim_{\mathbf{C}} \mathbf{C}[[x_1, \ldots, x_n]]/(f_1, \ldots, f_n);$$

$$\mu_0 = \mu(f|x_1 = 0) = \dim_{\mathbf{C}} \mathbf{C}[x_2, \ldots, x_n]]/(f_2, \ldots, f_n(0, x_2, \ldots, x_n)).$$

Thus, V_λ is homotopy equivalent to a bouquet of μ_1 of dimension $n-1$ (vanishing cycles) and its submanifold V'_λ is homotopy equivalent to a bouquet of μ_0 spheres of dimension $n-2$ (for $n = 2$, V'_λ reduces to $\mu_0 + 1$ points). It is easy to verify that $\mu = \mu_0 + \mu_1$ and that the μ spheres in Theorem 1 can be obtained from the μ_1 spheres representing the vanishing cycles of f and the μ_0 discs that cover the μ_0 vanishing spheres of the function $f|x_1 = 0$ inside the complement $V_\lambda - V'_\lambda$.

EXAMPLE. Let $n = 2$. Then (V_λ, V'_λ) is a Riemann surface with Betti number $b_1 = \mu_1$ and $\mu_0 + 1$ distinguished points on it. A basis of the one-dimensional relative cycles for (V_λ, V'_λ) can be chosen from the μ_1 circles that emanate from a distinguished point and determine vanishing cycles on V_λ and the μ_0 segments that join this distinguished point to the others. We can choose the system of circles and segments so that V_λ / V'_λ shrinks to its image in V_λ / V'_λ.

We call these generators vanishing cycles (for circles or spheres) and *vanishing half-cycles* (for segments or discs).

Furthermore, we can carry over to the relative case the idea of a distinguished basis of vanishing cycles. To this end we consider a line $\mathbf{C}^1$ in general position in the base of a versal deformation of f near the point 0 of the boundary $x_1 = 0$. This line intersects the bifurcation diagram of the zeros of $\Sigma(f \mid x_1)$ in μ points, of which μ_0 are on the intersection with the part of the bifurcation diagram where transversality is lost, and μ_1 on that where V_λ loses smoothness. To points of the second kind there correspond ordinary vanishing cycles. To points of the first kind there correspond singularities in a neighbourhood of which V_λ is given by an equation of the form $x_1 + x_2^2 + \ldots + x_n^2 = \varepsilon$. As $\varepsilon \to 0$, the cycle $\operatorname{Im} x = 0, x_2^2 + \ldots + x_n^2 = \varepsilon$, $x_1 = 0$ vanishes on V'_λ; a covering disc on V_λ is given by the equation $x_1 = \varepsilon - x_2^2 - \ldots - x_n^2$, $\operatorname{Im} x = 0$, $x_2^2 + \ldots + x_n^2 < \varepsilon$; this disc also determines the vanishing half-cycle. The basis of these μ vanishing cycles and half-cycles is also called *distinguished.*

The proofs of the theorem and the assertions are little different from those of the analogous assertions in the absolute case (see, for example, [17], [13], and [14]).

§4. The form of intersections

Although $\dim_{\mathbf{C}} H_{n-1}(V_\lambda, V'_\lambda) = \mu$, the most natural results are obtained by considering the action of the fundamental group of the complement of the bifurcation diagram not on this space, but on another space of the same dimension: the space of relative cycles with torsion coefficients. This new space is defined by means of a two-sheeted covering, and the introduction of it is a decisive factor in the present article (similar to the introduction of a two-sheeted covering of a plane, ramified along a curve, which is a decisive factor in [4] and subsequent articles on the geometry of real algebraic curves).

Let f be the germ of a holomorphic function of $x_1, \ldots, x_n$ at the point 0 of a manifold with boundary $x_1 = 0$. We consider a two-sheeted ramified covering $\varphi\colon \mathbf{C}^n \to \mathbf{C}^n$ with ramification on the boundary $x_1 = 0$. We denote by $\hat{f}$ the function f lifted to the covering, this is, $\hat{f} = \varphi^* f$. In other words,

$$\hat{f}(z;\ x_2,\ \ldots,\ x_n) = f(z^2;\ x_2,\ \ldots,\ x_n).$$

The local non-singular level manifold $\hat{V}\colon \hat{f} = \varepsilon$ doubly covers the local non-singular level manifold $V\colon f = \varepsilon$ with ramification along $V^{\cdot}\colon f = \varepsilon, x_1 = 0$. The mapping $z \mapsto -z$ determines a holomorphic involution of $\hat{V}$.

DEFINITION. The *form of intersections* of a singularity f at a point of the boundary $x_1 = 0$ is defined as the pair consisting of: 1) the space H^- of *anti-invariant* homology classes under the action of the involution on $H_{n-1}(\hat{V}, \mathbf{Z})$, and 2) the restriction to this space of the anti-invariant cycles H^- of the form of intersections of the covering manifold $\hat{V}$.

THEOREM 4. *The manifold $\hat{V}$ is homotopy equivalent to the bouquet of $\hat{\mu} = \mu_0 + 2\mu_1$ spheres of middle dimension (equal to $n - 1$).*

The involution $z \mapsto -z$ of the pair $(\hat{V}, \hat{V}')$ is homotopy equivalent to an involution of the bouquet of the $\hat{\mu}$ spheres under which $2\mu_1$ of the spheres intersect in pairs and on μ_0 of the spheres the involution is the reflection in the equator.

For the μ_0 anti-invariant spheres and the μ_1 pairs of symmetric spheres we can take the geometric complete inverse images of the vanishing cycle and half-cycles of §3, having oriented these circuits in a suitable way.

EXAMPLE. For $n = 2$, $\hat{V}$ is a two-sheeted ramified covering of V with ramification at points of V'. The segment joining two points of V' on V is lifted to $\hat{V}$ as a circle. This circle changes orientation under the involution.

Theorem 4 is proved by the standard arguments (see [13] and [14]), like Theorem 3.

REMARK. To the homologies described above there correspond in analysis integrals of the form

$$\int \frac{\alpha(x)}{\sqrt{x_1}} \frac{dx_1 \wedge \ldots \wedge dx_n}{df}$$

over cycles of H^-.

§5. The monodromy group

The fundamental group of the complement of the bifurcation diagram $\Sigma(f \,|\, x_1)$ acts on the space defined in §4, preserving the form of intersections. We describe the system of generators of the monodromy group defined in this way.

DEFINITION. The homology class of the inverse image of a vanishing half-cycle in H^- is called a *short cycle,* and an element of H^- that is the difference of two orthogonal symmetric vanishing cycles of $\hat{V}$ is called a *long cycle.*

A *distinguished basis* of H^- is defined as a basis consisting of μ_0 short and μ_1 long cycles, corresponding to the distinguished basis of the relative homology in $H_{n-1}(V, V')$ constructed in §3 from the μ_0 vanishing half-cycles and μ_1 cycles corresponding to the points of intersection of a line in general position with the bifurcation diagram $\Sigma(f \,|\, x_1)$.

Let V be the non-singular level manifold corresponding to a point $* \in \mathbf{C}^1 \setminus \Sigma$, and p_i the points of intersection of $\mathbf{C}^1$ with Σ. A distinguished basis $\{e_i\}$ in H^- is constructed with respect to the system of paths α_i that do not intersect outside $*$ and lead from $*$ in p_i to $\mathbf{C}^1$. (The cycle e_i vanishes along the path α_i.) We consider the system of generators γ_i of the fundamental group $\pi_1(\mathbf{C}^1 \setminus \Sigma, *)$ that corresponds to this system of paths (the loop γ_i goes along α_i to p_i, goes round p_i in the positive direction, and returns to the old path).

We denote by T_γ the monodromy transformation $H^- \to H^-$ corresponding to $\gamma \in \pi_1$.

THEOREM 5. *The generator γ_i acts on H^- according to the generalized Picard–Lefschetz formulae*: $T_{\gamma_i}(x) = x + (-1)^{n(n+1)/2}(x, e_i)e_i/2$ *if e_i is a long cycle*; $T_{\gamma_i}(x) = x + (-1)^{n(n+1)/2}(x, e_i)\, e_i$ *if e_i is a short cycle. Here, as before, n is the number of arguments of the functions.*

The proof is the same as for the usual formulae (see [14]); I am grateful to S. M. Husein-Zade for verifying this.

§6. Elliptic singularities

The Dynkin diagrams $A_{2\mu-1}$, $D_{\mu+1}$, and E_6 have non-trivial involutions. An involution of the diagram induces an involution of the corresponding Euclidean space which preserves the scalar product, the set of simple roots, and the lattice generated by them.

We consider a subspace of our Euclidean space that is fixed under the involution. In this subspace the lattice of integral combinations of the simple roots cuts out a sublattice, which consists of all integral combinations of the simple roots in which the roots transposed by the involution have equal coefficients. It is, therefore, generated by those simple roots that are fixed under the involution, and by the sums of transposed roots and their images. The following lemma is well known and easy to verify.

LEMMA. *A subspace of the space of the diagram $A_{2\mu-1}$, $D_{\mu+1}$, or E_6 that is fixed under the involution and is endowed with the scalar product, the lattice, and the basis of this lattice described above, is none other than the Euclidean space of one of the diagrams B_μ, C_μ, or F_4, respectively, endowed with the standard system of simple roots and the lattice of their integral combinations.*

REMARK. We normalize the system of roots so that the scalar square of each short root is 2 and that of each long root is 4 (this coincides with the usual normalization in the cases C_μ; in the remaining cases our squares are twice as large as usual).

We now turn to singularities of functions of n variables on a manifold with boundary. We assume that $n \equiv 3 \bmod 4$ (this can be achieved without leaving the stable equivalence class).

DEFINITION. A singularity is called *elliptic* (respectively, *parabolic* or *hyperbolic*) if its quadratic form is negative definite (respectively, semidefinite, or has 1 positive square).

THEOREM 6. 1) *Elliptic singularities are precisely the simple singularities B_μ, C_μ, and F_4.*

2) *We can obtain the form of intersections of an elliptic singularity from the corresponding Dynkin diagram by considering in the Euclidean space the system of simple roots* (*short roots with square* 2 *and long roots with square* 4); *the form is defined on the lattice generated by the roots, and differs from the scalar product of the surrounding Euclidean space only in sign.*

The action of the fundamental group of the complement of the bifurcation diagram on H^- is the natural representation of the Brieskorn braid group on the corresponding Weyl group.

3) *For simple singularities f of the types B_μ, C_μ, and F_4 the function $\hat{f}$ also has a simple singularity of the types $A_{2\mu-1}$, $D_{\mu+1}$, and E_6, respectively. Distinguished bases for $\hat{f}$ in $H_{n-1}(\hat{V})$ and for f in $H^-_{n-1}(\hat{V})$ can be chosen so that they are obtained from one another by the construction described in the lemma. The role of the involution of the Dynkin diagram is played by the composition of the involution of the covering (induced by the change of sign of z; see* §4) *and the involution* -1 *on* $H_{n-1}(\hat{V})$.

Theorem 6, 1) is proved in the same way as the analogous theorem of Tyurina [21] for the usual simple singularities; we need only calculate the quadratic forms of the first adjoining non-simple singularities. These singularities are considered in §§7–9.

2) and 3) are verified directly; this can be done very quickly by the method of Husein-Zade [14], which is easily adapted to the relative case (or the even case with respect to z).

§7. Classification of singularities

In this section we make a start on the classification of critical points of holomorphic functions $f(x_1, x_2, \ldots, x_n)$ with respect to diffeomorphisms that preserve the hyperplane $x_1 = 0$ and the critical point O; the critical value is also assumed to be zero. The classification is to within stable equivalence. For $n = 2$ the variables (x_1, x_2) are denoted by (x, y). Thus, the equation of the boundary in this case is $x = 0$, and y is the coordinate along the boundary.

LIST OF NORMAL FORMS

$$\begin{array}{llllllll} \textit{Series } F: & F_{k,0} & \leftarrow & F_{k,1} & \leftarrow & F_{k,2} & \leftarrow & \ldots \\ & & & \uparrow & & & & \\ & & & F_{6k+2} & \leftarrow & F_{6k+3} & \leftarrow & F_{6k+4} \leftarrow (F_{k+1,0}). \end{array}$$

Notation	Normal form	Restrictions	μ	cod
F_{6k+2}	$x^{3k+1}+y^3+\underset{\sim}{a}x^{2k+1}y$	$k\geqslant 1$	$6k+2$	$5k$
F_{6k+3}	$x^{2k+1}y+y^3+\underset{\sim}{a}x^{k+1}y^2$	$k\geqslant 1$	$6k+3$	$5k+1$
F_{6k+4}	$x^{3k+2}+y^3+\underset{\sim}{a}x^{2k+2}y$	$k\geqslant 1$	$6k+4$	$5k+2$
$F_{k,0}$	$x^{3k}+bx^ky^2+y^3+\underset{\sim}{c}x^{2k+1}y$	$k\geqslant 1,\ 4b^3+27\neq 0$	$6k$	$5k-2$
$F_{k,p}$	$\underset{\sim}{a}x^{3k+p}+x^ky^2+y^3$	$k\geqslant 1,\ a_0\neq 0$	$6k+p$	$5k-2+p$

Here $\underset{\sim}{a} = a_0 + \ldots + a_{k-1}^k x^{k-1}$ and $\underset{\sim}{c} = c_0 + \ldots + c_{k-2}x^{k-2}$ for $k \geqslant 2$; for $k = 1$, $\underset{\sim}{c} = 0$. Here and later, cod denotes the codimension of the class in the space of functions with critical point 0 and critical value zero.

We say that a function f *adjoins* a class of functions X if in any small neighbourhood of the jet of f in the space of jets of arbitrarily high order there are jets of functions of the class X.

THEOREM 7. *Every function with critical point of corank* <2 *on the* *boundary is either stably equivalent to B_μ, C_μ, or F_4, or to one of the functions of the series F or K, or it adjoins one of the classes K_0^*, K_0^{**}, or K_0^{***}.*

Every function with a critical point of corank $\geqslant 2$ on the boundary is either stably equivalent to one of the functions $f(x, y, z)$ of the class L_6 at the point 0 of the boundary $x = 0$

$$L_6 = y^3 + z^3 + axz + xy, \; a^3 \neq 1, \; \mu = 6, \; \text{cod} = 3,$$

or it adjoins L_6.

Series K.

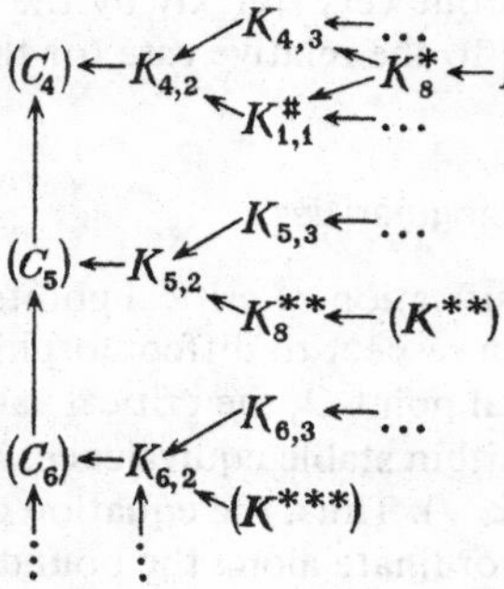

Notation	Normal form	Restrictions	μ	cod
$K_{4,2}$	$y^4+axy^2+x^2$	$a^2 \neq 4$	6	3
$K_{4,q}$	$y^4+xy^2+ax^q$	$a \neq 0,\ q>2$	$q+4$	$q+1$
$K_{p,q}$	$y^p+xy^2+ax^q$	$a \neq 0,\ p>4,\ q \geqslant 2$	$p+q$	$p+q-3$
$K^{\#}_{1,2p-3}$	$(y^2+x)^2+ax^py$	$a \neq 0$	$2p+3$	$2p$
$K^{\#}_{1,2p-4}$	$(y^2+x)^2+ax^p$	$a \neq 0$	$2p+2$	$2p-1$
K_8^*	$y^4+x^2y+ax^3$	—	8	5
K_9^*	$y^4+x^3+ax^2y^2$	—	9	6
K_0^*	$y^4+ax^2y^2+bx^3y+x^4+cx^3y^2$	$\Delta(a, b) \neq 0$	12	7
K_8^{**}	$y^5+x^2+axy^3$	—	8	5
K_0^{**}	$y^5+x^2y+(a+by)xy^3$	$a^2 \neq 4$	10	6
K_0^{***}	$y^6+(a+by)xy^3+x^2$	$a^2 \neq 4$	10	6

The parameters a, b, and c in all our normal forms are moduli, that is, a singularity equivalent to the given one occurs by changing these parameters finitely many times.

The proof of this theorem is sketched in §8.

§8. The determinant of singularities

We shall use the abridged notation of [5], the only difference being that all the equivalences preserve the boundary ($x = 0$). In the determinant, capital letters denote the stable equivalence classes defined above (for the definition of B_μ, C_μ and F_4, see §1). It is implied that each function depends on as many variables as are explicitly mentioned: $n = 2$ in Theorems **2–41**, and $n = 3$ in Theorems **43** and **44.** By F we always mean a function with critical point 0 and critical value zero; $\approx$ means quasihomogeneous equivalence of quasijets; $j_{(x^m)}f$ means the quasijet of f consisting of monomials whose exponents lie not above the hyperplane determined by the exponents of the m monomials in parentheses; $\Rightarrow$ means "implies", and $|\Rightarrow$ means "see".

THE DETERMINANT.

1. $\mu(f) < \infty \Rightarrow$ one of four:

the corank of the restriction of f to the boundary equal to $0 \mid\Rightarrow$ **2**,

$1 \mid\Rightarrow$ **3**,

$2 \mid\Rightarrow$ **42**,

$>2 \mid\Rightarrow$ **46**.

2. The corank of the restriction of f to the boundary equal to $0 \Rightarrow f \in B_k$, $k \geqslant 2$.

3. The corank of the restriction of f to the boundary equal to $1 \Rightarrow$ the restriction belongs to A_{p-1}, $p > 2$.

4. $f|_{x=0} \in A_{p-1}$, $p > 2 \Rightarrow f \sim y^p + x\varphi(x, y)$, $\varphi(0, 0) = 0$.

5. $f = y^p + x\varphi(x, y)$, $\varphi(0, 0) = 0$, $p > 2 \Rightarrow$ one of five:

$p = 3 \mid\Rightarrow$ **6**,

$p = 4 \mid\Rightarrow$ **16**,

$p = 5 \mid\Rightarrow$ **27**,

$p = 6 \mid\Rightarrow$ **33**,

$p > 6 \mid\Rightarrow$ **38**.

Series F.

6. $j_{y^3, x}f = y^3 \Rightarrow$ one of three:

$j_{y^3, xy}f \approx y^3 + xy \mid\Rightarrow$ **7**,

$j_{y^3, x^2}f \approx y^3 + x^2 \mid\Rightarrow$ **8**,

$j_{y^3, x^2}f \approx 0 \mid\Rightarrow$ **13**$_1$.

7. $j_{y^3, xy}f = y^3 + xy \Rightarrow f \in C_3$.

8. $j_{y^3, x^2}f = y^3 + x^2 \Rightarrow f \in F_4$.

In Theorems **9–15**, $k \geqslant 1$.

9$_k$. $j_{y^3, x^{3k}}f = y^3 \Rightarrow$ one of four:

$j_{y^3, x^{3k+1}}f \approx y^3 + x^{3k+1} \mid\Rightarrow$ **10**$_k$,

$j_{y^3, x^{2k+1}y}f \approx y^3 + x^{2k+1}y \mid\Rightarrow$ **11**$_k$,

$j_{y^3, x^{3k+2}}f \approx y^3 + x^{3k+2} \mid\Rightarrow$ **12**$_k$,

$j_{y^3, x^{3k+2}}f \approx y^3 \mid\Rightarrow$ **13**$_{k+1}$.

10_k. $j_{y^3,\,x^{3k+1}}f=y^3+x^{3k+1}\Longrightarrow f\in F_{6k+2}$.
11_k. $j_{y^3,\,x^{2k+1}y}f=y^3+x^{2k+1}y\Longrightarrow f\in F_{6k+3}$.
12_k. $j_{y^3,\,x^{3k+2}}f=y^3+x^{3k+2}\Longrightarrow f\in F_{6k+4}$.
13_k. $j_{y^3,\,x^{3k-1}}f\approx y^3\Longrightarrow$ one of three:

$$j_{y^3,\,x^{3k}}f\approx y^3+bx^ky^2+x^{3k},\ 4b^3+27\neq 0\ |\!\Longrightarrow 14_k,$$
$$j_{y^3,\,x^{3k}}f\approx y^3+x^ky^2\ |\!\Longrightarrow 15_k,$$
$$j_{y^3,\,x^{3k}}f\approx y^3\ |\!\Longrightarrow 9_k.$$

14_k. $j_{y^3,\,x^{3k}}f=y^3+bx^ky^2+x^{3k},\ 4b^3+27\neq 0\Longrightarrow f\in F_{k,\,0}$.
15_k. $j_{y^3,\,x^ky^2}f=y^3+x^ky^2\Longrightarrow f\in F_{k,\,p},\quad p>0$.

Series K_4.

16. $j_{y^4,\,xy}f\approx y^4+xy\Longrightarrow f\in C_4$.
17. $j_{y^4,\,xy}f=y^4\Longrightarrow$ one of four:

$$j_{y^4,\,x^2}f\approx y^4+axy^2+x^2,\ a^2\neq 4\ |\!\Longrightarrow 18,$$
$$j_{y^4,\,x^2}f\approx y^4+xy^2\ |\!\Longrightarrow 19,$$
$$j_{y^4,\,x^2}f\approx (y^2+x)^2\ |\!\Longrightarrow 20,$$
$$j_{y^4,\,x^2}f\approx y^4\ |\!\Longrightarrow 21.$$

18. $j_{y^4,\,x^2}f=y^4+axy^2+x^2,\ a^2\neq 4\Longrightarrow f\in K_{4,\,2}$.
19. $j_{y^4,\,x^2}f=y^4+xy^2\Longrightarrow f\in K_{4,\,q},\ q>2$.
20. $j_{y^4,\,x^2}f=(y^2+x)^2\Longrightarrow f\in K^{\#}_{1,\,q},\ q\geqslant 1$.
21. $j_{y^4,\,x^2}f=y^4\Longrightarrow$ one of three:

$$j_{y^4,\,x^2y}f\approx y^4+x^2y\ |\!\Longrightarrow 22,$$
$$j_{y^4,\,x^3}f\approx y^4+x^3\ |\!\Longrightarrow 23,$$
$$j_{y^4,\,x^3}f=y^4\ |\!\Longrightarrow 24.$$

22. $j_{y^4,\,x^2y}f=y^4+x^2y\Longrightarrow f\in K^*_8$.
23. $j_{y^4,\,x^3}f=y^4+x^3\Longrightarrow f\in K^*_9$.
24. $j_{y^4,\,x^3}f=y^4\Longrightarrow$ one of two:

$$j_{y^4,\,x^4}f\approx y^4+ax^2y^2+bx^3y+y^4,\ \Delta(a,b)\neq 0\ |\!\Longrightarrow 25,$$

f adjoins the preceding class $|\Rightarrow$ **26.**

25. $j_{y^4,\,x^4}f=y^4+ax^2y^2+bx^3y+x^4,\ \Delta(a,b)\neq 0\Longrightarrow f\in K^*_0$.
26. f adjoins K^*_0 and $f\in K^*$.

Series K_5.

27. $j_{y^5,\,xy}f\approx y^5+xy\Longrightarrow f\in C_5$.
28. $j_{y^5,\,xy}f=y^5\Longrightarrow$ one of four:

$$j_{y^5,\,xy^2}f\approx y^5+xy^2\ |\!\Longrightarrow 29,$$
$$j_{y^5,\,x^2}f\approx y^5+x^2\ |\!\Longrightarrow 30,$$
$$j_{y^5,\,x^2y}f\approx y^5+x^2y+axy^3,\quad a^2\neq 4\ |\!\Longrightarrow 31,$$

f adjoins the preceding class $|\Rightarrow$ **32.**

29. $j_{y^5, xy^2}f = y^5 + xy^2 \Longrightarrow f \in K_{5,q},\ q \geqslant 2.$
30. $j_{y^5, x^2}f = y^5 + x^2 \Longrightarrow f \in K_8^{**}.$
31. $j_{y^5, x^2y}f = y^5 + x^2y + axy^3,\ a^2 \neq 4 \Longrightarrow f \in K_0^{**}.$
32. f adjoins K_0^{**} and $f \in K^{**}$.

Series K_6.

33. $j_{y^6, xy}f \approx y^6 + xy \Longrightarrow f \in C_6.$
34. $j_{y^6, xy}f = y^6 \Longrightarrow$ one of three:

$$j_{y^6, xy^2}f \approx y^6 + xy^2 \mid\Longrightarrow \mathbf{35},$$
$$j_{y^6, x^2}f \approx y^6 + axy^3 + x^2,\quad a^2 \neq 4 \mid\Longrightarrow \mathbf{36},$$

f adjoins the preceding class $\mid\Rightarrow$ **37**.

35. $j_{y^6, xy^2}f = y^6 + xy^2 \Longrightarrow f \in K_{6,q},\ q \geqslant 2.$
36. $j_{y^6, x^2}f = y^6 + axy^3 + x^2,\ a^2 \neq 4 \Longrightarrow f \in K_0^{***}.$
37. f adjoins K_0^{***}, $f \in K^{***}$.

Series K_p, $p > 6$.

38$_p$. $j_{y^p, xy}f \approx y^p + xy \Longrightarrow f \in C_p$
39$_p$. $j_{y^p, xy}f = y^p \Longrightarrow$ one of two:

$$j_{y^p, xy^2}f \approx y^p + xy^2 \mid\Longrightarrow \mathbf{40}_p,$$
$$j_{y^p, xy^2}f \approx y^p \mid\Longrightarrow \mathbf{41}_p.$$

40$_p$. $j_{y^p, xy^2}f = y^p + xy^2 \Longrightarrow f \in K_{p,q},\ q \geqslant 2.$
41$_p$. $j_{y^p, xy^2}f = y^p \Longrightarrow f$ adjoins K_0^{***}.

The corank of the restriction to the boundary greater than 1.

42. The corank of the restriction of f to the boundary equal to 2 $\Rightarrow$ one of two:

the restriction to the boundary equivalent to $y^3 + z^3$ $\mid\Rightarrow$ **43**,
f adjoins the preceding class $\mid\Rightarrow$ **45**.

43. $f(0, y, z) = y^3 + z^3 \Longrightarrow$ one of two:
$j_{y^3, z^3, xy}f \approx y^3 + z^3 + axz + xy,\ a^3 \neq 1 \mid\Longrightarrow \mathbf{44}$,
f adjoins the preceding class $\mid\Rightarrow$ **45**.

44. $j_{y^3, z^3, xy}f = y^3 + z^3 + axz + xy,\ a^3 \neq 1 \Longrightarrow f \in L_6.$
45. f adjoins L_6, $f \in L^*$, cod $L^* = 5$.
46. The corank of the restriction to the boundary greater than 2 $\Rightarrow f$ adjoins L_6.

The theorem of §7 is an obvious consequence of Theorems **1–46**.

These theorems are proved by standard methods (see [3], [5], [6], and [7]), namely:

Number of theorem	Method of proof
2, 3, 4, 7, 8, 10, 11, 12, 13, 14, 16, 17, 18, 22, 23, 24, 25, 27, 30, 31, 33, 36, 38, 39, 42, 43, 44	(Semi)quasihomogeneous normal form
2, 3, 6, 9, 21, 28, 34, 42, 43	Rotation of the Newton ruler
15, 19, 20, 29, 35, 40,	Spectral sequence
1, 5, 26, 32, 37, 41, 45, 46	Tautology

§9. Some corollaries

Many corollaries follow from the theorems given above, of which we mention here only a few.

COROLLARY. 1. *Each singularity, except B_μ, C_μ, F_4, adjoins either $F_{1,0}$ or $K_{4,2}$ or L_6.*

Since singularities of these three classes are unimodal, Theorem 1 of §1 follows.

By the method of Husein-Zade it is easy to verify that singularities of these three classes are parabolic (two zero squares). This proves Theorem 6 of §6.

Husein-Zade has also calculated the signatures of a number of other singularities of functions of two variables in the list of §7. The singularities $F_{1,k}$, $K_{p,q}$, and $K_{1,q}^{\#}$ are hyperbolic (one zero and one positive square). The singularities $F_{2,0}$, K_8^*, K_9^*, K_0^*, K_8^{**}, K_0^{**}, K_0^{***}, F_8, F_9, and F_{10} have two positive squares.

COROLLARY 2. *The parabolic singularities of functions of two variables are precisely $F_{1,0}$ and $K_{4,2}$.*

Apparently all parabolic singularities are exhausted by the three families $F_{1,0}$, $K_{4,2}$, and L_6.

COROLLARY 3. *The hyperbolic singularities of functions of two variables are precisely $F_{1,k}$ with $k > 0$, $K_{p,q}$ with $(p, q) > (4, 2)$, and $K_{1,q}^{\#}$.*

COROLLARY 4. *All simple and unimodal singularities of functions of two variables are contained in the following diagram*:

$$\begin{array}{l}
C_2 = B_2 \leftarrow B_3 \leftarrow B_4 \leftarrow \ldots \\
C_3 \leftarrow F_4 \leftarrow F_{1,0} \leftarrow F_{1,1} \leftarrow \ldots \\
\quad F_8 \leftarrow F_9 \leftarrow F_{10} \\
C_4 \leftarrow K_{4,2} \leftarrow K_{4,3} \leftarrow \ldots \\
\quad K_8^* \leftarrow K_9^* \\
\quad K_{1,1}^{\#} \leftarrow K_{1,2}^{\#} \leftarrow \ldots \\
C_5 \leftarrow K_{5,2} \leftarrow K_{5,3} \leftarrow \ldots \\
\quad K_8^{**} \\
C_6 \leftarrow K_{6,2} \leftarrow \ldots
\end{array}$$

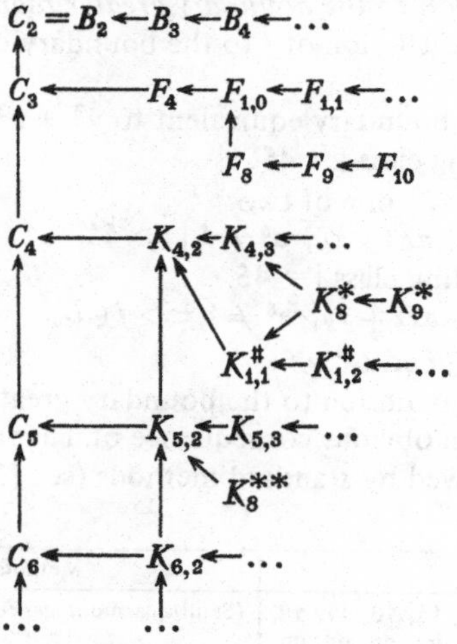

The classification of critical points of functions on a manifold with boundary is equivalent to that of critical points of functions that are even with respect to one of the coordinates (see [9]). All our results can, therefore, be interpreted as a classification of equivariant critical points of functions for the simplest action of a group of order 2. The initial part of this classification has been rediscovered several times by various authors (see, for example, [9], [18], [20], and [22]). Guts has even found a complete list of simple singularities of symmetric functions, but the main point – the connection between critical points of functions on a manifold with boundary and the simple groups B_μ, C_μ, and F_4 – has apparently gone unnoticed.

The only missing simple Lie group G_2 arises naturally in the classification of simple functions that are invariant under the action of the symmetric group of permutations of the coordinates (that is, under the classification of simple symmetric functions). The classification problem corresponding to G_2 is related to the group of permutations of three elements and can also be interpreted as the problem of classifying functions on a plane to within diffeomorphisms that preserve a semicubical parabola (see [9]).

§10. Singularities of a family of evolutes

Let γ be a plane curve, parametrized by the natural parameter s (arc length), that is, a smooth map $\gamma\colon \mathbf{R} \to \mathbf{R}^2$ of a line in the Euclidean plane with $|d\gamma/ds| = 1$.

The *family of evolutes* of γ is defined as the family of images of the lines $s + t = \text{const}$ in the plane with the coordinates (s, t) under the mapping

$$f\colon \mathbf{R}^2 \to \mathbf{R}^2, \quad f(s, t) = \gamma(s) + t\gamma'(s)$$

(s is the coordinate of the point of contact of the string with the curve, and t is the length of the free part of the string).

The family of evolutes of a plane curve arises as the family of level lines of time in the problem of the shortest path in a plane with an obstruction bounded by the curve.

THEOREM A. *In a neighbourhood of a point of non-zero curvature on a curve the family of evolutes is diffeomorphic to the family of plane semi-cubical parabolas $y = (x - c)^{2/3}$ in a neighbourhood of the axis $y = 0$.*

The proof is based on the following construction. We consider the mapping $\hat{f}\colon \mathbf{R}^2 \to PT^*\mathbf{R}^2$ that associates with the pair (s, t) a line element of the tangent to the evolute at $f(s, t)$. It is easy to verify that $\hat{f}$ is an immersion.

By contracting the domain of definition of f, we may suppose that the image of $\hat{f}$ is a smooth surface M in the three-dimensional space of line elements. We may also carry over the curves $s + t = \text{const}$ to M by means of $\hat{f}$; it remains to investigate the restriction to M of the projection $p\colon PT^*\mathbf{R}^2 \to \mathbf{R}^2$.

It is easy to compute that this restriction (and consequently the composition $f = p \circ \hat{f}$) has a fold whose image is the curve under investigation and whose inverse image is the set of its normal elements. Moreover, curves of the family $s + t = \text{const}$ touch the kernel of the derivative of our mapping. This guarantees the reduction to the indicated normal form (see [10]) under some non-degeneracy condition, which is satisfied because the curvature is non-zero.

We now suppose that the curvature of the curve at some point vanishes, but its first derivative is non-zero; we then talk of an ordinary point of inflexion (example: $y = x^3$).

DEFINITION. We call a line element in the plane *singular* with respect to γ if it either lies on the surface $M = \hat{f}(\mathbf{R}^2)$ described above or is applied at a point of $\gamma(\mathbf{R})$.

THEOREM B. *The set of singular elements in a neighbourhood of an element of the normal to the curve at an ordinary point of inflexion is locally diffeomorphic to the bifurcation diagram B_3 in* $\mathbf{R}^3$ (we have in mind diffeomorphisms of the ambient three-dimensional manifolds).

In a neighbourhood of an element of the normal at a point of non-zero curvature the set of singular elements is locally diffeomorphic to the direct product of the bifurcation diagram B_2 in $\mathbf{R}^2$ and a line.

PROOF. We fix Cartesian orthonormal coordinates in which the curve is given by the equation $y = x^3 + \ldots$. Through the point of application of each line element of the plane we draw the line perpendicular to it and consider the points of intersection of this line with our curve. For elements close to an element of the normal at the point of inflexion, there are three such points of intersection (generally speaking, complex and not necessarily distinct) close to $x = 0$. Let (x_1, x_2, x_3) be the abscissae of the points of intersection, and x_0 the abscissa of the point of application of the element. We associate with our element the polynomial of degree 3 with the roots $\xi_i = x_i - x_0$. It is easy to verify that this defines a local diffeomorphism $PT^*\mathbf{R}^2 \to \mathbf{R}^3$. The surface M goes over to the set of polynomials with multiple roots, and the surface of elements applied to the curve to the set of polynomials with zero roots. These two sets also constitute the bifurcation diagram B_3.

In the non-analytic case these arguments are placed by a reference to the Malgrange division theorem [16].

REMARK 1. This theorem makes it possible to split the investigation of singularities of the family of evolutes into two problems: the investigation of singularities of the family of curves on M and that of the additional singularities that arise by the projection of M onto $\mathbf{R}^2$.

The first problem is solved by applying the theorem of [9] on the normal form of a function to the space of the bifurcation diagram.

REMARK 2. My attention was drawn to the problem of the singularities of a family of evolutes by V. I. Borzenko. Having drawn the surface M, I recognized in it part of the bifurcation diagram B_3 studied by V. V. Goryunov. An analysis of the causes of this coincidence also led to the results of §§1–9.

References

[1] V. I. Arnol'd, Normal forms of functions near degenerate critical points, the Weyl groups A_k, D_k, E_k, and Lagrangian singularities, Funktsional. Anal. i Prilozhen. **6**:4 (1972), 3–25. MR **50** # 8595.
= Functional Anal. Appl. **6** (1972), 254–272.

[2] ———, Remarks on the stationary phase method and the Coxeter numbers, Uspekhi Mat. Nauk **28**:5 (1973), 17–44. MR **53** # 1635.
= Russian Math. Surveys **28**:5 (1973), 19–48.

[3] ———, Normal forms of functions in a neighbourhood of degenerate critical points, Uspekhi Mat. Nauk **29**:2 (1974), 11–49.
= Russian Math. Surveys **29**:2 (1974), 10–50.

[4] ———, The situation of ovals of real plane algebraic curves, the involutions of four-dimensional smooth manifolds, and the arithmetic of integral quadratic forms, Funktsional. Anal. i Prilozhen. **5**:3 (1971), 1–9. MR **44** # 3999.
= Functional Anal. Appl. **5** (1971), 169–176.

[5] ———, Critical points of smooth functions and their normal forms, Uspekhi Mat. Nauk **30**:5 (1975), 3–65. MR **54** # 8701.
= Russian Math. Surveys **30**:5 (1975), 1–75.

[6] ———, A spectral sequence for the reduction of functions to normal forms, in the collection "Problems of mechanics and mathematical physics", Nauka, Moscow 1976, 7–20.

[7] ———, A spectral sequence for the reduction of functions to normal forms, Funktsional. Anal. i Prilozhen. **9**:3 (1975), 81–82. MR **52** # 4332.
= Functional Anal. Appl. **9** (1975), 251–253.

[8] ———, Braids of algebraic functions, and cohomology of swallow tails, Uspekhi Mat. Nauk **23**:4 (1968), 247–248. MR **38** # 156.

[9] ———, Wave front evolution and equivariant Morse lemma. Comm. Pure Appl. Math. **29** (1976), 557–582 (in English). MR **55** # 9148.

[10] ———, *Dopolnitel'nye glavy teorii obyknovennykh differentsial'nykh uravnenii* (Additional chapters of the theory of ordinary differential equations), Nauka, Moscow 1978, §4.

[11] E. V. Brieskorn, Singular elements of semisimple algebraic groups, Proc. Internat. Congress Mathematicians, Nice 1970, vol. 2, Gauthier–Villars, Paris 1971, 279–284.

[12] ———, Sur les groupes de tresses (d'après V.I. Arnold), Sém. N. Bourbaki **24** (1971/72), exp. 401.

[13] ———, Die Monodromie der isolierten Singularitäten von Hyperflächen, Manuscripta Math. **2** (1970), 103–161. MR **42** # 2509.

[14] S. M. Husein-Zade, Monodromy groups of isolated singularities of hypersurfaces, Uspekhi Mat. Nauk **32**:2 (1977), 23–65.
= Russian Math. Surveys **32**:2 (1977, 23–69.

[15] P. Deligne, Les immeubles des groupes de tresses généralisés, Invent. Math. **17** (1972), 273–302. MR **54** # 10659.

[16] B. Malgrange, Ideals of differentiable functions, Tata Inst. Fund. Research, Bombay; Oxford Univ. Press, London 1967. MR **35** # 3445.
Translation: *Idealy differentsiruemykh funktsii,* Mir, Moscow 1968.

[17] J. Milnor, Singular points of complex hypersurfaces, Annals of Math. Studies, **61**, Princeton Univ. Press, Princeton, N.J., 1968. MR **39** # 969.
Translation: *Osobye tochki kompleksnykh giperpoverkhnostei,* Mir, Moscow 1971.

[18] T. Poston and I. N. Stewart, Catastrophe theory and its applications, Pitman, London 1978.

[19] P. Slodowy, Einfache Singularitäten und einfache algebraische Gruppen, Regensburger Math. Schriften 2, Univ. Regensburg 1978.

[20] ———, Einige Bemerkungen zur Entfaltung symmetrischer Funktionen, Math. Z. **158** (1978), 157–170.

[21] G. N. Tyurina, The topological properties of isolated singularities of complex spaces of codimension 1, Izv. Akad. Nauk SSSR Ser. Mat. **32** (1968), 605–620. MR **37** # 3053. = Math. USSR-Izv. 2 (1968), 557–571.

[22] G. Wassermann, Classification of singularities with compact Abelian symmetry, Regensberger Math. Schriften 1, Univ. Regensburg 1977.

Received by the Editors 5 June 1978

Translated by E. J. F. Primrose

INDICES OF SINGULAR POINTS OF 1–FORMS ON A MANIFOLD WITH BOUNDARY, CONVOLUTION OF INVARIANTS OF REFLECTION GROUPS, AND SINGULAR PROJECTIONS OF SMOOTH SURFACES

V. I. Arnol'd

Contents

Introduction
§1. Singularities and their indices 226
§2. A theorem on the sum of indices 228
§3. Indices of singular points of adapted vector fields 229
§4. Connection between the indices of fields and forms 230
§5. Covering indices 231
§6. Algebraic definitions 233
§7. Indices of critical points of functions on the boundary of a manifold 238
§8. Indices of simple boundary singularities 238
§9. Duality 245
§10. Some applications 252
References 266

Introduction

The notion of the index of a singular point of a vector field does not generalize directly to the case of manifolds with boundary. However, it turns out that a generalization to the case of boundary singularities becomes possible if vector fields are replaced by differential 1-forms. For a 1-form on a manifold with boundary we define singular points and their indices so that the sum of the indices of all (interior and boundary) singular points of any 1-form with isolated singularities on a compact manifold with boundary is equal to the Euler characteristic of the manifold. For boundary singularities we define several integral invariants (indices) and we establish the equivalence of their geometric definitions with algebraic definitions in terms of the signatures of quadratic forms on local rings.

In the special case when the 1-form is the differential of a function and the boundary singularity of the function is simple, the resulting index may also be described as the signature of a quadratic form, constructed using the special operation of "convolving" the invariants of the corresponding reflection group. Moreover, the family of quadratic forms on a local ring that arises through the operation of multiplication and the family of quadratic forms on the dual space that arises through the operation of convolution, are dual to each other in a certain sense. One of the results of studying this duality is the assignment of a Lie algebra whose dimension is equal to the dimension of the space to each irreducible reflection group in Euclidean space: the corresponding Lie group is formed by the linearizations of those diffeomorphisms of the orbit manifold of the reflection group that map the set of non-regular orbits into itself.

This technique leads to the solution of a number of problems on normal forms of various geometric objects. In particular, we consider the projection of smooth surfaces from 3-space onto the plane. The singularities of projections in general position were already classified in a classical paper of Whitney. However, by choosing the direction of projection in a special way we can add more complicated singularities. There turn out to be ten singularities such that for surfaces in general position all the singularities of the projections with respect to all the directions are equivalent to these ten.

Precise definitions and statements of these and other results and conjectures are contained in §§1–7 (indices), §§8–9 (convolutions) and §10 (projections).
The author thanks V. V. Goryunov, V. I. Danilov, V. M. Zakalyukin, A. G. Kushnirenko, O. V. Lyashko, O. A. Platonova, A. G. Khovanskii, A. N. Shoshitaishvili, M. A. Shubin, and D. B. Fuks for useful discussions.

§1. Singularities and their indices

We consider a 1-form on a manifold with boundary. A boundary point is called a *boundary singularity* if this form vanishes on the whole tangent plane to the boundary at the point.

We assume that this singularity is isolated, and we define its index. We fix a non-degenerate equation for the boundary, that is, a function f that vanishes on the boundary, has no critical points on the boundary, and is positive inside the manifold.

DEFINITION. The *index* i_+ of a boundary singularity is the integer indicating the number of points on a small hemisphere with centre at the singularity at which points the form is positively proportional to the differential of the equation of the boundary.

Here the points on the hemisphere are included with their orientations. More precisely, we consider the cosphere bundle of the manifold restricted to this hemisphere. Our 1-form ω and the differential df of the equation of the boundary determine a pair of sections of this bundle. The number i_+ is the intersection index of these sections. (The orientations are chosen so that the index of the singularity at 0 for the form $\omega = x_1 dx_1 + \ldots + x_n dx_n$ is 1.)

EXAMPLE. We consider a function of two variables $F(x_1, x_2)$ on the half-plane $x_1 \geqslant 0$. The 1-form dF has as boundary singularities the critical points of the restriction of F to the boundary. The part of the half-plane where the values of F are less than that at the critical point approaches the critical point through several sectors. It is not hard to show that $i_+ = 1 - k_+$, where k_+ is the number of sectors of the set of smaller values in the half-plane.

REMARK 1. The index i_+ does not depend on the non-degenerate function f in its definition. In fact, any two such functions can be joined by a homotopy in the class of non-degenerate functions. The section generated by df on the boundary of our hemisphere does not vary under homotopy, so that its intersection index with the section generated by ω does not vary.

REMARK 2. The index of a 1-form ω can be defined for any transversally oriented hypersurface with boundary lying in the domain of regularity of f, under the assumptions that the boundary of the hypersurface lies on the boundary of the manifold in question and that the hypersurface does not pass through the singularities of the form (neither the interior nor the boundary ones).

With this aim we consider the cosphere bundle over the domain of regularity of f, excluding from the base small neighbourhoods around the singularities of ω. This is an oriented manifold of dimension $2n - 1$ (if the original manifold M is n-dimensional). The form df defines a submanifold of dimension n, and the restriction of ω to the hypersurface in question defines a submanifold of dimension $n - 1$. A (local) orientation of M defines (local) orientations of both these submanifolds. Under a change of the local orientation of M, both these induced orientations vary. Therefore, the intersection index is well-defined; it is also the index of the form along a transversally oriented hypersurface.

If the hypersurface is the boundary of a domain in which f is regular, then the index of ω along a hypersurface transversally oriented along an outward normal is equal to the sum of the indices i_+ of the boundary singularities of the form in this domain and the usual indices of the interior singular points of the form in the domain. (This follows from standard properties of the intersection index.)

The detailed description of the homology group on which the index is defined is left to the reader.

REMARK 3. On a Riemannian manifold with boundary one can define the boundary singularities of a vector field as the points where the field is orthogonal to the boundary. The index i_+ of a boundary singularity of the field indicates the number of points (including signs) on a small hemisphere with centre at the singular point at which the field is directed along the normal to the boundary. However, in contrast to the index of a boundary singularity of a form, the index of a boundary singularity of a field is not a diffeomorphism-invariant: it depends on the Riemannian metric.

§2. A theorem on the sum of indices

We consider a 1-form on a compact manifold with boundary and assume that it has only isolated (both interior and boundary) singularities.

THEOREM. *The sum of the indices of all singularities of 1-forms on a compact manifold with boundary (that is, the sum of the ordinary indices for the interior singular points and the indices i_+ for the boundary singularities) is equal to the Euler characteristic of the manifold.*

PROOF. We consider the double $\widetilde{M}$ of M. By definition,

$$\widetilde{M} = \{(x, z) \in M \times \mathbf{R}\colon z^2 = f(x)\},$$

where $f = 0$ is a non-degenerate equation for the boundary ∂M. $\widetilde{M}$ is a closed manifold.

The natural projection $\pi\colon \widetilde{M} \to M$ allows us to lift ω to $\widetilde{M}$: we obtain a 1-form $\widetilde{\omega} = \pi^*\omega$ on $\widetilde{M}$. Every interior singular point of ω on M corresponds to a pair of singular points of $\widetilde{\omega}$ on $\widetilde{M}$, each of them having the same index. The remaining singular points of $\widetilde{\omega}$ are precisely the points $(x, 0)$, where x is a boundary singularity of ω. We denote by $\widetilde{i}$ the index of this singularity for $\widetilde{\omega}$ and by i_0 the index of the restriction of ω to ∂M at x.

LEMMA. $\widetilde{i} = 2i_+ - i_0$.

This lemma will be proved in §5.

END OF THE PROOF OF THE THEOREM. The sum $\widetilde{I}$ of the indices of all the singular points of $\widetilde{\omega}$ on $\widetilde{M}$ is equal to the Euler characteristic $\chi(\widetilde{M})$, and the sum I_0 of the indices of all the singular points of the restriction of ω to ∂M is equal to $\chi(\partial M)$. We denote by I_+ the sum of the indices of all the boundary singularities of ω, and by I the sum of the indices of all its interior singularities. By the lemma,

$$\widetilde{I} = 2I + 2I_+ - I_0.$$

But $\chi(\widetilde{M}) = 2\chi(M) - \chi(\partial M)$. Comparing this with the preceding formula we obtain $I + I_+ = \chi(M)$, as was required.[1]

The proof given later for the lemma proceeds in the following way. With a boundary singularity of a form we associate a singularity of a vector field tangent to the boundary (§4). For such fields we define a number of indices and establish relations between them (§3). Next we verify that the corresponding indices for forms and fields are equal (§§4–5), and we derive a relation between the indices of forms from the corresponding relation for fields.

[1] The generalized Lefschetz theorem is proved similarly: the alternating sum of the traces of the automorphisms of the homology groups under a diffeomorphism of a compact manifold with boundary to itself is equal to the sum of the ordinary indices of the interior fixed points and the indices i_+ of the boundary fixed points defined above.

§3. Indices of singular points of adapted vector fields

A vector field on a manifold with boundary is said to be *adapted* if the vector of the field at every boundary point is tangent to the boundary. Several numbers (indices) are associated with a boundary singular point of an adapted vector field.

In Euclidean n-space $\mathbf{R}^n$ with the coordinates $(x_1, \ldots, x_n)$ we consider the half-space $x_1 \geqslant 0$. Let v be an adapted field with an isolated singular point O. We denote by S the unit sphere $S^{n-1} = \{x: |x| = 1\}$, and for sufficiently small $\varepsilon > 0$ we construct the usual spherical mapping $f: S \to S$, $f(x) = v(\varepsilon x)/|v(\varepsilon x)|$. If v is adapted, then f takes an equator of the sphere $(x_1 = 0)$ into itself. This equator, which we denote by S_0, divides S into a right and a left hemisphere $(x_1 \geqslant 0$ and $x_1 \leqslant 0)$, which we denote by S^+ and S^-, respectively. We denote by σ the involution of reflection in the equator (changing the sign of x_1).

DEFINITION. The following four numbers are the *partial indices* of a singular point of an adapted field:

$$\left.\begin{aligned} i_+ &= \deg[(S^+/S_0) \to (S/S^-)] && \text{(right-right index)} \\ i_- &= \deg[(S^-/S_0) \to (S/S^+)] && \text{(left-left index)} \end{aligned}\right\} \text{direct,}$$

$$\left.\begin{aligned} j_+ &= \deg[(S^+/S_0) \to (S/S^+)] && \text{(right-left index)} \\ j_- &= \deg[(S^-/S_0) \to (S/S^-)] && \text{(left-right index)} \end{aligned}\right\} \text{crossed.}$$

The three numbers below are called the *complete indices*:

$$\begin{aligned} i &= \deg[(f \mid S^+) \vee (\sigma f \mid S^-)] && \text{(complete relative index),} \\ i_1 &= \deg(S \to S) && \text{(usual index),} \\ i_0 &= \deg(S_0 \to S_0) && \text{(restricted index).} \end{aligned}$$

The two numbers below are called the *covering indices*:

$$\tilde{i}_+ = \deg[(f \mid S^+) \vee (\sigma f \sigma \mid S^-)], \quad \tilde{i}_- = \deg[(f \mid S^-) \vee (\sigma f \sigma \mid S^+)].$$

Here the arrows denote the mappings of spheres that are induced by the spherical mapping f of v.

THEOREM. *The following relations hold between the indices of an adapted vector field*:

$$\begin{aligned} i_+ + i_- &= i_1 + i_0, \\ i_+ - i_- &= i, \\ 2i_+ &= \tilde{i}_+ + i_0, \\ 2i_- &= \tilde{i}_- + i_0. \end{aligned}$$

LEMMA A. *The complete indices can be expressed by the partial indexes*: $i = i_+ - i_- = j_+ - j_-$, $i_1 = i_+ + j_- = i_- + j_+$, $i_0 = i_+ - j_+ = i_- - j_-$.

PROOF. The formulae for i_1 and for i follow from the definition of degree as the number of inverse images of a point of S^+ (first formulae) or of S^- (second formulae). The formulae for i_0 are obtained from the following commutative diagrams induced by $f \mid S^+$:

$$\begin{array}{ccccc} S_0 \to & S^+ \to & S^+/S_0 \\ \downarrow & \downarrow_{f|S^+} & \downarrow \\ S_0 \to & S \to & S/S_0, \end{array} \qquad \begin{array}{ccccccccc} 0 \to & \pi_{n-1}(S^+) \to & \pi_{n-1}(S^+, S_0) & \xrightarrow{\alpha} & \pi_{n-2}(S_0) \to 0 \\ & \downarrow & \beta\downarrow & & \downarrow\gamma \\ 0 \to & \pi_{n-1}(S) \to & \pi_{n-1}(S, S_0) & \xrightarrow{\delta} & \pi_{n-2}(S_0) \to 0. \end{array}$$

Under the usual orientation for spheroids ($\partial[S^+] = [S_0] = -\partial[S^-]$) we have:

$$\alpha([S^+, S_0]) = [S_0], \quad \gamma([S_0]) = i_0[S_0],$$
$$\beta([S^+, S_0]) = (i_+[S^+, S_0] + j_+[S^-, S_0]),$$
$$\delta([S^+, S_0]) = [S_0], \quad \delta([S^-, S_0]) = -[S_0].$$

It follows from the commutativity that $i_0 = i_+ - j_+$, as required.

LEMMA B. *The covering indices can be expressed by the partial indexes*:

$$\tilde{i}_+ = i_+ + j_+, \quad \tilde{i}_- = i_- + j_-.$$

PROOF. Considering the inverse images of a point of S^+ under the mapping $(f \mid S^+) \vee (\sigma f \sigma \mid S^-)$, we obtain i_+ inverse images in S^+ and j_+ inverse images in S^-; similarly for $\tilde{i}_-$.

The theorem now follows from the formulae of Lemmas A and B.

COROLLARY. *All nine indices can be expressed in terms of the three complete indexes*:

$$\tilde{i}_+ = i_1 + i, \quad \tilde{i}_+ + \tilde{i}_- = 2i_1, \quad 2i_+ = i_1 + i_0 + i,$$
$$\tilde{i}_- = i_1 - i, \quad \tilde{i}_+ - \tilde{i}_- = 2i, \quad 2i_- = i_1 + i_0 - i.$$

§4. Connection between the indices of fields and forms

In Euclidean space $\mathbf{R}^n$ with the coordinates $(x_1, \ldots, x_n)$ we consider the right and left half-spaces $\mathbf{R}^n_+$ $(x_1 \geqslant 0)$ and $\mathbf{R}^n_-$ $(x_1 \leqslant 0)$. Let $\Omega = \omega_1 dx_1 + \ldots + \omega_n dx_n$ be a differential form in $\mathbf{R}^n$ with an isolated boundary singularity at the origin. This singularity can be regarded as a boundary singularity for both half-spaces, so that right and left indices arise; for them we introduce the notation

$i_+(\Omega, \mathbf{R}^n_+) = i_+(\Omega)$ (right index of the form),
$i_+(\Omega, \mathbf{R}^n_-) = i_-(\Omega)$ (left index of the form).

LEMMA A. *The right and left indices of Ω are the same as those of the form*

$$\Omega_1 = |x_1| \omega_1 dx_1 + \omega_2 dx_2 + \ldots + \omega_n dx_n.$$

PROOF. We construct a family of forms $(0 \leqslant \lambda \leqslant 1)$:

$$\Omega_\lambda = [1 - \lambda(1 - |x_1|)]\omega_1 dx_1 + \omega_2 dx_2 + \ldots + \omega_n dx_n.$$

The singular points of the forms Ω_λ (both interior and boundary) do not depend on λ. Moreover, the condition of proportionality of Ω_λ with dx_1 has the form $\omega_2 = \ldots = \omega_n = 0$, hence, does not depend on λ.

The coefficient of proportionality $[1 - \lambda(1 - |x_1|)]\omega_1$ does not change sign as λ varies (if $x_1 \neq 0$), that is, its sign is the same as that of ω_1. Therefore, the indices of Ω_λ along a small right (left) hemisphere with centre at the origin do not change as λ varies from 0 to 1, that is, they are the same for Ω and for Ω_1, as required.

LEMMA B. *The right and left indices of Ω_1 are equal to the indices i_+ and i_-, as defined in* §3, *for the adapted field*

$$W = |x_1| \omega_1 \frac{\partial}{\partial x_1} + \omega_2 \frac{\partial}{\partial x_2} + \ldots + \omega_n \frac{\partial}{\partial x_n},$$

that is, $i_+(\Omega_1) = i_+(W)$, $i_-(\Omega_1) = i_-(W)$.

PROOF. Under the spherical mapping of W the number of (oriented) inverse images in the right hemisphere of the pole of the right hemisphere S^+ is equal to the number of points on a small right hemisphere with centre at zero at which the form Ω_1 is positively proportional to dx, and similarly for the left hemisphere, as required.

We introduce three more vector fields

$$V = x_1\omega_1 \frac{\partial}{\partial x_1} + \omega_2 \frac{\partial}{\partial x_2} + \ldots + \omega_n \frac{\partial}{\partial x_n},$$

$$V_1 = \omega_1 \frac{\partial}{\partial x_1} + \omega_2 \frac{\partial}{\partial x_2} + \ldots + \omega_n \frac{\partial}{\partial x_n},$$

$$V_0 = \omega_2 \frac{\partial}{\partial x_2} + \ldots + \omega_n \frac{\partial}{\partial x_n} \Big|_{\mathbf{R}^{n-1}} \quad (x_1 = 0).$$

LEMMA C. *The complete indices of zero of the adapted field W is equal to the usual indices at zero of the constructed fields, namely,*

$$i(W) = \operatorname{ind} V, \quad i_1(W) = \operatorname{ind} V_1, \quad i_0(W) = \operatorname{ind} V_0.$$

PROOF. For i_0 this is the definition, for i_1 the proof is as in Lemma A. The formula for i follows from the fact that V is the same as W on S^+ and as σW on S^-.

THEOREM. *The right and left indices i_+ and i_- of Ω at zero are connected with the indices i, i_1, and i_0 of the singular points of V, V_1, and V_0 by the relations*

$$i_+(\Omega) + i_-(\Omega) = i_1 + i_0,$$
$$i_+(\Omega) - i_-(\Omega) = i.$$

PROOF. Lemmas A, B, and C express all five of these indices in terms of the corresponding indices of W, and the theorem in §3 contains the required relations between them.

§5. Covering indices

We regard the double $\widetilde{\mathbf{R}}^n_+$ of the right half-space $x_1 \geqslant 0$ as the space with coordinates $(z, x_2, \ldots, x_n)$ equipped with the projection $\pi_+ : \widetilde{\mathbf{R}}^n_+ \to \mathbf{R}^n_+$, taking the point $(z, x_2, \ldots, x_n)$ into $(z^2, x_2, \ldots, x_n)$. The form Ω is pulled back to

$$\pi_+^*\Omega = 2z\tilde{\omega}_1\, dz + \tilde{\omega}_2\, dx_2 + \ldots + \tilde{\omega}_n\, dx_n, \qquad \tilde{\omega} = \pi_+^*\omega.$$

LEMMA. *The index of the singular point zero of the form $\pi_+^*\Omega$ on the double $\widetilde{\mathbf{R}}^n_+$ is equal to the covering index $\tilde{i}_+$ (defined in* §3) *of the adapted field*

$$W = |x_1| \omega_1 \frac{\partial}{\partial x_1} + \omega_2 \frac{\partial}{\partial x_2} + \ldots + \omega_n \frac{\partial}{\partial x_n}.$$

PROOF. In calculating the index we can replace a small sphere in the double by the surface $z^4 + x_2^2 + \ldots + x_n^2 = \varepsilon^2$, which is the inverse image of the small sphere $|x| = \varepsilon$. We consider the spherical mapping of the field

$$\widetilde{V}_+ = 2z\widetilde{\omega}_1\frac{\partial}{\partial z} + \widetilde{\omega}_2\frac{\partial}{\partial x_2} + \ldots + \widetilde{\omega}_n\frac{\partial}{\partial x_n}.$$

The inverse image of the centre $(\partial/\partial z)$ of the right hemisphere under this spherical mapping is obtained from the points at which $\widetilde{\omega}_2 = \ldots = \widetilde{\omega}_n = 0$, $z\widetilde{\omega}_1 > 0$. The inverse images in the right half-space ($z > 0$) correspond to the inverse image S^+ of the pole of the right hemisphere S^+, under the spherical mapping f of W. The inverse images of the left half-space ($z < 0$) in S^+ under f correspond to the pole of the left hemisphere S^-. This relation (completed by a check of the orientations) leads to the formula

$$\operatorname{ind} \widetilde{V}_+ = \deg\,[(f \mid S^+) \vee (\sigma f \sigma \mid S^-)] = \widetilde{\imath}_+(W),$$

as was required.

In just the same way we can prove the analogous formulae for the double of the left half-space: for $\pi_-(z, x_2, \ldots, x_n) = (-z^2, x_2, \ldots, x_n)$

$$\pi_-^*(\Omega) = -2z\widetilde{\omega}_1\,dz + \widetilde{\omega}_2\,dx_2 + \ldots + \widetilde{\omega}_n\,dx_n, \qquad \widetilde{\omega} = \pi_-^*\omega,$$

$$\widetilde{V}_- = -2z\widetilde{\omega}_1\frac{\partial}{\partial z} + \widetilde{\omega}_2\frac{\partial}{\partial x_2} + \ldots + \widetilde{\omega}_n\frac{\partial}{\partial x_n}, \qquad \operatorname{ind} \widetilde{V}_- = \widetilde{\imath}_-(W).$$

THEOREM. *The indices $\widetilde{\imath}_+$ and $\widetilde{\imath}_-$ of the singular points of the lift of Ω to the doubles of the left and right half-spaces are associated with the left, right, and restricted indices i_+, i_-, and i_0 of the form by the relations*

$$\widetilde{\imath}_+ = 2i_+ - i_0, \qquad \widetilde{\imath}_- = 2i_- - i_0.$$

PROOF. The preceding lemma and the lemmas of §4 identify the covering and the other indices of Ω with the corresponding indices of the adapted field W, and for them the assertion was already proved in §3.

These theorems imply that all the indices of Ω are equal to the corresponding indices of W and satisfy the same relations. We give to each of the indices of the form the name of the corresponding index of the field. For example:

$i(\Omega) = i_+(\Omega) - i_-(\Omega)$ is the complete relative index,

$i_1(\Omega)$ is the usual index,

$i_0(\Omega)$ is the index of the restriction of the form to the boundary,

$\widetilde{\imath}_+(\Omega)$ is the index of the form lifted to the double of the right half-space.

REMARK. In [6] Khovanskii considers a vector field P in $\mathbf{R}^n$ whose components are homogeneous polynomials $P_1, \ldots, P_n$ of degrees $m_1, \ldots, m_n$, satisfying a non-degeneracy condition that is valid almost everywhere. In $\mathbf{R}^{n-1} = \{x\colon x_1 = 1\}$ he constructs a vector field with the components

$$p_k(x_2, \ldots, x_n) = P_k(1, x_2, \ldots, x_n) \qquad (k = 2, \ldots, n).$$

The sum of the indices of all the singular points of this field at which $p_1 > 0$ is denoted by ind^+, and those at which $p_1 < 0$ by ind^-. Khovanskii found best possible estimates for ind^+, for ind^-, for $\text{ind}^+ - \text{ind}^-$, and for $\text{ind}^+ + \text{ind}^-$ in terms of the number of lattice points in certain polyhedra. He also noted that in the case

$$m_1 + \ldots + m_n \equiv n \bmod 2,$$

the index of the singular point O of the field P in $\mathbf{R}^n$ is equal to the characteristic $\text{ind}^+ - \text{ind}^-$ and he raised the question of the meaning of the characteristic $\text{ind}^+ - \text{ind}^-$ in terms of the singular point O of P for a different parity of the sum of the exponents. We can now answer this question. The following proposition is easy to prove.

PROPOSITION. *In the notation introduced above,* $\text{ind}^+ = i_+$, $\text{ind}^- = j_+$, *and the Kronecker characteristic* $\text{ind}^+ - \text{ind}^-$ *is equal to the covering index* $\tilde{i}_+$ *of the form* $\Sigma P_k dx_k$ *at* O *on the manifold with boundary* $x_1 \geqslant 0$.

The covering index $\tilde{i}_+$ is also equal to the sum $i_1 + i$, by the formula of the corollary to Lemma B in §3. For a field with homogeneous components, depending on the parity of the sum $m_1 + \ldots + m_n$, one of the terms i_1 or i vanishes (this is easily obtained from a comparison of the vectors of the field $(P_1, \ldots, P_n)$ and $(x_1 P_1, P_2, \ldots, P_n)$, respectively, at antipodal points). The sums of the degrees for both fields are different from 1. Thus, the *characteristic* $\text{ind}^+ - \text{ind}^-$, depending *on the parity of the sum of the degrees, is equal either to the usual index* i_1 *of the singular point* O *of the field* $P = (P_1, \ldots, P_n)$ *(the case* $m_1 + \ldots + m_n \equiv n \bmod 2$*), or to our complete index* $i(P)$*, which is equal to the usual index of* O *of the field* $(x_1 P_1, P_2, \ldots, P_n)$ *(the case* $m_1 + \ldots + m_n \equiv n + 1 \bmod 2$*).*

§6. Algebraic definitions

Let

$$\Omega = \omega_1\, dx_1 + \ldots + \omega_n\, dx_n$$

be a 1-form with an isolated boundary singularity O in the half-space $\mathbf{R}^n_+$ $(x_1 \geqslant 0)$.

We consider the following three local algebras over $\mathbf{R}$:

$$Q = \mathbf{R}\,[[x_1, x_2, \ldots, x_n]]/(x_1\omega_1, \omega_2, \ldots, \omega_n),$$
$$Q_1 = \mathbf{R}\,[[x_1, x_2, \ldots, x_n]]/(\omega_1, \omega_2, \ldots, \omega_n),$$
$$\tilde{Q} = \mathbf{R}\,[[z, x_2, \ldots, x_n]]/(z\tilde{\omega}_1, \tilde{\omega}_2, \ldots, \tilde{\omega}_n).$$

Here $\tilde{\omega}(z, x_2, \ldots, x_n) = \omega(z^2, x_2, \ldots, x_n)$.

We assume that the dimensions μ, μ_1, and $\tilde{\mu}$ of these algebras over $\mathbf{R}$ are finite and consider the corresponding Jacobians

$$J = \partial(x_1\omega_1, \omega_2, \ldots, \omega_n)/\partial(x_1, x_2, \ldots, x_n),$$
$$J_1 = \partial(\omega_1, \omega_2, \ldots, \omega_n)/\partial(x_1, x_2, \ldots, x_n),$$
$$\tilde{\tilde{J}} = \partial(z\tilde{\omega}_1, \tilde{\omega}_2, \ldots, \tilde{\omega}_n)/\partial(z, x_2, \ldots, x_n).$$

Let

$$q: Q \to \mathbf{R}, \quad q_1: Q_1 \to \mathbf{R} \text{ and } \tilde{\tilde{q}}: \tilde{Q} \to \mathbf{R}$$

be any linear forms taking positive values on the corresponding Jacobians. The multiplication in the local algebra then defines a *non-degenerate* symmetric bilinear form:

$$\varphi(a, b) = q(ab) \quad \text{for } a, b \in Q,$$
$$\varphi_1(a, b) = q_1(ab) \quad \text{for } a, b \in Q_1,$$
$$\tilde{\tilde{\varphi}}(a, b) = \tilde{\tilde{q}}(ab) \quad \text{for } a, b \in \tilde{Q},$$

respectively.

THEOREM. *The signatures of these three forms are equal, respectively, to the complete relative index i, the usual index i_1, and the covering index $\tilde{i}_+$ of the boundary singularity of Ω at zero.*

COROLLARY. *The three signatures defined above are connected by the relation $\tilde{i}_+ = i_1 + i$ (the third signature is equal to the sum of the first two).*

PROOF OF THE THEOREM. By a theorem of Eisenbud–Levine [3] and Khimshiashvili [5], these signatures are equal to the usual indices of singular points of the corresponding fields V, V_1, and $\tilde{V}$. According to §§4 and 5, these usual indices are equal to the corresponding indices i, i_1, and $\tilde{i}_+$ of the boundary singularity.

Although the corollary follows immediately from the theorem and formulae in §3, we give here a direct algebraic proof.

PROOF OF THE COROLLARY. A) *The linear space $\tilde{Q}$ splits into the direct sum $\tilde{Q} = \tilde{Q}_{\text{even}} + \tilde{Q}_{\text{odd}}$ of even and odd constituents* mod z.

For we split $\mathbf{R}[[z, x_2, \ldots, x_n]]$ into the direct sum of elements that are odd and even mod z. The generators of the ideal $(z\tilde{\omega}_1, \tilde{\omega}_2, \ldots, \tilde{\omega}_n)$ are purely even $(\tilde{\omega}_2, \ldots, \tilde{\omega}_n)$ or purely odd $(z\tilde{\omega}_1)$. Therefore, the ideal splits into the direct sum of the odd and even parts. Consequently, the factor-algebra

$$\tilde{Q} = \mathbf{R}[[z, x_2, \ldots, x_n]]/(z\tilde{\omega}_1, \tilde{\omega}_2, \ldots, \tilde{\omega}_n)$$

also splits into a direct sum.

B) *The space $\tilde{Q}_{\text{even}}$ is a subalgebra of $\tilde{Q}$ and is naturally isomorphic to Q.*

In fact, the even elements mod z of the ideal $(z\tilde{\omega}_1, \tilde{\omega}_2, \ldots, \tilde{\omega}_n)$ in $\mathbf{R}[[z, x_2, \ldots, x_n]]$ belong to the ideal $(x_1\omega_1, \omega_2, \ldots, \omega_n)$ in $\mathbf{R}[[x_1, x_2, \ldots, x_n]]$ (we identify the latter algebra of even elements mod z in $\mathbf{R}[[z, x_2, \ldots, x_n]]$, assuming that $x_1 = z^2$).

C) *The mapping $\mathbf{R}[[x_1, \ldots, x_n]] \to \mathbf{R}[[z, x_2, \ldots, x_n]]$ that associates with $a(x)$ the element $z\tilde{a}(x)$ defines an isomorphism of linear spaces $Q_1 \to \tilde{Q}_{\text{odd}}$.*

For the elements of the ideal $(\omega_1, \ldots, \omega_n)$ in $\mathbf{R}[[x_1, \ldots, x_n]]$ are mapped to elements of the ideal

$$(z\tilde{\omega}_1, \ldots, z\tilde{\omega}_n) \subset (z\tilde{\omega}_1, \tilde{\omega}_2, \ldots, \tilde{\omega}_n),$$

so that multiplication by z gives a linear mapping $Q_1 \to \tilde{Q}$. The image of this mapping is exactly $\tilde{Q}_{\text{odd}}$. We claim that this mapping is an isomorphism. Suppose that the image of a in $\tilde{Q}$ is zero, that is,

$$z\tilde{a}(x) = \alpha_1 z\tilde{\omega}_1 + \alpha_2\tilde{\omega}_2 + \ldots + \alpha_n\tilde{\omega}_n$$

($\alpha_k \in \mathbf{R}[[z, x_2, \ldots, x_n]]$). Then

$$z(\tilde{a} - \alpha_1\tilde{\omega}_1) = \alpha_2\tilde{\omega}_2 + \ldots + \alpha_n\tilde{\omega}_n,$$

that is, $z(\tilde{a} - \alpha_1\tilde{\omega}_1)$ belongs to the ideal $(\tilde{\omega}_2, \ldots, \tilde{\omega}_n)$ in $\mathbf{R}[[z, x_2, \ldots, x_n]]$. But z is not a zero-divisor in

$$\mathbf{R}[[z, x_2, \ldots, x_n]]/(\tilde{\omega}_2, \ldots, \tilde{\omega}_n),$$

because the sequence $(z, \tilde{\omega}_2, \ldots, \tilde{\omega}_n)$ is regular, by the Cohen–Macaulay theorem, $\tilde{Q}$ being finite-dimensional. Hence, we have the expansion

$$\tilde{a} - \alpha_1\tilde{\omega}_1 = \beta_2\tilde{\omega}_2 + \ldots + \beta_n\tilde{\omega}_n$$

with coefficients $\beta \in \mathbf{R}[[z, x_2, \ldots, x_n]]$. Replacing each coefficient α_1, β_k by its even part, we find that

$$a = \gamma_1\omega_1 + \ldots + \gamma_n\omega_n$$

with coefficients $\gamma_k \in \mathbf{R}[[x_1, \ldots, x_n]]$. Consequently, the class of a in Q_1 is zero, and our mapping is an isomorphism.

D) *The mapping*

$$\mathbf{R}[[x_1, \ldots, x_n]] \to \mathbf{R}[[x_1, \ldots, x_n]]$$

taking $a(x)$ to $x_1 a(x)$ defines an isomorphism of the linear space Q_1 with the linear space of the ideal (x_1) in Q.

The elements of the ideal $(\omega_1, \ldots, \omega_n)$ go into elements of the ideal $(x_1\omega_1, \omega_2, \ldots, \omega_n)$, so that the mapping is defined. It is clear that it is "onto"; we verify that the kernel is (0). If

$$x_1 a(x) = \alpha_1 x_1\omega_1 + \alpha_2\omega_2 + \ldots + \alpha_n\omega_n,$$

then $x_1(a - \alpha_1\omega_1)$ belongs to the ideal $(\omega_2, \ldots, \omega_n)$ in $\mathbf{R}[[x_1, \ldots, x_n]]$. But the sequence $(x_1, \omega_2, \ldots, \omega_n)$ is regular by sense of the Cohen–Macaulay theorem (since Q is finite-dimensional). Consequently, there is an expansion

$$a - \alpha_1\omega_1 = \beta_2\omega_2 + \ldots + \beta_n\omega_n,$$

that is, the class of a in Q_1 is zero.

E) *The Jacobians corresponding to the rings Q, Q_1, and $\tilde{Q}$ are equal to*

$$J = \omega_1 J_2 + x_1 J_1, \quad J_1, \quad \text{and} \quad \tilde{J} = \tilde{\omega}_1\tilde{J}_2 + 2z^2\tilde{J}_1,$$

where

$$J_k = \det(\partial\omega_i/\partial x_j), \qquad i, j \geqslant k.$$

F) LEMMA. *The classes* $\omega_1 J_2$ *and* $x_1 J_1$ *in* Q *are proportional, and*

$$\omega_1 J_2 / x_1 J_1 = \mu_0/\mu_1,$$

where

$$\mu_0 = \dim_{\mathbf{R}} \mathbf{R}[[x_1, x_2, \ldots, x_n]]/(x_1, \omega_2, \ldots, \omega_n).$$

REMARK. More generally, *we consider the rings*

$$Q_{ab} = \mathbf{R}[[x]]/(ab, \omega_2, \ldots, \omega_n),\ Q_a = \mathbf{R}[[x]]/(a, \omega_2, \ldots, \omega_n),$$
$$\text{and} \quad Q_b = \mathbf{R}[[x]]/(b, \omega_2, \ldots, \omega_n),$$

of finite dimensions μ_{ab}, μ_a, *and* μ_b, *respectively, with the Jacobians*

$$J(ab) = aJ(b) + bJ(a),$$
$$J(a) = \partial(a, \omega_2, \ldots, \omega_n)/\partial(x_1, \ldots, x_n),$$
$$J(b) = \partial(b, \omega_2, \ldots, \omega_n)/\partial(x_1, \ldots, x_n).$$

Then the classes of both the terms of the Jacobian $J(ab)$ *in* Q_{ab} *are proportional, and*

$$[aJ(b)]/[bJ(a)] = \mu_b/\mu_a.$$

PROOF. The line spanned by the Jacobian is the annihilator of the maximal ideal. Let us consider one of the terms, say $[aJ(b)]$. We claim that it belongs to the annihilator of the maximal ideal. The products of $aJ(b)$ by the elements of the maximal ideal of Q_{ab} are generated by the elements $x_i aJ(b)$ in $\mathbf{R}[[x]]$. But $x_i J(b)$ lies in the ideal $(b, \omega_2, \ldots, \omega_n)$. Consequently, $x_i aJ(b)$ lies in the ideal $(ab, \omega_2, \ldots, \omega_n)$, that is, $[aJ(b)]$ belongs to the annihilator. It remains to compute the coefficient of proportionality. We assume that the various ω are holomorphic (the general case reduces easily to this one). We use the standard formula for the multiplicity of a holomorphic mapping: for small ρ

$$\left(\frac{1}{2\pi i}\right)^n \oint\limits_{|f_k| = \rho_k} \frac{J \prod dx}{\prod f_k} = \mu,$$

where

$$J = \partial(f_1, \ldots, f_n)/\partial(x_1, \ldots, x_n),$$
$$\mu = \dim_{\mathbf{C}} \mathbf{C}[[x_1, \ldots, x_n]]/(f_1, \ldots, f_n).$$

Applying this formula in the case

$$f_1 = ab,\ f_2 = \omega_2, \ldots, f_n = \omega_n,$$

we find

$$(2\pi i)^n \mu_{ab} = \oint\limits_{|f_k| = \rho_k} \frac{J(b)\, dx}{b\omega_2 \ldots \omega_n} + \oint\limits_{|f_k| = \rho_k} \frac{J(a)\, dx}{a\omega_2 \ldots \omega_n}.$$

In the first integral we can replace the chain $|ab| = \rho_1$, $|\omega_k| = \rho_k$ by a homologous chain in the complement of the zeros of the denominator, namely,

$|b| = \rho_1$, $|\omega_k| = \rho_k$, and in the second by $|a| = \rho_1$, $|\omega_k| = \rho_k$. Again using the multiplicity formula we find

$$\oint_{|f_k|=\rho_k} \frac{aJ(b)\,dx}{ab\omega_2 \ldots \omega_n} = (2\pi i)^n \mu_b, \quad \oint_{|f_k|=\rho_k} \frac{bJ(a)\,dx}{ab\omega_2 \ldots \omega_n} = (2\pi i)^n \mu_a.$$

The mapping

$$\alpha \mapsto \left(\frac{1}{2\pi i}\right)^n \oint_{|f_k|=\rho_k} \frac{\alpha\,dx}{\prod f_k}$$

gives a linear form on the local algebra

$$Q = \mathbf{C}[[x_1, \ldots, x_n]]/(f_1, \ldots, f_n).$$

We have shown above that the values of this form on the elements $[J(ab)]$, $[aJ(b)]$, and $[bJ(a)]$ proportional in Q are, respectively, equal to μ, μ_b, and μ_a, as was required.

G) *The choice of the quadratic forms on* Q_1, *on* Q, *and on* $\widetilde{Q}$. Let $q_1 : Q_1 \to \mathbf{R}$ be a linear form with positive values on J_1. We denote by $\pi: \widetilde{Q} \to Q$ the projection of $\widetilde{Q}$ onto $\widetilde{Q}_{\text{even}} \approx Q$, which annihilates $\widetilde{Q}_{\text{odd}}$. Let $P: Q \to x_1 Q$ be any projection onto the ideal in Q generated by x_1. Let $\rho: Q_1 \to x_1 Q$ be the isomorphism of part D). We define linear forms $q: Q \to \mathbf{R}$ and $\widetilde{\widetilde{q}} : \widetilde{Q} \to \mathbf{R}$ by

$$q = q_1 \circ \rho^{-1} \circ P, \quad \widetilde{\widetilde{q}} = q_1 \circ \rho^{-1} \circ P \circ \pi.$$

We consider the corresponding symmetric bilinear forms on Q_1, Q, and $\widetilde{Q}$: $\varphi_1(a, b) = q_1(ab)$ for a, b in Q_1, $\varphi(a, b) = q(ab)$ for a, b in Q, and $\widetilde{\widetilde{\varphi}}(a, b) = \widetilde{\widetilde{q}}(ab)$ for a, b in $\widetilde{Q}$. We note that

$$q_1([J_1]) > 0, \quad q([J]) > 0, \quad \widetilde{\widetilde{q}}([\widetilde{\widetilde{J}}]) > 0.$$

For it follows from Lemma F that

$$q([J]) = \frac{\mu_0 + \mu_1}{\mu_1} q_1([J_1]), \quad \widetilde{\widetilde{q}}([\widetilde{\widetilde{J}}]) = \frac{\mu_0 + 2\mu_1}{\mu_1} q_1([J_1]).$$

H) *The subspaces* $\widetilde{Q}_{\text{even}}$ *and* $\widetilde{Q}_{\text{odd}}$ *are* $\widetilde{\widetilde{\varphi}}$*-orthogonal.*

For, $\pi(a_{\text{even}}, b_{\text{odd}}) = 0$.

I) *The restriction of* $\widetilde{\widetilde{\varphi}}$ *to* $\widetilde{Q}_{\text{even}}$ *is* φ *(under the identification of* Q *and* $\widetilde{Q}_{\text{even}}$ *in part* B)),

For $\widetilde{Q}_{\text{even}}$ is a subalgebra of $\widetilde{Q}$, and $\widetilde{\widetilde{q}} = q \circ \pi$.

J) *The restriction of* $\widetilde{\widetilde{\varphi}}$ *to* $\widetilde{Q}_{\text{odd}}$ *is* φ_1 *(under the identification of* Q_1 *and* $\widetilde{Q}_{\text{odd}}$ *in part* C)),

For using C), G), and D) in succession, we find that

$$\widetilde{\widetilde{\varphi}}(z\widetilde{a}, z\widetilde{b}) = q_1 \circ \rho^{-1} \circ P \circ \pi\,(z^2 \widetilde{a}\widetilde{b}) = q_1 \circ \rho^{-1} \circ P\,(x_1 ab) =$$
$$= q_1 \circ \rho^{-1}(x_1 ab) = q_1(ab) = \varphi_1(a, b).$$

K) *It follows from* H), I), *and* J) *that the signature of* $\widetilde{\widetilde{\varphi}}$ *is equal to the sum of the signatures of* φ *and* φ_1, *as required.*

§7. Indices of critical points of functions on the boundary of a manifold

By the indices of a critical point of a function we mean the corresponding indices of its differential. Let $f\colon (\mathbf{R}^n, O) \to (\mathbf{R}, 0)$ be a smooth function with an isolated boundary singularity O on the half-space $\mathbf{R}^n_+$ $(x_1 > 0)$. We fix a sufficiently small ρ-neighbourhood of O and then we choose a sufficiently small $\varepsilon > 0$. We consider the manifold with boundary

$$M_\varepsilon = \{x\colon\ f(x) = -\varepsilon,\ |x| \leqslant \rho\}.$$

We divide it into the parts in the half-spaces $\mathbf{R}^n_+$ $(x_1 \geqslant 0)$ and $\mathbf{R}^n_-$ $(x_1 \leqslant 0)$ and in the hypersurface $\mathbf{R}^{n-1}$ $(x_1 = 0)$:

$$M_\varepsilon = M^+ \cup M^-,\quad M^+ \cap M^- = M^0.$$

Let $\widetilde{M}_+$ and $\widetilde{M}_-$ denote the corresponding doubles.

THEOREM. *The indices of the boundary singularity of the differential of f are connected with the Euler characteristics of the corresponding level sets by the relations*

$$i_+ = 1 - \chi(M^+),\qquad i_- = 1 - \chi(M^-),$$

$$\widetilde{i}_+ = 1 - \chi(\widetilde{M}^+),\qquad \widetilde{i}_- = 1 - \chi(\widetilde{M}^-),$$

$$i_1 = 1 - \chi(M_\varepsilon),\qquad i_0 = 1 - \chi(M^0).$$

PROOF. The last four formulae are obvious (see, for example, [5]), and the first two follow from them by the relations between the indices (§3) and between the Euler characteristics

$$\chi(\widetilde{M}_+) = 2\chi(M^+) - \chi(M^0).$$

REMARK. Actually, the definitions of the indices i_+ and i_- were suggested by these formulae on the one hand, and by the equality of the signatures of the two families of quadratic forms connected with the so-called simple singularities on the other. These two families of quadratic forms and the connections between them are discussed in §§8 and 9.

§8. Indices of simple boundary singularities

The simplest boundary singularities of functions are connected with the simple Lie algebras A_μ, B_μ, C_μ, D_μ, E_6, E_7, E_8, F_4 (only G_2 is not included):

A_μ: $f = x_1 \pm x_2^{\mu+1} + \ldots$,
B_μ: $f = \pm x_1^\mu + \ldots$,
C_μ: $f = x_1x_2 \pm x_2^\mu + \ldots$,
D_μ: $f = x_1 + x_2^2x_3 \pm x_3^{\mu-1} + \ldots$,
E_6: $f = x_1 + x_2^3 \pm x_3^4 + \ldots$,
E_7: $f = x_1 + x_2^3 + x_2x_3^3 + \ldots$,
E_8: $f = x_1 + x_2^3 + x_3^5 + \ldots$,
F_4: $f = \pm x_1^2 + x_2^3 + \ldots$

Here the dots denote a non-degenerate quadratic form in the remaining variables (that is, $\pm x_k^2 \pm \ldots \pm x_n^2$, where $k = 2$ for B, 3 for A, C, and F, and 4 for D and E); the boundary equation is $x_1 = 0$.

Apart from the quadratic forms on the local algebras

$$Q = \mathbf{R}[[x_1, \ldots, x_n]]/(x_1\partial f/\partial x_1, \partial f/\partial x_2, \ldots, \partial f/\partial x_n),$$

which were described in §6, there are also quadratic forms on the dual spaces that are connected with the simple singularities; the signatures of these forms also turn out to be equal (in modulus) to the indices $i_+ - i_-$ of the correspond-boundary singularities. These new quadratic forms are defined immediately for all reflection groups.

Let G be an finite irreducible reflection group in the Euclidean space $\mathbf{R}^\mu$. The G also acts on the complexified space $\mathbf{C}^\mu$. The orbit *variety* $B = \mathbf{C}^\mu/G$ is also diffeomorphic to $\mathbf{C}^\mu$. A coordinate system in B is given by a basis of the ring of symmetric (G-invariant) functions (polynomials) in $\mathbf{C}^\mu$. The natural projection π: $\mathbf{C}^\mu \to B$ takes all the reflecting planes into some (in general, reducible) hypersurface $\Sigma \subset B$. This hypersurface is called the *variety of non-regular orbits* (or *discriminant variety*).

We denote by $\mathcal{O}$ the algebra of germs of holomorphic functions on B at O. We define a bilinear operation of "*convolution of invariants*" Φ: $\mathcal{O} \times \mathcal{O} \to \mathcal{O}$ by the relation

$$\Phi(a, b) = \pi_* \langle \nabla \pi^* a, \nabla \pi^* b \rangle,$$

where $\langle\, ,\rangle$ and ∇ are the Euclidean scalar product and gradient.

Explicit formulae are given in [1] for the convolution Φ in the case of the series A.

We denote by T the tangent space to B at zero and by T^* the dual space. The linearization of the convolution leads to the operation of "*linearized convolution*" φ: $T^* \times T^* \to T^*$, which is defined by the relation

$$\varphi(\alpha, \beta) = d(\Phi(a, b)),$$

where $\alpha = da$ and $\beta = db$ (d is the differential of a function at zero).

Let q be a point of T. We define a symmetric bilinear form on T^* by

$$\psi_q(\alpha, \beta) = \varphi(\alpha, \beta)|\, q.$$

THEOREM A. *If q does not belong to the tangent plane to the discriminant* *variety at zero, then the form ψ_q is non-degenerate.*

Before we prove Theorem A we state it in other terms.

DEFINITION. A *vector field with potential* $a \in \mathcal{O}$ is a field

$$v_a = \pi_* \nabla \pi^* a$$

on B. Obviously, the fields v_a can be expressed in terms of the convolution operation:

$$v_a b = \Phi(a, b)$$

THEOREM B. *Let $a_1, \ldots, a_\mu$ be a coordinate system on B. Then the vectors $v_{a_1}, \ldots, v_{a_\mu}$ are linearly independent at every point of $B \setminus \Sigma$, and their determinant has a first-order zero on Σ.*

DEFINITION. A linear field with potential $\alpha \in T^*$ is a field on T:

w_α = linear part of v_a at O, where α is the linear part of a at O.

The fields w_α can, obviously, be expressed in terms of the linearized convolution operation: $w_\alpha \beta = \varphi(\alpha, \beta)$.

THEOREM C. *Let $\alpha_1, \ldots, \alpha_\mu$ be a basis of T^*. Then the vectors $w_{\alpha_1}, \ldots, w_{\alpha_\mu}$ are linearly independent at each point of T that does not belong to the tangent plane τ to the discriminant variety Σ, and the determinant of the components of these vectors has a zero of order μ on τ.*

PROOF OF THEOREM B. The differentials of the functions a_i at each point of B are linearly independent. Hence, the differentials of $\pi^* a_i$ are independent at each point of $\mathbf{C}^\mu$ that does not lie on a reflecting hyperplane. Consequently, the vectors $\nabla \pi^* a_i$ are independent outside such a hyperplane. Therefore, the vectors $v_{a_i} = \pi_* \nabla \pi^* a_i$ are independent at each point of B outside the discriminant variety Σ.

We consider a *generic* point of the discriminant variety. We choose local coordinates $\Lambda_1, \ldots, \Lambda_\mu$ in a neighbourhood of this point such that the equation of the discriminant variety takes the form $\Lambda_1 = 0$ (and so that all the coordinates of this point are zero).

We consider one of the inverse images of this point under the mapping

$$\pi \colon \mathbf{C}^\mu \to B.$$

The functions

$$z_2 = \pi^* \Lambda_2, \ldots, z_\mu = \pi^* \Lambda_\mu \text{ and } z_1 = \sqrt{(\pi^* \Lambda_1)}$$

form a system of local coordinates in a neighbourhood of this point in $\mathbf{C}^\mu$; here the equation of the reflecting hyperplane has the form $z_1 = 0$, and the reflection corresponding to it changes the sign of the coordinate z_1 without changing the values of the remaining coordinates of the point being reflected.

We compute the Jacobian J of the components of the fields v_{Λ_i} at a point with a small non-zero Λ_1-coordinate. We have

$$J = (\rho d\Lambda_1 \wedge \ldots \wedge d\Lambda_\mu)|(v_1 \wedge \ldots \wedge v_\mu) =$$
$$= \rho \det(\partial\Lambda/\partial z)[(dz_1 \wedge \ldots \wedge dz_\mu)|(\nabla\pi^*\Lambda_1 \wedge \ldots \wedge \nabla\pi^*\Lambda_\mu)],$$

where $\rho \neq 0$. Note that $\det(\partial\Lambda/\partial z) = 2z_1$. Further, the vectors $\nabla\pi^*\Lambda_k$ are obtained from $\Sigma(\partial\Lambda_k/\partial z_i)\partial/\partial z_i$ by a non-degenerate linear transformation whose determinant g also does not vanish on the reflecting hyperplane (for $z_1 = 0$). Therefore,

$$J = \rho g 2z_1[(dz_1 \wedge \ldots \wedge dz_\mu)|(2z_1\partial/\partial z_1 \wedge \partial/\partial z_2 \wedge \ldots \wedge \partial/\partial z_\mu)] =$$
$$= 4\rho g z_1^2 = 4\rho g \Lambda_1.$$

Hence, the determinant of the components of the fields v_{Λ_i} has a first-order zero on Σ. For any coordinate system $a_1, \ldots, a_\mu$ the v_{a_i} can be expressed linearly and non-degenerately in terms of the v_{Λ_i}; therefore, the determinant of their components has a first-order zero at the generic point of Σ.

PROOF OF THEOREM C. Let $a_1, \ldots, a_\mu$ be a coordinate system such that $da_i = \alpha_i$. We consider the Jacobian J formed from the components of the v_{a_i}. This function on B has a first-order zero on Σ (Theorem B). Hence, the algebraic-geometric multiplicity of the singular point O (the order of the zero of J on a general line passing through O) is equal to μ. But Σ has a tangent plane at zero. Consequently, the initial form of J has the shape $c\lambda^\mu$, where $\lambda = 0$ is the equation of the tangent plane to Σ and $c \neq 0$.

Now we note that the determinant of the components of the v_{a_i} differs from the determinant of the components of the vectors w_{α_i} at the corresponding point by terms of order $\mu + 1$ (since w_{α_i} is the linearization of v_{a_i} at zero). Hence, the last determinant is *equal* to $c\lambda^\mu$, as required.

PROOF OF THEOREM A. The fact that the form ψ_q is non-degenerate indicates that the defining form of the linear operator

$$\Psi_q\colon T^* \to T, \; \Psi_q(\alpha)|\beta = \varphi(\alpha, \beta)|q, \quad \alpha, \beta \in T^*,$$

is non-degenerate. But the operator ψ_q takes α to the value of w_α at q (see the definition of the w_α). Hence, ψ_q being non-degenerate is equivalent to the linear independence of the system of values of the basis fields w_{α_i} at q. Thus, Theorem A follows from Theorem C, which we have just proved.

We return to the simple boundary singularity of a real function on the manifold with boundary $x_1 = 0$. The complexification of the corresponding local ring

$$Q = \mathbf{R}[[x_1, \ldots, x_n]]/(x_1\partial f/\partial x_1, \partial f/\partial x_2, \ldots, \partial f/\partial x_n)$$

is identified with the tangent space T at O to the variety B of orbits of the corresponding reflection group. For real q the form ψ_q is real on Q^*. By Theorem A it is non-degenerate if q is an invertible element of Q. Consequently, the signature of this form does not change under a continuous variation of q, and the modulus of the signature also does not depend at all on the choice of the invertible element $q \in Q$.

THEOREM D. *The signature of the form ψ_q defined above is equal in modulus to the index $i = i_+ - i_-$ of the corresponding boundary singularity.*

PROOF. Computing the indices i for all real forms of boundary singularities, we obtain

Type	A_{2k}	A_{2k+1}	B_{2k}	B_{2k+1}	C_{2k}	C_{2k+1}	D^+_{2k}	D^-_{2k}	D_{2k+1}	E_6	E_7	E_8	F_4
$\|i\|$	1	0	0	1	0	1	0	2	1	0	1	0	0

Here the sign in $D^\pm_\mu$ corresponds to the sign in the normal form $x_2^2 x_3 \pm x_3^{\mu-1}$.

Calculating the signatures of the ψ_q in these cases, we obtain the same answers. For in the cases A_μ, B_μ, C_μ, and D_μ it is fairly easy to compute the linearized convolution operation φ (see the remarks below), and the computations of the signature of ψ_q can be done explicitly. In the cases E_μ and F_4 the moduli of the signatures can be found without calculating φ, but at the expense of considerations of quasi-homogeneity (this is also applicable in the cases A_μ, B_μ, C_μ, and D_{2k+1}).

We carry out this computation completely for the case E_6. The normal form is $x_1 + x_2^3 \pm x_3^4$. The local algebra Q is the same as for the usual (non-boundary) singularity E_6 of the function $x^3 \pm y^4$.

Q is generated (over **R**) by the basis $e_1 = 1, e_2 = y, e_3 = x, e_4 = y^2, e_5 = xy, e_6 = xy^2$. The corresponding coordinates $\lambda_1, \dots, \lambda_6$ in Q form a basis of Q^*.

These elements of Q^* determine the basis invariants of the corresponding reflection group. Their degrees are, respectively, equal to 12, 9, 8, 6, 5, and 2. We equip the ring of polynomials in λ with a "quasi-homogeneous grading", in which $\deg \lambda_1 = 12$, $\deg \lambda_2 = 9, \dots, \deg \lambda_6 = 2$. If a is a homogeneous invariant of degree k, and b is of degree l, then $\langle \nabla a, \nabla b \rangle$ is a homogeneous invariant of degree $k + l - 2$. This allows us to control the action of the convolution operation Φ on the grading.

We form a matrix from the functions $\Phi(\lambda_p, \lambda_q) = \pi_* \langle \pi^* \lambda_p, \pi^* \lambda_q \rangle$ on Q. Its elements are (quasi-)homogeneous polynomials in λ whose degrees are given by the following table:

	λ_6	λ_5	λ_4	λ_3	λ_2	λ_1
λ_6	2	5	6	8	9	12
λ_5	5	8	9	11	12	
λ_4	6	9	10	12		
λ_3	8	11	12			
λ_2	9	12				
λ_1	12					

(the unoccupied spaces are filled with numbers larger than 12). From this table it is clear that the functions in the lower right (unoccupied) corner of the matrix of Φ are decomposable invariants (they lie in the square of the maximal ideal of $\mathcal{O}$); they do not contribute to the linearized convolution φ.

For the calculation of the signature of ψ_q we take as q the element e_1. Each element $\varphi(\lambda_p, \lambda_q)$ is a linear combination of the functions λ_r, and its value on e_1 is equal to the coefficient of λ_1 in this linear combination. But the degree of λ_1 is equal to 12, which is the largest of all the numbers in the left upper corner of our matrix.

Therefore, $\psi_{e_1}(\lambda_p, \lambda_q) \neq 0$ only for dual p and q (such that $p + q = 7$). Consequently, the quadratic form on Q^* corresponding to ψ_{e_1} reduces to $\xi_1\xi_6 + \xi_2\xi_5 + \xi_3\xi_4$ and hence, has signature zero.

The same argument allows us to compute the modulus of the signature in all the cases except D_{2k} (this is the only case when the set of degrees contains a repetition). We give here only the result of a direct computation of the linearized convolution operation φ for D_μ.

The normal form is $x_1 + x_2^2 \pm x^{\mu-1}$. The local algebra Q is the same as for the usual (non-boundary) singularity $D_\mu^\pm$ of the function $x^2y \pm y^{\mu-1}$. Q is generated (over $\mathbf{R}$) by the basis $e_1 = y^{\mu-2}, \ldots, e_{\mu-1} = 1; e_\mu = x$. The corresponding coordinates $\lambda_1, \ldots, \lambda_\mu$ in Q form a basis of Q^*.

Let $(z_1, \ldots, z_\mu)$ be coordinates in $\mathbf{R}^\mu$, where the reflection group D_μ acts by permuting the coordinates and changing the signs of an even number of coordinates. For the μ basis invariants we can take, for example,

$$s_2 = \sum z_i^2, \quad s_4 = \sum z_i^4, \quad \ldots, \quad s_{2\mu-2} = \sum z_i^{2\mu-2}, \quad P_\mu = z_1 \ldots z_\mu.$$

We compute the linearized convolution operation φ on the basis invariants. We denote the field $\partial/\partial z_i$ by ∂_i. In succession we find

$$\nabla s_{2k} = 2k \sum z_i^{2k-1}\partial_i, \quad \langle \nabla s_{2k}, \nabla s_{2l}\rangle = 4kls_{2k+2l-2},$$
$$\nabla P_\mu = Pz_1^{-1}\partial_1 + \ldots + Pz_\mu^{-1}\partial_\mu, \quad \langle \nabla s_{2k}, \nabla P_\mu\rangle = 2k \sum z_i^{2k-2}P_\mu,$$
$$\langle \nabla P_\mu, \nabla P_\mu\rangle = \sum (P_\mu/z_i)^2 = \sigma_{\mu-1}(z^2) = (-1)^\mu(\mu-1)^{-1}s_{2\mu-2} + \ldots$$

Here the dots denote decomposable invariants (contained in the square of the ideal generated by $s_2, \ldots, s_{2\mu-2}, P_\mu$). We note that $\langle \nabla s_{2k}, \nabla s_{2l}\rangle$ is also decomposable for $2k + 2l > 2\mu$, and that $\langle \nabla s_{2k}, \nabla P_\mu\rangle$ is decomposable for $k \neq 1$; in the last case $\langle \nabla s_2, \nabla P_\mu\rangle = 2\mu P_\mu$. Finally, the linearized convolution φ is given by the table

	s_2	s_4	$\ldots$	$s_{2\mu-2}$	P_μ
s_2	$4s_2$	$8s_4$	$\ldots$	$4(\mu-1)s_{2\mu-2}$	$2\mu P_\mu$
s_4	$8s_4$		$\cdot^{\cdot^{\cdot}}$	0	0
$\vdots$	$\vdots$			$\vdots$	$\vdots$
$s_{2\mu-2}$	$4(\mu-1)s_{2\mu-2}$	0	$\ldots$	0	0
P_μ	$2\mu P_\mu$	0	$\ldots$	0	$\dfrac{(-1)^\mu s_{2\mu-2}}{\mu-1}$

.

The isomorphism of the complexification of Q with the tangent space T to the orbit space of D_μ on $\mathbf{C}^\mu$ at O is constructed in the following way. We consider the function $x^2 y \pm y^{\mu-1} + \Sigma\lambda_k e_k$. A point $\Sigma\lambda_k e_k \in \mathbf{C}Q$ is singular if zero is a critical value of this function of x and y. The set of all singular points is called the *discriminant variety. Under the mapping of* $\mathbf{C}Q$ *into* B *the discriminant variety is mapped to the variety of non-regular orbits.*

Computing the singular points $\Sigma\lambda_k e_k$ we obtain

$$2xy + \lambda_\mu = 0, \quad x^2 + p'(y) = 0, \quad x^2 y + p(y) + \lambda_\mu x = 0,$$

where

$$p(y) = \pm y^{\mu-1} + \lambda_1 y^{\mu-2} + \ldots + \lambda_{\mu-1}.$$

Eliminating x from the first equation, we conclude that the point $\Sigma\lambda_k e_k$ is singular if and only if the polynomial $4yp(y) - \lambda_\mu^2$ has a multiple root.

On the other hand, the variety of non-regular orbits Σ in B can be described in the following way. Let $\sigma_1, \ldots, \sigma_\mu$ be the fundamental symmetric functions of $z_1^2, \ldots, z_\mu^2$. The invariants $\sigma_1, \ldots, \sigma_{\mu-1}, P_\mu$ can be taken as coordinates on B. We form the equation

$$Z^\mu - \sigma_1 Z^{\mu-1} + \ldots + (-1)^{\mu-1}\sigma_{\mu-1} Z + (-1)^\mu P_\mu^2 = 0.$$

The point with the coordinates $(\sigma_1, \ldots, \sigma_{\mu-1}, P_\mu)$ belongs to the variety of non-regular orbits if and only if this equation has a multiple root (since the reflecting planes are given by the equations $z_i^2 = z_j^2$, that is, $Z_i = Z_j$).

Comparing the coefficients of this polynomial with the coefficients of the polynomial

$$4yp - \lambda_\mu^2 = \pm 4y^\mu + 4\lambda_1 y^{\mu-1} + \ldots + 4\lambda_{\mu-1} y - \lambda_\mu^2,$$

we find

$$\sigma_k = \pm(-1)^k \lambda_k \quad (k = 1, \ldots, \mu - 1); \; P_\mu = \lambda_\mu \sqrt{(\mp(-1)^\mu/2)}.$$

But by Newton's formula,

$$k\sigma_k = (-1)^{k-1} s_{2k} + \text{decomposable elements.}$$

Consequently, up to decomposable elements,

$$s_{2k} = \mp k\lambda_k, \quad 2P_\mu = \lambda_\mu \sqrt{(\mp(-1)^\mu)}.$$

These relations allow us to calculate the linearized convolution operation φ in the basis $\lambda_1, \ldots, \lambda_\mu$. The resulting table for $\varphi(\lambda_k, \lambda_l)$ has the form

	λ_1	$\lambda_2 \ldots$	$\lambda_{\mu-1}$	λ_μ
λ_1	$\mp 4\lambda_1$	$\mp 8\lambda_2 \ldots$	$\mp 4(\mu-1)\lambda_{\mu-1}$	$\mp 2\mu\lambda_\mu$
λ_2	$\mp 8\lambda_2$	$\cdot$	0	0
$\vdots$	$\vdots$	$\cdot \; \cdot$	$\vdots$	$\vdots$
$\lambda_{\mu-1}$	$\mp 4(\mu-1)\lambda_{\mu-1}$	$0 \ldots$	0	0
λ_μ	$\mp 2\mu\lambda_\mu$	$0 \ldots$	0	$4\lambda_{\mu-1}$

To compute the signature of the form ψ_q we take $q = e_{\mu-1}$ (that is, q is a unit of the ring). Then we obtain a quadratic form ψ_q of the kind $\mp(\xi_1\xi_{\mu-1} + \xi_2\xi_{\mu-2} + \ldots) + \xi_\mu^2$. If μ is odd, then this form has signature 1. But if μ is even, the signature of the form in parentheses is equal to 1, so that

$$\operatorname{sgn} \psi_q(D_{2k}^{+}) = 0, \quad \operatorname{sgn} \psi_q(D_{2k}^{-}) = 2,$$

as required.

§9. Duality

From the theorems proved above it follows that the (moduli of the) signatures of the quadratic forms are equal, of which one form is defined on the local algebra Q and the other on the dual space Q^*. This suggests that these forms are themselves related. Such a connection in fact exists. It is one of the manifestations of the astonishing duality between the multiplication $Q \times Q \to Q$ and the linearized convolution operation φ: $T^* \times T^* \to T^*$. This duality seems surprising to me because it is verified separately for each singularity, while the general reasons for it are not clear.

Let $\alpha \in Q^*$ be a general element of the dual space to Q. We define a bilinear form f_α on Q by the relation $f_\alpha(u, v) = \alpha \mid (uv)$. From the multiplication in Q we obtain a family of bilinear forms on Q, parametrized by the elements of Q^*.

Precisely the same construction transforms the linearized convolution operation φ: $Q^* \times Q^* \to Q^*$ (defined in §8) into a family of bilinear forms on Q^*, parametrized by the elements of Q.

In §8 these forms were denoted by ψ_q ($q \in Q$) (strictly speaking, they also depend on the choice of identification of the space of the local algebra Q with the tangent space T to the orbit space of a reflection group; we discuss the influence of this identification below in Remark 3).

We consider a non-degenerate quadratic form on Q. It is given by a non-degenerate operator $Q \to Q^*$. The inverse operator gives a non-degenerate form on Q^*. This form is said to be *dual* to the original one (or the *Legendre transform* of the original form).

PROPOSITION 1. *The choices of the forms* $\{f_\alpha\}$ *and* $\{\psi_q\}$ *are dual in the sense that for each non-degenerate form* f_α *the dual is one of the* ψ_q, *and vice versa.*

This (surprising) proposition is proved by direct calculations in the cases A_μ, B_μ, C_μ, and D_μ; apparently, it also holds in general.[1]

The question arises of the relation between the parameters α and q of dual forms. This relation is non-linear, but can be expressed linearly in terms of the inversion operation in Q. More accurately, the linearized displacement φ can be expressed in terms of multiplication in Q in the following way.

Let D be a quasi-homogeneous derivation in the graded local algebra Q, $D = \Sigma d_k \lambda_k \partial/\partial\lambda_k$, where $d_k = \deg e_k$ (we are concerned with a grading in which the function f from which the local algebra

[1] Translator's note: In a letter dated 18 November 1979 the author states that A. B. Guivental has proved this conjecture.

$$Q = \mathbf{C}[[x]]/(x_1 f_1,\ f_2,\ \ldots,\ f_n),$$

is constructed is homogeneous of degree 1). We define a **C**-linear operator $R = E - D$: $Q \to Q$. We denote by $M_p : Q \to Q$ the operator of multiplication by $p \in Q$, so that $M_p q = pq$. The dual operator M_p^*: $Q^* \to Q^*$ gives, for each $\alpha \in Q^*$, a linear mapping $N_\alpha : Q \to Q^*$ by the formula

$$N_\alpha p = M_p^* \alpha \qquad \text{for every} \qquad p \in Q.$$

For an α in general position the mapping N_α is invertible.

Let G be an irreducible reflection group in $\mathbf{C}^\mu$, B the orbit variety, T the tangent space to B at O, T^* the dual space, and φ: $T^* \times T^* \to T^*$ the linearized convolution operation, defined by the Euclidean structure in $\mathbf{C}^\mu$ (see §8).

PROPOSITION 2. *Let α be an element of Q^* in general position (so that N_α is an invertible operator). Then the operation $\varphi_\alpha : Q^* \times Q^* \to Q^*$, given by the relation*

$$\varphi_\alpha(a,\ b) = R^* N_\alpha (N_\alpha^{-1} a \cdot N_\alpha^{-1} b),$$

goes over into the linearized convolution operation φ: $T^* \times T^* \to T^*$ under some isomorphism of the linear spaces T^* and Q^*.

This (surprising) assertion is proved in the cases, A, B, C, and D by a direct calculation if α is taken to be a basis element (the coefficient of a basis monomial of highest degree, proportional to the Jacobian). But apparently it also holds in the general case,[1] and for any (general) α. In particular, the operations φ_α with different α are equivalent (they go over into each other under some linear mapping of Q^* into itself), which is not at all obvious from the formula for φ_α.

PROOF OF PROPOSITION 2 FOR THE SINGULARITIES A_μ, B_μ, C_μ, D_μ AND A SPECIAL α. Since in the cases B_μ and C_μ both the reflection groups and the graded local rings are isomorphic, the case C_μ is not considered further. The versal deformations of the singularities A_μ, B_μ, and D_μ are chosen in the following way:

$$\begin{aligned} A_\mu &: x^{\mu+1} + \lambda_1 x^{\mu-1} + \ldots + \lambda_\mu \cdot 1, \\ B_\mu &: x^\mu + \lambda_1 x^{\mu-1} + \ldots + \lambda_\mu \cdot 1 \ \text{(boundary: } x = 0\text{)}, \\ D_\mu &: x^2 y \pm y^{\mu-1} + \lambda_1 y^{\mu-2} + \ldots + \lambda_{\mu-1} \cdot 1 + \lambda_\mu x. \end{aligned}$$

We denote by $e_0, e_1, \ldots, e_{\mu-1}$ the basis elements of Q over **C** (or over **R**) corresponding to the monomials in these formulae. In the dual space Q^* we choose a basis $f_0, f_1, \ldots, f_{\mu-1}$. These bases are connected with the monomials of the versal deformation and the coefficients λ_k for these monomials in the following way:

	$e_0 \ldots e_{\mu-2}\ \ e_{\mu-1}$	$f_0 \ldots f_{\mu-2}\ \ f_{\mu-1}$
A_μ, B_μ	$1 \ldots x^{\mu-2}\ \ x^{\mu-1}$	$\lambda_1 \ldots \lambda_{\mu-1}\ \ \lambda_\mu$
D_μ	$1 \ldots y^{\mu-2}\ \ x$	$\lambda_1 \ldots \lambda_{\mu-1}\ \ \lambda_\mu$

[1] This has also been proved since by A. B. Guivental (letter from author, 18 November 1979). (Transl.)

The multiplication tables in Q in these bases have the following form:

A_μ, B_μ	e_0	$e_1 \ldots e_{\mu-1}$
e_0	e_0	$e_1 \ldots e_{\mu-1}$
e_1	e_1	0
$\vdots$		$\vdots$
$e_{\mu-1}$	$e_{\mu-1}$	$0 \ldots 0$

D_μ	e_0	$e_1 \ldots e_{\mu-2}$	$e_{\mu-1}$
e_0	e_0	$e_1 \ldots e_{\mu-2}$	$e_{\mu-1}$
e_1	e_1	0	0
$\vdots$		$\vdots$	$\vdots$
$e_{\mu-2}$	$e_{\mu-2}$	$0 \ldots 0$	0
$e_{\mu-1}$	$e_{\mu-1}$	$0 \ldots 0$	$\mp(\mu-1)e_{\mu-2}$

In all cases we take $\alpha = f_0$. It is not hard to check that the operator $N_\alpha : Q^* \to Q^*$ then acts on the above bases according to the formula $N_\alpha e_k = f_k$ in all cases except one, namely, for D_μ the last basis element is transformed into $N_\alpha e_{\mu-1} = \mp(\mu-1)f_{\mu-1}$.

The gradings in Q are defined by the conditions

$$\deg x^{\mu+1} = 1 \ (\text{for } A_\mu)$$
$$\deg x^{\mu} = 1 \ (\text{for } B_\mu),$$
$$\deg(x^2y,\ y^{\mu-1}) = 1 \ (\text{for } D_\mu);$$

consequently, $\deg e_k = k/(\mu+1)$ for A_μ, $\deg e_k = 2k/2\mu$ for B_μ; $\deg e_k = 2k/(2\mu-2)$ for D_μ when $k < \mu - 1$, $\deg e_{\mu-1} = (\mu-2)/(2\mu-2)$. In place of R it is convenient to consider the proportional operator hR, where $h(A_\mu) = \mu+1$, $h(B_\mu) = 2\mu$, $h(D_\mu) = 2\mu-2$. The eigenvectors of the operator hR^* are f_k, and the corresponding eigenvalues are given, in accordance with the $\deg e_k$ found above, by the table

Type	f_0	f_1	$\ldots$	$f_{\mu-2}$	$f_{\mu-1}$
A_μ	2	3	$\ldots$	μ	$\mu+1$
B_μ	2	4	$\ldots$	$2\mu-2$	2μ
D_μ	2	4	$\ldots$	$2\mu-2$	μ

Finally, the operations

$$h\varphi_\alpha = hR^*N_\alpha(N_\alpha^{-1}\cdot N_\alpha^{-1}\cdot)$$

are given by the tables

A_μ	f_0	$f_1 \ldots f_{\mu-1}$
f_0	$2f_0$	$3f_1 \ldots (\mu+1)f_{\mu-1}$
f_1	$3f_1$	0
$\vdots$		$\vdots$
$f_{\mu-1}$	$(\mu+1)f_{\mu-1}$	$0 \ldots 0$

B_μ	f_0	$f_1 \ldots f_{\mu-1}$
f_0	$2f_0$	$4f_1 \ldots 2\mu f_{\mu-1}$
f_1	$4f_1$	0
$\vdots$	$\vdots$	$\vdots$
$f_{\mu-1}$	$2\mu f_{\mu-1}$	$0 \ldots 0$

D_μ	f_0	$f_1 \ldots f_{\mu-2}$	$f_{\mu-1}$
f_0	$2f_0$	$4f_1 \ldots (2\mu-2)f_{\mu-2}$	$\mu f_{\mu-1}$
f_1	$4f_1$	0	0
$\vdots$	$\vdots$	$\vdots$	$\vdots$
$f_{\mu-2}$	$(2\mu-2)f_{\mu-2}$	$0 \ldots 0$	0
$f_{\mu-1}$	$\mu f_{\mu-1}$	$0 \ldots 0$	$\mp 2f_{\mu-2}$

The linearized convolution operation φ is computed in §8 for the case D_μ. In the remaining cases the computations are similar but simpler. In the case A_μ the group acts by permutations on the roots of the equation $x^{\mu+1} + f_0 x^{\mu-1} + \dots + f_{\mu-1} = 0$; in the case B_μ by permutations and changes of sign of the roots of the equation $z^{2\mu} + f_0 z^{2\mu-2} + \dots + f_{\mu-1} = 0$. The calculation of $\langle \nabla f_k, \nabla f_l \rangle$ modulo decomposable elements is easily carried out by means of Newton's formula $\sigma_k = (-1)^{k-1} s_k/k$ mod (decomposable elements). Finally, the operations φ are given by the tables

A_μ	f_0	f_1	$\dots$	$f_{\mu-1}$
f_0	$-2f_0$	$-3f_1$	$\dots$	$-(\mu+1)f_{\mu-1}$
f_1	$-3f_1$			0
$\vdots$	$\vdots$		$\ddots$	$\vdots$
$f_{\mu-1}$	$-(\mu+1)f_{\mu-1}$	0	$\dots$	0

B_μ	f_0	f_1	$\dots$	$f_{\mu-1}$
f_0	$-4f_0$	$-8f_1$	$\dots$	$-4\mu f_{\mu-1}$
f_1	$-8f_1$			0
$\vdots$	$\vdots$		$\ddots$	$\vdots$
$f_{\mu-1}$	$-4\mu f_{\mu-1}$	0	$\dots$	0

D_μ	f_0	f_1	$\dots$	$f_{\mu-2}$	$f_{\mu-1}$
f_0	$\mp 4f_0$	$\mp 8f_1$	$\dots$	$\mp 4(\mu-1)f_{\mu-2}$	$\mp 2\mu f_{\mu-1}$
f_1	$\mp 8f_1$			0	0
$\vdots$	$\vdots$		$\ddots$	$\vdots$	$\vdots$
$f_{\mu-2}$	$\mp 4(\mu-1)f_{\mu-2}$	0	$\dots$	0	0
$f_{\mu-1}$	$\mp 2\mu f_{\mu-1}$	0	$\dots$	0	$4f_{\mu-2}$

Comparing these with the preceding tables, we obtain in all cases

$$\varphi = ch\varphi_\alpha,$$

where $c(A_\mu) = -1$, $c(B_\mu) = -2$, and $c(D_\mu^\pm) = \mp 2$.

Upon multiplication by a constant φ goes into an equivalent operation, because

$$\varkappa\varphi(a, b) = \varkappa^{-1}\varphi(\varkappa a, \varkappa b).$$

Consequently, the operations φ and φ_α are equivalent, as required.

REMARK 1. Proposition 1 follows from Proposition 2. For it is not difficult to calculate that N_γ is an operator of the form f_γ on Q. With an element $\gamma \in Q^*$ we associate a $q \in Q$ such that $\gamma = N_\alpha (Rq)^{-1}$. Then N_γ^{-1} is an operator of the form $\psi_{a,q} = \varphi_\alpha | q$ on Q^*. Thus, a form of the family $\{f_\gamma\}$ on Q corresponding to the parameter value γ and a form of the family $\{\psi_{\alpha,q}\}$ on Q^* corresponding to the parameter value q are dual to each other, provided that the element Rq is linearly independent of q, the element N_γ^{-1} of Q is linearly independent of, and these two elements are inverses of each other.

REMARK 2. Apart from the operation φ_α we can consider more general operations $\varphi_{\alpha,p}$ (respectively, $\varphi_{\alpha,p,q}$), in which instead of R we take $R_p = M_p R$ (respectively, $R_{p,q} = M_p R M_q$; here p and q are elements of Q).

However, there are identities linking these new operations with $\varphi_\alpha = \varphi_{\alpha,1,1}$:

1) $\varphi_{M_\pi^* \alpha, \pi p, q} = \varphi_{\alpha, p, q}$,

2) $\varphi_{\alpha, \pi^{-2} p, \pi q} \sim \varphi_{\alpha, p, q}$.

In particular,

$$\varphi_{\alpha, p} = \varphi_{M_{p^{-1}}^* \alpha}, \quad \varphi_{\alpha, p, q} = \varphi_{M_{q^{-2}p^{-1}}^* \alpha}.$$

The calculations leading to these identities are based on the following easily proved identities:

$$N_{N_\alpha p}q = N_\alpha(pq), \quad N_\alpha M_q = M_q^* N_\alpha, \quad DM_q = M_q D + M_{D[q]}.$$

REMARK 3. The fact that instead of a single operation φ on T we obtain a whole family φ_α of operations on Q^* is not surprising, since the identification of the spaces T and Q is not canonical. For when we identify the base of the versal deformation with the variety B of orbits of a reflection group, we watch only that the discriminant variety is identified with the variety of non-regular orbits.[1] The group of diffeomorphisms $B \to B$ that preserve the variety of non-regular orbits acts on the space of such identifications.

The group indicated above can also be defined as the group of germs of diffeomorphisms of B at O, lifted by the mapping π: $\mathbf{C}^\mu \to B$ (see [1]). Thus, it is naturally isomorphic to the group of equivariant (germs of) diffeomorphisms in $\mathbf{C}^\mu$.

The corresponding Lie algebra consists of germs of vector fields tangent to the variety of non-regular orbits. This algebra is a free module with μ generators over the ring $\mathcal{O}$ of germs of holomorphic functions on B at O. As generators we can take the fields $v_{\alpha_i} = \pi_* \nabla \pi^* a_i$, where the potentials a_i form a system of local coordinates on B in a neighbourhood of zero.

From this it is clear that on the space of identifications of T with Q it is natural to consider the action of the group of linear parts of π-liftable diffeomorphisms of T at O. The corresponding Lie algebra is of dimension μ and consists of the linearizations of the fields v_a at zero. These fields were denoted above by w_β ($\beta \in T^*$). Thus, with each reflection group in $\mathbf{C}^\mu$ there is associated a μ-dimensional Lie algebra, the Lie algebra of the fields w_β. In terms of the linearized convolution operation φ in T^* these fields can be expressed as follows: for any γ of T^* we have $w_\beta \gamma = \varphi(\beta, \gamma)$.

From Proposition 2 we can extract a description of these fields on T in terms of the algebra Q. We fix an element α from Q^* that determine the operation φ_α, and we fix an identification of T with Q that takes φ_α to φ. It is easy to prove the following proposition without using special properties of Q.

PROPOSITION 3. *Let D be a derivation of the commutative algebra Q. For any $p \in Q$ we consider* **C**-*linear operators W_p taking Q into Q that are given by the formula $W_p = M_p R$, where M_p is the operator of multiplication by p and $R = E - D$. The linear operators W_p corresponding to all the possible $p \in Q$ form a μ-dimensional Lie algebra.*

REMARK 4. The operators

$$W_{p,q} = M_p R M_q \ (p, q \in Q).$$

also form a Lie algebra. This algebra can also be described as the Lie algebra of operators of the form

[1] The corresponding restriction is not imposed on the isomorphism of T^* and Q^* in Proposition 2, but it holds for a suitable normalization of α.

$$M_u + M_v D(u, v \in Q).$$

PROPOSITION 4. *If Q is the local algebra of a simple singularity and D is the quasi-homogeneous derivation introduced above, then under the identification of Q with T indicated above the linear vector fields on Q corresponding to the linear operators W_p of Proposition 3 are taken to the linear fields w_β on T defined above. The parameters p (of Q) and β (of Q^*) are related linearly: $\beta = N_\alpha p$.*

PROOF. We have to verify that $w_\beta \gamma \equiv (W_p)^* \gamma$. According to Propositions 1 and 3, this means that

$$R^* N_\alpha (N_\alpha^{-1} N_\alpha p \cdot N_\alpha^{-1} \gamma) \equiv R^* M_p^* \gamma,$$

that is, that

$$N_\alpha (p \cdot N_\alpha^{-1} \gamma) \equiv M_p^* \gamma.$$

But by the definition of N_α we have

$$N_\alpha (p \cdot N_\alpha^{-1} \gamma) \equiv M_p^* M_{N_\alpha^{-1}\gamma}^* \equiv M_p^* N_\alpha N_\alpha^{-1} \gamma,$$

which proves the required identity.

PROPOSITION 5. *The operations $\varphi_\alpha : Q^* \times Q^* \to Q^*$ corresponding to different α are mapped to each other by linear mappings of Q^* into itself.*

PROOF. 1°. The isomorphism $N_\alpha : Q \to Q^*$ allows us to transfer the operation φ_α from Q^* to Q. We define an operation $\Phi_\alpha : Q \times Q \to Q$ by the relation

$$\Phi_\alpha (p, q) = N_\alpha^{-1} \varphi_\alpha (N_\alpha p, N_\alpha q).$$

By the definition of φ_α,

$$\Phi_\alpha (p, q) = S_\alpha (pq),$$

where $S_\alpha = N_\alpha^{-1} R^* N_\alpha$.

2°. Let $\alpha_0 \in Q^*$ be a form equal to 1 on the Jacobian J and to 0 on all monomials of smaller degree. Then, as is not hard to check,

$$N_{\alpha_0}^{-1} R^* N_{\alpha_0} = D + (2E/K),$$

where K is the Coxeter number defined by

$$\deg J = 1 - (2/K).$$

3°. We denote the operator $S_{\alpha_0} = D + (2E/K)$ simply by S. If $\alpha = N_{\alpha_0} a$, then $N_{\alpha_0}^{-1} N_\alpha : Q \to Q$ is multiplication by a. Therefore, $S_\alpha = M_a^{-1} S M_a$, where a is the class in Q of a function that is non-zero at the origin.

4°. To prove that the operations Φ_α and Φ_{α_0} are similar, it suffices for any function a with $a(0) = 1$ to construct an automorphism h in the algebra Q and a function u with $u(0) = 1$ for which

$$M_a^{-1} S M_a = M_u^{-1} (h^{-1} S h) M_u^2.$$

5°. We look for a (semi)quasi-homogeneous formal diffeomorphism h

inducing an automorphism of Q such that

(1) $$h^{-1}Dh = \lambda D,$$

where λ is a (formal) function. To solve the equation in 4° it suffices to find a (formal) solution h, λ, u of the system (1)–(2), where

(2) $$(2/K) + a^{-1}Da = (2/K)u + u^{-1}\lambda Du^2.$$

6°. The equation (2) is equivalent to the system

(3) $$(2/K) + b = (2/K)u + u^{-1}\lambda D[u^2],$$
(4) $$D = \lambda u D,$$

where $b = a^{-1}D[a]$ is a known function.

7°. The system (3)–(4) reduces to the equation

(5) $$D[u] = u\left(\frac{1-u}{K}+\frac{b}{2}\right).$$

8°. The equation (5) is solved by expanding in quasi-homogeneous terms: if $u = 1 + \Sigma u_d$, where $\deg u_d = d > 0$ (that is, $Du_d = du_d$) and $b = \Sigma b_d$, $d > 0$, then

$$(K^{-1} + d)u_d = 2^{-1}b_d + M(u_{<d}), \qquad d > 0,$$

where M is a polynomial of terms of lower degree.

9°. Having found a series u as indicated above and the series $\lambda = 1/u$, we solve (1) relative to the formal diffeomorphism $h = \exp(\gamma D)$, where γ is a formal function, by the usual method of Poincaré.

REMARK 5. The proof is applicable not only to simple singularities, but to all quasi-homogeneous (usual or boundary) singularities for which the number $-1/K$ does not occur among the weights of the basis monomials.

REMARK 6. The operator taking φ_α to φ_β has the form $\exp W_p^*$, where the operator $W_p^*\colon Q^* \to Q^*$ is dual to $W_p = M_p R\colon Q \to Q$, at least for A_μ, $\mu < 4$.[1]

REMARK 7. Proposition 4 can be used to recover the local algebra Q from the linearized convolution operation φ. For φ determines the Lie algebra of the fields w_β. These fields can be regarded as linear operators $w_\beta\colon T \to T$. We choose an invertible element w_{β_0} and form the operator $w_\beta w_{\beta_0}^{-1} = u_\beta$. If we began with the operation φ for a reflection group associated with a simple singularity, then these operators u_β form an algebra isomorphic to Q.

Thus, for any reflection group we obtain a linear family of operators u_β. Calculations show that these families are always commutative algebras: for the groups $I_2(n)$, H_3, and H_4 we obtain the algebras A_2, A_3, and D_4, respectively, but with unusual gradings.

REMARK 8. Proposition 4 can be used for the construction of (generalized) linear convolution operation φ from local algebra of a not necessarily simple

[1] This conjecture has also been proved by A. B. Guivental (letter from the author, 18 November 1979) (Transl.)

quasi-homogeneous singularity. It would be interesting to find out whether these operations are connected with some generalized reflection groups in Euclidean spaces.

§ 10. Some applications

The Lie algebras of liftable fields and their linearizations w_β are to the utmost extent a useful instrument for the reduction of various geometric objects to a normal form. This method was already used in [1]. Below we state some new results of applying this method.

10.1. Normal forms of curves in the swallow-tail space. We recall that B denotes the variety of orbits of a reflection group in $\mathbf{C}^\mu$. For the group A_μ of permutations of coordinates in the plane $z_0 + \ldots + z_\mu = 0$ of $\mathbf{C}^{\mu+1}$ the variety B is identified with the base of the versal deformation $z^{\mu+1} + \lambda_1 z^{\mu-1} + \ldots + \lambda_\mu$. The basis invariants $\lambda_1, \ldots, \lambda_\mu$ determine elements of the ring $\mathcal{O}$ of germs of holomorphic functions on B at O.

The differentials of these functions at O are denoted by $d\lambda_i$ and form a basis in the space T^* dual to the tangent space T to B at O. The corresponding vector fields v_{λ_i} form a free basis of the $\mathcal{O}$-module of germs of vector fields tangent to the variety of non-regular orbits Σ (the image of the reflecting hyperplane under the factorization π: $\mathbf{C}^\mu \to B$). An explicit form of the components of these fields, that is, of the convolution operation Φ, is indicated in [1]. In particular, the linearized convolution is given by the formula

$$\varphi(d\lambda_i, d\lambda_j) = (i+j)d\lambda_{i+j-1}$$

(or 0 for $i + j - 1 > \mu$). The tangent plane to the discriminant variety Σ at O is denoted by τ. In this notation the equation of τ has the form $d\lambda_\mu = 0$.

Turning to the general case of an arbitrary irreducible reflection group, we consider a curve γ: $(\mathbf{C}, 0) \to (B, O)$. We call it *transversal* to the variety of non-regular orbits Σ at O if the velocity vector $\dot{\gamma}(0)$ does not belong to τ.

THEOREM 1. *All the curves transversal to Σ at O are mapped into each other by diffeomorphisms of B that preserve Σ (so that $\gamma_1(t) \equiv h(\gamma_0(t))$).*

In particular, in the case A_μ every such curve reduces to the form $\lambda_1 = \ldots = \lambda_{\mu-1} = 0, \lambda_\mu = t$.

We consider the germ of the holomorphic vector field u at O in B. We say that *u is transversal* to Σ at zero if the field vector at O does not belong to τ.

THEOREM 2. *All the vector fields transversal to Σ at O are mapped into each other by diffeomorphisms of the base B that preserve Σ (at least in the cases A_μ, D_μ, E_μ, B_μ, C_μ, F_4, G_2, and $I_2(n)$).*

In particular, in the case A_μ every such field is equivalent to the field $\partial/\partial\lambda_\mu$.

A simple proof of Theorem 2 can probably be obtained from the following arguments. We consider a general complex phase curve of the given field. It intersects Σ in μ points. The passage times of these points are determined up to an initial reference time. Consequently, our curve corresponds uniquely to an (unordered) set of μ complex numbers with sum zero. This mapping of the (local) manifold of phase curves (which is diffeomorphic to a neighbourhood of O in $\mathbf{C}^{\mu-1}$) into the variety of sets (which can be identified with the variety $\mathbf{C}^{\mu-1}$ of the polynomials $z^\mu + a_2 z^{\mu-2} + \ldots + a_\mu$) is proper and finite-to-one (this mapping was used by O. V. Lyashko and E. Looijenga in [4] in the cases A_μ, D_μ, and E_μ, but it can also be constructed for the other reflection groups).

Apart from the given field we consider the standard "constant" field ($\partial/\partial\lambda_\mu$ in the case A_μ). The preceding construction determines for each phase curve of our field a phase line of the constant field, up to finitely many possibilities. With our phase curve we associate the nearest line. The required diffeomorphism of the ambient spaces is now constructed from the mappings of the phase curves into the phase lines that take the (complex) time on the phase curve into time on the phase line. (The origin of reference in both cases is chosen so that the sum of the times of intersection with Σ is zero.)

Apparently, Theorem 2 remains true also in the real C^∞ (not necessarily analytic) case if we allow the field to change sign.[1]

Theorem 1 follows from Theorem 2, but we give a direct proof.

PROOF OF THEOREM 1. It follows from Theorem C of §8 that each connected component of the space $T \setminus \tau$ is an orbit of the group of the linear parts of diffeomorphisms of the base that preserve Σ (in the real case there are two components, and in the complex case one). Therefore, each vector transversal to τ is carried by this linear group into any other such vector (in the real case it is perhaps necessary to permit sign changes in the vector). Thus, it suffices to prove the equivalence of curves of the form

$$\gamma_0(t) = et, \quad \gamma_1(t) = et + \delta(t),$$

where $\delta(t) = t^2\beta(t)$ and e is a vector transversal to τ. (We have identified our space B with its tangent space at zero, after fixing the basis invariants, say.)

Now we form the family of curves $\gamma_\theta(t) = \gamma_0(t) + \theta\delta(t)$, $0 \leqslant \theta \leqslant 1$, and we look for a family of (local) diffeomorphisms $g_\theta : (B, \Sigma) \to (B, \Sigma)$ from the condition

$$\gamma_0(t) + \theta\delta(t) \equiv g_\theta(\gamma_0(t)).$$

For g we obtain the equation $\partial g/\partial\theta\,|_{\varphi(t),\theta} = \delta(t)$. We then consider a vector field h, depending on the "time" θ and corresponding to the family of diffeomorphisms g_θ : by definition,

$$\partial g/\partial\theta\,|_{x,\theta} = h(g_\theta(x), \theta).$$

In particular, for $x = \gamma_0(t)$ the following relation must hold:

$$\delta(t) = h(et + \theta\delta(t), \theta),$$

from which we can find h.

We fix a basis of liftable fields v_{a_i}. We look for a liftable field h in the form of a linear combination of basis fields

[1] As we have learned during the preparation of the manuscript, O. V. Lyashko and V. M. Zakalyukin have also proved Theorem 2 and a number of generalizations of it (see [7]).

$$h(x, \theta) = \sum h_i(x, \theta)v_{a_i}(x).$$

We have

$$v_{a_i}(et + t^2\theta\beta(t)) = t[w_{\alpha_i}(e) + tu_i(t, \theta)].$$

Here $\alpha_i = da_i$ is a vector of the basis of T^* corresponding to the coordinate system $\{a_i\}$ in B; the u_i are smooth functions.

For the coefficients h_i we obtain the equation

$$t\beta(t) = \sum h_i(et + \theta\delta(t), \theta)[w_{\alpha_i}(e) + tu_i(t, \theta)].$$

By Theorem C of §8, the vectors $w_{\alpha_i}(e)$ are linearly independent and form a basis of T. For sufficiently small $|t|$ the vectors in square brackets also form a basis. Therefore, the coefficients h_i, as functions of t and θ, are uniquely determined and regular (smooth, holomorphic, . . .). For $t = 0$ these coefficients vanish for any θ. We extend the functions h_i for each θ from the curve γ_θ, where they are already defined, to a neighbourhood of the origin in B. This extension can be chosen to be regular in θ. The resulting family of vector fields h defines a one-parameter family of local liftable diffeomorphisms g_θ for all θ from 0 to 1, because $h(0, \theta) \equiv 0$. The diffeomorphism g_1 takes γ_0 into γ_1, and Theorem 1 is proved.

10.2. Normal forms of projections of surfaces in 3-space. DEFINITION. A *projection* of a surface S is a diagram $S \to \mathbf{R}^3 \to \mathbf{R}^2$, where the left arrow is an embedding of the smooth surface S and the right arrow is the linear projection. An *equivalence* of projections of surfaces is a commutative diagram.

$$\begin{array}{ccccc} S_1 & \to & \mathbf{R}^3 & \to & \mathbf{R}^2 \\ \downarrow & & \downarrow & & \downarrow \\ S_2 & \to & \mathbf{R}^3 & \to & \mathbf{R}^2, \end{array}$$

where the vertical arrows are diffeomorphisms. Similar definitions can be given for the germs (S, O).

THEOREM 3. *In the space of compact smooth hypersurfaces in* $\mathbf{R}^3$ *there is an open everywhere dense set formed by those surfaces whose projection at any point and in any direction is locally equivalent to the projection of one of the* 10 *surfaces of the following list at* (0, 0, 0) *along the x-axis*:

$$z = x,\ z = x^2,\ z = x^3 + xy,\ z = x^3 \pm xy^2,\ z = x^3 + xy^3,$$
$$z = x^4 + xy,\ z = x^4 + x^2y + xy^2,\ z = x^5 + x^3y \pm xy.$$

These projections are pairwise inequivalent. (*In the complex case the projections with the minus signs in the equations for the surfaces are equivalent to the projections with the plus signs, so that the list can be shortened to eight surfaces.*) *The singularities of the projections of the list are not removable by a small perturbation of the surface.*

For a surface in general position the first normal form in the list describes the projection in any direction at almost every point, the second at the points of some curve, the third at the remaining points. The normal forms given by the fourth and sixth equations of the list describe the projection at the remaining points in the directions belonging to some curve in the space of directions. The fifth, seventh, and eighth equations give projections that occur for a general surface at the remaining points in the remaining directions.

The proof of Theorem 3 is based on the fact that we can associate with a projection a two-parameter family of "vertical" lines (inverse images in $\mathbf{R}^3$ of points from $\mathbf{R}^2$). A surface S cuts out several points on each vertical line. Thus, we obtain a two-parameter family of zero-dimensional hypersurfaces on the vertical lines.

The points of intersection of S with the verticals can be combined into several verticals. Suppose that μ points merge at the relevant point of S on the vertical (assuming also complex points; in other words, μ is the dimension over $\mathbf{R}$ of the local ring corresponding to the intersection). From the Weierstrass preparation theorem (in the real C^∞ case, the Malgrange preparation theorem) it follows that our two-parameter family of zero-dimensional hypersurfaces is induced from the universal $(\mu - 1)$-parameter family

$$x^\mu + \lambda_1 x^{\mu-2} + \ldots + \lambda_{\mu-1} = 0. \tag{1}$$

In the universal family each $\lambda \in \mathbf{R}^{\mu-1}$ (in the complex case $\mathbf{C}^{\mu-1}$) corresponds to the zero-dimensional hypersurface $\{x\}$. Thus, the equation of S has (locally) the form

$$x^\mu + \lambda_1(y, z)x^{\mu-2} + \ldots + \lambda_{\mu-1}(y, z) = 0. \tag{2}$$

We may assume that all the coordinates of the point in question (the vertical coordinate x and the horizontal ones y and z) vanish and that $\lambda(0) = 0$. Here y and z are coordinates in the plane $\mathbf{R}^2$, which is the base of our family of vertical lines.

LEMMA 1. *The projection is equivalent to the one for which the equation of S has the form*

$$z = x^\mu + \lambda_1(y)x^{\mu-2} + \ldots + \lambda_{\mu-2}(y)x.$$

PROOF. Since S is smooth, $d\lambda_{\mu-1}(0) \neq 0$. By choosing coordinates in $\mathbf{R}^2$ (that is, a diffeomorphism of the base) we can achieve that $\lambda_{\mu-1} = -z$. For $y = 0$ the equation of S gives

$$z = x^\mu + \lambda_1(0, z)x^{\mu-2} + \ldots + \lambda_{\mu-2}(0, z)x, \qquad \lambda_k(0, 0) = 0.$$

By the implicit function theorem, $z = f(x)$, where f has a zero of order μ at the origin. A local diffeomorphism of the x-axis reduces f to the form x^μ.

Now the equation of S can be regarded as a deformation of the function $z = x^\mu$ with parameter y. By the versal deformation theorem a diffeomorphism of the x-axis depending on y reduces the equation of S to the form

$$z = x^\mu + \lambda_1(y)x^{\mu-2} + \ldots + \lambda_{\mu-1}(y).$$

This in turn takes the form indicated in the lemma by choosing the function $z - \lambda_{\mu-1}(y)$ as the new z-coordinate on the base.

LEMMA 2. *A hypersurface in general position in* $\mathbf{R}^n$ *does not admit tangent lines of order* $\mu > 2n - 1$.

PROOF. In $\mathbf{R}^n$ we fix a point, a hyperplane passing through it, and a line in the hyperplane also passing through this point. We consider the space of jets of the hypersurfaces for which this plane is tangent at the distinguished point. The set of jets of surfaces having a tangent of order μ at the distinguished point (say, the x-axis) has codimension $\mu - 2$ in this jet space. For, if the equation of the surface is $z = f(x, y, \ldots)$, then the coefficients of the expansion

$$f(x, 0) = f_2 x^2 + \ldots + f_{\mu-1} x^{\mu-1} \bmod x^\mu,$$

must vanish, and there are $\mu - 2$ of them. The dimension of the variety of lines passing through a point and lying in a hyperplane is $n - 2$. Therefore, the variety of jets of surfaces admitting a tangent line of order μ at a fixed point has codimension $\mu - n$ in the variety of jets of hypersurfaces at a point with a fixed tangent plane. By the transversality theorem, by a small perturbation of an $(n-1)$-dimensional hypersurface we can get rid of all the tangents with lines of order μ if $n - 1 < \mu - n$, as required.

COROLLARY. *For surfaces in general position in* $\mathbf{R}^3$ *we have* $\mu \leqslant 5$ *for projection in any direction and at any point.*

For by Lemma 2, a surface in $\mathbf{R}^3$ does not have contact of order more than five with a line.

Now we consider the cases $\mu = 1, 2, 3, 4, 5$ in turn. In the first two cases we obtain immediately the normal forms $z = x$ and $z = x^2$ from Lemma 1.

LEMMA 3. *For surfaces in general position the singularities of the projections with* $\mu = 3$ *are equivalent to those of the projection of the surfaces* $z = x^3 \pm xy^k$, $k \leqslant 3$, *along the x-axis at the origin.*

PROOF. Lemma 1 gives $z = x^3 + \lambda(y)x$. If λ has a zero of order k at $y = 0$, then a change of the variable y reduces the equation of the surface to the equivalent form $z = x^3 \pm y^k x$. For surfaces in general position $k = 1$ occurs for the common directions of projection (Whitney's theorem), $k = 2$ for directions on some curve in the space of directions, and $k = 3$ for individual directions; $k > 3$ does not occur for S in general position.

Before we consider the cases $\mu = 4$ and $\mu = 5$, we return to the formula (2), which associates our projection $S \to \mathbf{R}^3 \to \mathbf{R}^2$ with the mapping $\lambda: \mathbf{R}^2 \to \mathbf{R}^{\mu-1}$ of the base of our family of vertical lines into the base of the universal family.

The universal family is given by the diagram $\Gamma \to \mathbf{R}^\mu \to \mathbf{R}^{\mu-1}$, where the left arrow is the embedding of the hypersurface given by (1) into $\mathbf{R}^\mu$ with the coordinates (x, λ), and the right arrow is the canonical projection $(x, \lambda) \mapsto (\lambda)$. The hypersurface Γ itself is diffeomorphic to $\mathbf{R}^{\mu-1}$ (we can take $(x, \lambda_1, \ldots, \lambda_{\mu-2})$ as coordinates on it). The composition of the two arrows gives the Whitney mapping $w: \Gamma \to \mathbf{R}^{\mu-1}$. The formula (2) gives a mapping $\lambda: \mathbf{R}^2 \to \mathbf{R}^{\mu-1}$, which induces a commutative diagram

$$\begin{array}{ccccc} S & \to & \mathbf{R}^3 & \to & \mathbf{R}^2 \\ \downarrow & & \downarrow & & \downarrow \lambda \\ \Gamma & \to & \mathbf{R}^\mu & \to & \mathbf{R}^{\mu-1} \end{array}$$

whose upper row defines the projection in question, while the lower row is the diagram of the universal family. We call this the *λ-diagram* of the projection.

An *equivalence* of a λ-diagram and a λ'-diagram is a commutative diagram shaped like a prism, whose horizontal faces are a λ-diagram and a λ'-diagram, and the six vertical edges are diffeomorphisms. The following result is obvious.

LEMMA 4. *If a λ-diagram and a λ'-diagram are equivalent, then the corresponding projections are also equivalent.*

It is useful to clarify in what case the mappings λ and λ' give equivalent diagrams, that is, one is taken into the other by an automorphism of the diagram of the universal family.

An *infinitesimal automorphism* of a diagram $\Gamma \to \mathbf{R}^\mu \to \mathbf{R}^{\mu-1}$ is a vector field in the ambient space $\mathbf{R}^\mu$ that is tangent to the hypersurface Γ and projects onto the base $\mathbf{R}^{\mu-1}$.

The projection of an infinitesimal automorphism onto the base is a vector field that is liftable relative to the Whitney mapping $w\colon \Gamma \to \mathbf{R}^{\mu-1}$.

LEMMA 5. *Every Whitney-liftable* (*that is, liftable to* Γ) *vector field on* $\mathbf{R}^{\mu-1}$ *is the projection of some infinitesimal automorphism* (*that is, can be lifted to* $\mathbf{R}^\mu$).

PROOF. Suppose that the field $v = \Sigma v_k(\lambda)\, \partial/\partial\lambda_k$ is lifted to Γ from a field which in the coordinates $(x, \lambda_1, \ldots, \lambda_{\mu-2})$ has the form

$$V = V_0(x, \lambda_1, \ldots, \lambda_{\mu-2})\partial/\partial x + \ldots .$$

The required field on $\mathbf{R}^\mu$ is

$$V_0 \partial/\partial x + \sum v_k(\lambda)\partial/\partial\lambda_k.$$

Now we pass to the problem of the classification of the diagrams $\mathbf{R}^2 \xrightarrow{\lambda} \mathbf{R}^{\mu-1} \xleftarrow{w} \Gamma$, where w is the Whitney mapping. The equivalences here are taken to be commutative diagrams

$$\begin{array}{ccccc} \mathbf{R}^2 & \xrightarrow{\lambda} & \mathbf{R}^{\mu-1} & \xleftarrow{w} & \Gamma \\ \downarrow & & \downarrow & & \downarrow \\ \mathbf{R}^2 & \xrightarrow{\lambda(t)} & \mathbf{R}^{\mu-1} & \xleftarrow{w} & \Gamma, \end{array}$$

where the vertical arrows are diffeomorphisms depending continuously on t. In other words, we must classify the mappings of the plane into the swallow-tail space relative to the group of diffeomorphisms of the ambient space that preserve the swallow-tail (more precisely, relative to the identity component of this group). Here the specific equivalence of the two mappings λ and $\lambda' = \lambda(1)$ is called a *w-equivalence.*

We consider the space Ω of germs of smooth mappings $\lambda\colon (\mathbf{R}^2, 0) \to (\mathbf{R}^{\mu-1}, 0)$.

LEMMA 6. *Let* $\mu = 4$. *Then the germ* λ *is either w-equivalent to the germ of the embedding of the plane* $\lambda_1 = 0$ *into* $\mathbf{R}^3$ *or it belongs to a certain set of co-*

dimension 1 *in* Ω. *In the last case, the germ of* λ *is either w-equivalent to the germ of the embedding of the cylinder* $\lambda_2 = \pm \lambda_1^2$ *in* $\mathbf{R}^3$ *or it belongs to a certain set of codimension* 2 *in* Ω.

PROOF. The set of germs whose rank at O is less than 2 has codimension 2 in Ω. Therefore, the image of λ can be assumed to be the smooth surface of a swallow-tail in $\mathbf{R}^3$. Now the first assertion of the lemma follows from the equivariant Morse lemma proved in [1]. The second assertion is also contained in [1], where the beginnings are given of a classification of degenerate critical points of S_4-invariant functions (in [1] we were concerned with the whole group of diffeomorphisms that preserve the swallow-tail, but only the identity component was used in the proof).

The λ-diagrams corresponding to the two cylinders $\lambda_2 = \pm \lambda_1^2$ are equivalent. The corresponding planes and the cylinder of projection are equivalent to the projection of the surfaces $z = x^4 + xy$ and $z = x^4 + x^2y + xy^2$ along the x-axis at O. From Lemmas 5 and 6 it follows that for surfaces of general position in $\mathbf{R}^3$ the projections with $\mu = 4$ are locally equivalent to those described above. (The case of a plane occurs only for some curve in the space of projection directions, and the case of the cylinder only for isolated directions).

LEMMA 7. *Let* $\mu = 5$. *Then the mapping* $\lambda\colon (\mathbf{R}^2, O) \to (\mathbf{R}^{\mu-1}, O)$ *is either w-equivalent to the embedding of one of the two planes* $\lambda_2 = 0$, $\lambda_1 = \pm\lambda_3$ *into* $\mathbf{R}^4$, *or it belongs to a certain set of codimension* 1 *in* Ω.

PROOF. The image of λ may be assumed to be a surface in $\mathbf{R}^4$ that is smooth at zero (otherwise λ belongs to a set of codimension 3 in Ω). We consider the tangent plane to this surface at zero. The group $\{\exp W\}$, where the W are the linear operators of the fields w_β in §8 for the case of A_4, acts on the variety of two-dimensional subspaces in $T = \mathbf{R}^4$.

We introduce the notation $\partial_k = \partial/\partial\lambda_k$ $(k = 1, \ldots, 4)$. The basis fields w_k are then given by the formulae

$$\begin{aligned} w_1 &= 2\lambda_1\partial_1 + 3\lambda_2\partial_2 + 4\lambda_3\partial_3 + 5\lambda_4\partial_4, \\ w_2 &= 3\lambda_2\partial_1 + 4\lambda_3\partial_2 + 5\lambda_4\partial_3, \\ w_3 &= 4\lambda_3\partial_1 + 5\lambda_4\partial_2, \\ w_4 &= 5\lambda_4\partial_1. \end{aligned}$$

LEMMA 8. *The orbit of a plane* $\mathbf{R}^2$ *in general position in* $\mathbf{R}^4$ *under the action of the group* $\{\exp W\}$ *contains one of the two planes generated by the vectors* ∂_4 *and* $\partial_1 \pm \partial_3$; *the union of the remaining orbits is a proper subvariety of the space* $G_{4,2}$ *of planes in* $\mathbf{R}^4$.

PROOF. It is easy to prove the following proposition.

PROPOSITION 1. *The orbit of the basis vector* ∂_k *under the action of* $\{\exp W\}$ *is the half-space* $\lambda_k > 0$ *of the plane* $\lambda_{k+1} = \lambda_{k+2} = \ldots = 0$.

A plane $\mathbf{R}^2$ in general position in $\mathbf{R}^4$ intersects the orbit of ∂_4 (that is, the half-space $\lambda_4 > 0$). Therefore, the action of the group can reduce this plane to the plane π generated by the vectors ∂_4 and $e = a\partial_1 + b\partial_2 + c\partial_3$.

A further reduction will be carried out so that the plane always contains ∂_4.

PROPOSITION 2. *Suppose that a smooth one-parameter family of linear transformations taking π into planes containing ∂_4 is given. Then the velocity of the motion at ∂_4 belongs to π.*

For let us denote the positions of the vectors ∂_4 and e at the moment ε by $d(\varepsilon) = \partial_4 + \alpha\varepsilon + \ldots$ and $e(\varepsilon) = e + \beta\varepsilon + \ldots$ The coefficients of the expansion of ∂_4 in the basis $d(\varepsilon)$ and $e(\varepsilon)$ of the plane $\pi(\varepsilon)$ are denoted by $p(\varepsilon) = 1 + \gamma\varepsilon + \ldots$ and $q(\varepsilon) = \delta\varepsilon + \ldots$ $(\partial_4 = p(\varepsilon)d(\varepsilon) + q(\varepsilon)e(\varepsilon))$. Equating the coefficients for ε, we find that $\alpha = -\gamma\partial_4 - \delta e$, as required.

PROPOSITION 3. *Conversely, if we are given a one-parameter family of linear transformations taking π into $\pi(\varepsilon)$, and if the velocity at ∂_4 at each moment of time ε belongs to $\pi(\varepsilon)$, then all the planes $\pi(\varepsilon)$ contain ∂_4.*

For in this case the velocity with which the distance of ∂_4 from $\pi(\varepsilon)$ changes is identically zero.

The values of the basis fields w at ∂_4 are $w_1 = 5\partial_4$, $w_2 = 5\partial_3$, $w_3 = 5\partial_2$, and $w_4 = 5\partial_1$. Thus, the field

$$kw_1 + K(cw_2 + bw_3 + aw_4)$$

from any real k and K defines a motion whose velocity at ∂_4 at the initial moment belongs to π. The velocity of e is then equal to

$$k[2a\partial_1 + 3b\partial_2 + 4c\partial_3)] + K[c(3b\partial_1 + 4c\partial_2) + b(4c\partial_1)].$$

According to Proposition 3, we obtain equivalent planes π (belonging to the same orbit of $\{\exp W\}$) if we transfer e with respect to the phase curve of each of the two systems

$$\dot{a} = 2a, \quad \dot{b} = 3b, \quad \dot{c} = 4c; \tag{3}$$

$$\dot{a} = 7bc, \quad \dot{b} = 4c^2, \quad \dot{c} = 0. \tag{4}$$

For π in general position $c \neq 0$. We make $c = 1$ by dilating e. Now from (4) we obtain

$$\dot{a} = 7b, \; \dot{b} = 4; \; b = b_0 + 4t, \; a = a_0 + 7b_0t + 14t^2.$$

We make $b = 0$ by choosing t suitably. For $b = 0$ the system (3) allows us to vary $|a|/|c|$. For π in general position we have $a \neq 0$ and $c \neq 0$. By transferring along the phase curve of the first system (3) we can achieve that $|a| = |c|$. By dilating e we make $|a| = |c| = 1$. This proves Lemma 8.

PROPOSITION 4. *Each of the two planes $\lambda_2 = 0$, $\lambda_3 = \pm\lambda_1$ in $\mathbf{R}^4$ is w-stable, in the sense that the germ of every surface with this tangent plane at O is w-equivalent to the germ of the corresponding plane at O.*

PROOF. We verify the relevant infinitesimal condition. With this aim we compute the tangent space to the variety of systems of equations $f_1 = 0$, $f_2 = 0$, which yield surfaces equivalent to the given one.

This tangent space may be assumed to be the space of vectors $(\delta f_1, \delta f_2) \in A^2$, where A is a ring of functions of the class in question (say, $A = \mathbf{R}[[\lambda]]$). The system goes over into an equivalent one if we add to one equation the other one multiplied by an arbitrary (small) function. Therefore, this tangent space contains the submodule of A^2 generated by the vectors $(f_1, 0)$, $(f_2, 0)$, $(0, f_1)$, and $(0, f_2)$. The system goes over to an equivalent one also under the action of the one-parameter group of diffeomorphisms given by a w-liftable vector field in $\mathbf{R}^4$. Therefore, the tangent space also contains all the vectors of the form (vf_1, vf_2), where v is a liftable field. The vectors of this kind form a submodule of A^2 with four generators $(v_k f_1, v_k f_2)$, where $v_1, \ldots, v_4$ are liftable fields of a basis.

Now we consider the linear parts of our four generators at zero, which have the form $(w_k f_1, w_k f_2)$, where w_k are the basic linear fields described above. Substituting $f_1 = \lambda_2$, $f_2 = \lambda_3 \pm \lambda_1$, we obtain eight linear parts of the generators of the submodule of A^2 in the form

$$\begin{pmatrix}\lambda_2\\0\end{pmatrix}\begin{pmatrix}0\\\lambda_2\end{pmatrix}\begin{pmatrix}\lambda_3 \mp \lambda_1\\0\end{pmatrix}\begin{pmatrix}0\\\lambda_3 \mp \lambda_1\end{pmatrix}\begin{pmatrix}3\lambda_2\\4\lambda_3 \mp 2\lambda_1\end{pmatrix}\begin{pmatrix}4\lambda_3\\5\lambda_4 \mp 3\lambda_2\end{pmatrix}\begin{pmatrix}5\lambda_4\\\mp 4\lambda_3\end{pmatrix}\begin{pmatrix}0\\\mp 5\lambda_4\end{pmatrix}.$$

Comparing the first column with the fifth, and the result with the fourth, we see that all the $(0, \lambda_k)$ occur among the linear parts of the vectors of the submodule. Now by considering columns 3, 5, 6 and 7 we see that all the $(\lambda_k, 0)$ also occur (this result can be seen from the fact that the orbit of π in the variety of planes is open). Since all linear vector-functions of λ occur among the linear parts of the vectors of the submodule, our submodule contains all the variations $(\delta f_1, \delta f_2)$ for which the functions δf_1 and δf_2 vanish at zero (Hadamard's lemma). This proves the infinitesimal w-stability; the actual proof of w-stability is now carried out by any of the usual methods in the theory of singularities of differentiable mappings. This proves Proposition 4.

Lemma 7 follows from Lemma 8 and Proposition 4.

Theorem 3 follows from Lemmas 2, 3, 6 and 7.

REMARK 1. The mutual inequivalence of the 10 singularities listed in the theorem is proved by comparing their sets of critical points.

For example, the projections of the surfaces $z = x^5 + x^3y + \varepsilon xy$ along the x-axis have curves of critical values for $\varepsilon = \pm 1$, and these are diffeomorphic to the curves $Y = t^4$, $Z = t^5 + \varepsilon t^7 + \ldots$ These two curves (with $\varepsilon = +1$ and -1) are not diffeomorphic. Hence, the corresponding two projections are not equivalent.

The irremovability is verified by changing y into $y + \alpha x$ and z into $z + \beta x$, which corresponds to a small change in the direction of projection. After these changes we can verify the independence of the equations of the system relative to α, β, y, and z, which determine that the projection belongs to the given type.

For example, the equation $z = x^5 + x^3y + xy$ is transformed to

$z = x^5 + \alpha x^4 + yx^3 + \alpha x^2 + (y - \beta)x$. The change of variables $x = \xi - \alpha/5$ takes the equation to $\xi^5 + p\xi^3 + Q\xi^2 + R\xi + S = 0$, where $P = y + \ldots$, $Q = \alpha + \ldots$, $R = y - \beta + \ldots$, $S = -z + \ldots$ (the dots denote terms of order higher than 1 in α, β, y, z). The system of equations $P = Q = R = S = 0$ has the solution $\alpha = \beta = y = z = 0$, which satisfies the conditions of the implicit function theorem. This proves the transversality of the mapping of the parameter spaces (α, β, y, z) into the variety of jets of projections to the subvariety of jets with $\mu = 5$, and hence, the irremovability of this class of singularities; on the other hand, it was proved above that the singularities of this class reduce to the indicated normal form.

REMARK 2. The above classification of mappings of the plane into the base of the universal family can also be carried out for $\mu = 3$ modulo w-equivalences. The answer has the following form: in two-parameter families in general position of mappings of the (y, z)-plane into the base ($\mathbf{R}^2 = (\lambda_1, \lambda_2)$ of the universal family $x^3 + \lambda_1 x + \lambda_2 = 0$, taking O into O, only those singularities occur that are w-equivalent to the singularities at O of the following four mappings:

$$\begin{cases} \lambda_1 = y, \\ \lambda_2 = z, \end{cases} \quad \begin{cases} \lambda_1 = \pm y^2, \\ \lambda_2 = z, \end{cases} \quad \begin{cases} \lambda_1 = y^3 + yz, \\ \lambda_2 = z. \end{cases}$$

These four singularities are mutually inequivalent and are irremovable in the sense that a small perturbation of the two-parameter family in general position of the mappings $(\mathbf{R}^2, O) \to (\mathbf{R}^2, O)$ cannot change the first singularity for a general value of the parameters, the two middle ones on a curve in the parameter plane, and the last one for individual values of the parameter.

The proof proceeds by standard methods [1]; calculations can be avoided by noting that the pair consisting of a semicubical parabola and a line passing through its vertex and transversal to the tangent, or a pair of two semicubical parabolas with transverse tangents at the vertex, has no moduli (which follows immediately from the normal forms of the singularities D_5 and $Y_{5,5}$ of functions of two variables).

The projections of surfaces corresponding to these for singularities of λ-mappings, are equivalent to the four projections indicated in Theorem 3, with $\mu = 3$ ($z = x^3 + yx$, $z = x^3 \pm y^2 x$, $z = x^3 + y^3 x$). However, the λ-mapping constructed from the last formula is not equivalent to the mapping $\lambda_1 = y^3 + yz$, $\lambda_2 = z$ in the sense of w-equivalence. Thus, the converse assertion of Lemma 4 is not true: *equivalent projections can have inequivalent λ-mappings.*

REMARK 3. Apparently, Lemma 5 generalizes to diffeomorphisms: each diffeomorphism of the base that can be lifted to Γ seems to be the projection onto the base of some diffeomorphism of the ambient space $\mathbf{R}^\mu$ that preserves the hypersurface Γ. In this context it would be useful to compute the group of connected components of the group of w-liftable diffeomorphisms (in the real and in the complex case). Besides, the analogous proposition generalizing Lemma 5 apparently holds for a wide class of hypersurfaces Γ.

When the manuscript of this paper was completed, I learned that O. V. Lyashko has solved the problems raised here (see [7]), and that V. V. Goryunov has found all the simple singularities of projections of hypersurfaces onto spaces of any codimension.

10.3. Vector fields tangent to projections and sections of the discriminant varieties. The theorem on w-stability of curves can be regarded as a dual (in the sense of the informal duality of §9) to the theorem proved in [1] on the w-stability of hypersurfaces. This duality also appears in the more delicate properties of liftable fields described below.

We consider the variety of orbits B of an irreducible reflection group in Euclidean space, acting on its complexification $\mathbf{C}^\mu$. This variety of orbits B is diffeomorphic to $\mathbf{C}^\mu$.

We fix a general line $\mathbf{C}^1$ passing through O in B (transversal to the variety of non-regular orbits of Σ at O). We consider the projection $\pi: B \to \widetilde{B}$ of B along $\mathbf{C}^1$ onto $\widetilde{B}$ (diffeomorphic to $\mathbf{C}^{\mu-1}$). We denote by $\widetilde{\Sigma}$ the projection of the set of singular points of Σ onto $\widetilde{B}$. The complex hypersurface $\widetilde{\Sigma}$ in $\widetilde{B}$ is, in terms of singularity theory, the bifurcation diagram of the functions. As a rule, it is reducible. From Theorem 2 we obtain the very strong stability of $\widetilde{\Sigma}$ relative to the choice of the $\mathbf{C}^1$ (the projection may even be curvilinear).

THEOREM 4.[1] *The germs of vector fields on $\widetilde{B}$ tangent to $\widetilde{\Sigma}$ are the projections of germs of vector fields on B tangent to Σ, at least in the cases A_2, A_3, B_2, B_3.*

The dual theorem consists in the following. We fix a general hyperplane passing through O in B (transversal to the "worst direction" on Σ at O). We denote it by $\approx{B}$, the embedding of $\approx{B}$ in B by i, and the intersection of $\approx{B}$ and Σ by $\approx{\Sigma}$. From [1] we obtain the very strong stability of $\approx{\Sigma}$ relative to the choice of hyperplane (or hypersurface) $\approx{B}$.

THEOREM 5. *The germs of vector fields on $\approx{B}$ tangent to $\approx{\Sigma}$ can be extended to germs of fields on B tangent to Σ, at least in the cases A_2, A_3, B_2, and B_3.*

REMARK. Theorems 4 and 5 use the special properties of the discriminant varieties. It is easy to construct a plane curve and a general projection of it onto a line so that not every field on the line that (the field) vanishes at the projection of a singular point lifts to a field on the plane tangent to the curve. However, in the case of curves it suffices for fields to be liftable that the curve is quasi-homogeneous and that the projection is along an axis. In the case of surfaces quasi-homogeneity is already insufficient, as is shown by the example of the surface

$$z^4 - x^2z^2 + x^3y = 0,$$

[1] While the manuscript of this paper was being prepared, I learned that O. V. Lyashko (see [7]) and V. M. Zakalyukin have found far-reaching generalizations of Theorem 4. In particular, the theorem is true for all simple Lie algebras ($A_k, B_k, C_k, D_k, E_k, F_4, G_2$).

which can be projected onto the (x, y)-plane along the z-axis. The field

$$(4xy - x^2)\partial/\partial x$$

on the (x, y)-plane is tangent to the discriminant variety $xy(4y - x) = 0$, but is not liftable to a field tangent to a hypersurface in 3-space.[1]

For a section of a quasi-homogeneous hypersurface of the coordinate hyperplane the vector fields in the hyperplane tangent to the intersection extend to fields in the ambient space tangent to the hypersurface whenever the intersection has an isolated singularity at the origin. For let

$$F \equiv \sum \alpha_k x_k \partial F/\partial x_k = 0$$

be the equation of the hypersurface, $x_1 = 0$ the equation of the hyperplane, $f \equiv F\,|_{x_1 = 0} = 0$ the equation of the section, and $v_2 \partial_2 + \ldots + v_n \partial_n$ the field tangent to a section. Then

$$\sum v_k\, \partial f/\partial x_k = hf,$$

that is,

$$\sum (v_k - h\alpha_k x_k)\, \partial f/\partial x_k = 0.$$

If the singularity of f at O is isolated, the Koszul complex for $\{\partial f/\partial x_k\}$ is acyclic. From the acyclicity in dimension 1 it follows that the module of fields in the hyperplane tangent to the section is generated by the fields

$$\omega = \sum \alpha_k x_k \partial/\partial x_k \text{ and } \omega_{k,\,l} = (\partial f/\partial x_k)\partial_l - (\partial f/\partial x_l)\partial_k \qquad (k,\ l \geqslant 2).$$

These fields extend to the fields

$$\Omega = \sum \alpha_k x_k \partial_k \quad (k \geqslant 1) \text{ and } \Omega_{k,\,l} = (\partial F/\partial x_k)\partial_l - (\partial F/\partial x_l)\partial_k \quad (k,\ l \geqslant 1),$$

tangent to the hypersurface $F = 0$ in the ambient space. Hence, in particular, we can obtain Theorem 5 for the cases A_2, A_3, B_2, and B_3. However, already in the case A_4 the module of fields tangent to the section $\tilde{\tilde{\Sigma}}$ is not generated by ω and $\omega_{k,l}$ (example: the liftable fields $i^* v_{k,l}$, $v_{k,l} = d_l \lambda_l v_k - d_k \lambda_k v_l$, where $d_p = p + 1$, and $v_k = v_{\lambda_k}$ are basis liftable fields).

PROOF OF THEOREMS 4 and 5. The bases of the modules of vector fields tangent to Σ, $\tilde{\Sigma}$, and $\tilde{\tilde{\Sigma}}$ in the cases A_2, A_3, and B_3, and also the equations of $\tilde{\Sigma}$ and $\tilde{\tilde{\Sigma}}$ are written down in the following Table 1. We use coordinates in B such that

$\Sigma(A_2) = \{\lambda_1, \lambda_2\colon x^3 + \lambda_1 x + \lambda_2 = 0$ has a multiple root$\}$,
$\Sigma(A_3) = \{\lambda_1, \lambda_2, \lambda_3\colon x^4 + \lambda_1 x^2 + \lambda_2 x + \lambda_3 = 0$ has a multiple root$\}$,
$\Sigma(B_3) = \{\lambda_1, \lambda_2, \lambda_3\colon x^3 + \lambda_1 x^2 + \lambda_2 x + \lambda_3 = 0$ has a multiple or a zero root$\}$.

The direction of projection of π is the λ_2-axis for A_2, the λ_3-axis for A_3 and B_3. The coordinates in $\tilde{B}$ are λ_1 for A_2, (λ_1, λ_2) for A_3 and B_3. The plane $\tilde{\tilde{B}}$

[1] O. V. Lyashko (see [7]) has proved liftability under the condition that the set of points of the discriminant variety corresponding to roots of multiplicity greater than 2 has codimension at least 2.

is $\lambda_1 = 0$, the coordinates in $\widetilde{B}$ are λ_2 for A_2, (λ_2, λ_3) for A_3 and B_3. We denote the field $\partial/\partial\lambda_k$ by ∂_k. The projection of B onto $\widetilde{B}$ is denoted by π, and the embedding of $\widetilde{\widetilde{B}}$ in B by i.

The bases $\{v_k\}$ of the modules of fields tangent to Σ are obtained by the method of [1] (see §8). The bases of the modules of fields tangent to $\widetilde{\Sigma}$ and $\widetilde{\widetilde{\Sigma}}$ can be found by direct computations. With the use of the natural gradings the calculations become very short (the degree of λ_k is equal to the coefficient of $\lambda_k \partial_k$ in v_1).

Each of the basis fields in Table 1 tangent to $\widetilde{\Sigma}$ can be represented as the projection $\pi_* v$ of a vector field v on B tangent to Σ, and each of the basis fields in Table 1 that is tangent to $\widetilde{\widetilde{\Sigma}}$ can be represented as the restriction i^*v of a vector field v on B and tangent to Σ. For fields tangent to $\widetilde{\Sigma}$ (respectively, $\widetilde{\widetilde{\Sigma}}$), which are indicated first in Table 1, we have $v = v_1$. For A_3 and B_3 there is a second field. For it v has the form of a linear combination of basis fields v_k with coefficients p_k (for tangency with $\widetilde{\Sigma}$) or q_k (for tangency with $\widetilde{\widetilde{\Sigma}}$), as given in Table 2:

TABLE 1

Type	Tangent to Σ	Tangent to $\widetilde{\Sigma}$	Tangent to $\widetilde{\widetilde{\Sigma}}$
A_2	$v_1 = 2\lambda_1\partial_1 + 3\lambda_2\partial_2$ $v_2 = 3\lambda_2\partial_1 - \frac{2}{3}\lambda_1^2\partial_2$	$2\lambda_1\partial_1$ $\widetilde{\Sigma}$: $\lambda_1 = 0$	$3\lambda_2\partial_2$ $\widetilde{\widetilde{\Sigma}}$: $\lambda_2 = 0$
A_3	$v_1 = 2\lambda_1\partial_1 + 3\lambda_2\partial_2 + 4\lambda_3\partial_3$ $v_2 = 3\lambda_2\partial_1 + (4\lambda_3 - \lambda_1^2)\partial_2 - \frac{1}{2}\lambda_1\lambda_2\partial_3$ $v_3 = 4\lambda_3\partial_1 - \frac{1}{2}\lambda_1\lambda_2\partial_2 + \left(2\lambda_1\lambda_3 - \frac{3}{4}\lambda_2^2\right)\partial_3$	$2\lambda_1\partial_1 + 3\lambda_2\partial_2$ $(8\lambda_1^3 + 27\lambda_2^2)\partial_1$ $\widetilde{\Sigma}$: $(8\lambda_1^3 + 27\lambda_2^2)\lambda_2 = 0$	$3\lambda_2\partial_2 + 4\lambda_3\partial_3$ $64\lambda_3^2\partial_2 + 9\lambda_2^3\partial_3$ $\widetilde{\widetilde{\Sigma}}$: $27\lambda_2^4 = 256\lambda_3^3$
B_3	$v_1 = \lambda_1\partial_1 + 2\lambda_2\partial_2 + 3\lambda_3\partial_3$ $v_2 = 2\lambda_2\partial_1 + (3\lambda_3 + \lambda_1\lambda_2)\partial_2 + 2\lambda_1\lambda_3\partial_3$ $v_3 = 3\lambda_3\partial_1 + 2\lambda_1\lambda_3\partial_2 + \lambda_2\lambda_3\partial_3$	$\lambda_1\partial_1 + 2\lambda_2\partial_2$ $\lambda_2(4\lambda_2 - \lambda_1^2)(3\partial_1 + 2\lambda_1\partial_2)$ $\widetilde{\Sigma}$: $\lambda_2(4\lambda_2 - \lambda_1^2)(3\lambda_2 - \lambda_1^2) = 0$	$2\lambda_2\partial_2 + 3\lambda_3\partial_3$ $(4\lambda_2^3 + 27\lambda_3^2)\partial_2$ $\widetilde{\widetilde{\Sigma}}$: $\lambda_3(4\lambda_2^3 + 27\lambda_3^2) = 0$

TABLE 2

Type	$\{p_k\}$	$\{q_k\}$
A_3	$4\lambda_1^2-12\lambda_3$, $9\lambda_2$, $6\lambda_1$	0, $16\lambda_3$, $-12\lambda_2$
B_3	$\lambda_1\lambda_2-9\lambda_3$, $6\lambda_2-2\lambda_1^2$, $3\lambda_1$	$2\lambda_2^2$, $9\lambda_3$, $-6\lambda_2$

The coefficients p_k and q_k in Table 2 were calculated by means of the grading in the ring of polynomials in λ described above. However, these calculations are not needed in the proof: it can be checked directly that for the fields v constructed from these coefficients the fields $\pi_* v$ and i^*v coincide with those in Table 1. This proves Theorems 4 and 5.

REMARK. Theorems 4 and 5 were discovered as by-products of The proof of Theorem 3 (which they do not use).

Warning: In spite of the fact that the surfaces in B that are of interest to us in Lemmas 6 and 7 of Theorem 3, and also their normal forms, turn out to be vertical (cylindrical with generating direction $\partial/\partial\lambda_\mu$), their reduction to normal forms, generally speaking, requires the whole group of Σ-preserving diffeomorphisms: the groups of diffeomorphisms preserving Σ when projected onto $\widetilde{B}$ (that is, taking vertical lines into vertical lines) need not be sufficient.

Theorems 1 and 2 were proved as preliminaries to Theorem 3. The question of the normal form of singular projections of smooth surfaces, which is solved by Theorem 3, arose in O. A. Platonova's study of the singularities in general position in the problem of the fastest way of avoiding obstructions. The proofs of Theorems 1–5 required the study of the Lie algebras of the fields w_β, and hence the operation of linearized convolution φ introduced in §8. We arrived at the proof of non-degeneracy (Theorem B of §8), and for this, Theorem C. Theorem A, which is equivalent to Theorem B, in combination with the formula of Levine–Eisenbud [3] and Khimshiashvili [5] suggests an interpretation of the signatures of the quadratic forms ψ_q generated by the displacement as the indices of singular points of vector fields. For the singularities A_μ, D_μ and E_μ this immediately reduces to Theorem D of §8, and then, as for the boundary singularities B_μ, C_μ, F_4 introduced in [2], the resulting index turned out not to be entirely usual. An analysis of this index led to all the results of §§1–8 on the boundary singularities of 1-forms. A comparison of both signature formulae for the index required the study of the duality between quadratic forms on Q and on T^*, which is carried out in §9. Thus, all the results of this paper arose from the problem of removing obstructions.

References

[1] V. I. Arnold, Wave fronts evolution and equivariant Morse lemma, Comm. Pure Appl. Math. **29** (1976), 557–582 (in English).

[2] V. I. Arnol'd, Critical points of functions on a manifold with boundary, the simple Lie groups B_k, C_k, F_4 and singularities of evolutes, Uspekhi Mat. Nauk **33**:5 (1978), 91–107.
= Russian Math. Surveys **33**:5 (1978), 99–116.

[3] D. Eisenbud and H. I. Levine, The topological degree of a finite C^∞ map germ, Ann. of Math. (2) **106** (1977), 19–39. MR **57** # 7651.

[4] E. Lloijenga, The complement of the bifurcation variety of a simple singularity, Inventiones Math. **32** (1974), 105–116.

[5] G. N. Khimshiashvili, The local degree of a smooth mapping, Sakharth. SSR Mecn. Akad. Moambe **85** (1977), no. 2, 309–311.

[6] A. G. Khovanskii, The index of a polynomial vector field, Funktsional. Anal. i Prilozhen. **13** (1979), 49–58.
= Functional. Anal. Appl. **13** (1979).

[7] O. V. Lyashko, The geometry of bifurcation diagrams, Uspekhi Mat. Nauk **34**:3 (1979), 205–206 (added by author, 18 November 1979).
= Russian Math. Surveys **34**:3 (1979).

Received by the Editors 15 October 1978

Translated by J. S. Joel